EmQM15: Emergent Quantum Mechanics 2015

Journal of Physics: Conference Series Volume 701

Vienna, Austria
23 – 25 October 2015

Editors:

Gerhard Grossing
Hans-Thomas Elze
Johannes Mesa Pascasio
Jan Walleczek

ISBN: 978-1-5108-2239-9
ISSN: 1742-6588

Printed from e-media with permission by:

Curran Associates, Inc.
57 Morehouse Lane
Red Hook, NY 12571

Some format issues inherent in the e-media version may also appear in this print version.

Printed by Curran Associates, Inc. (2016)

For permission requests, please contact the Institute of Physics
at the address below.

Institute of Physics
Dirac House, Temple Back
Bristol BS1 6BE UK

Phone: 44 1 17 929 7481
Fax: 44 1 17 920 0979

techtracking@iop.org

Additional copies of this publication are available from:

Curran Associates, Inc.
57 Morehouse Lane
Red Hook, NY 12571 USA
Phone: 845-758-0400
Fax: 845-758-2633
Email: curran@proceedings.com
Web: www.proceedings.com

Table of contents

Volume 701

EmQM15: Emergent Quantum Mechanics 2015
23–25 October 2015, Vienna, Austria

Accepted papers received: 2 March 2016
Published online: 30 March 2016

Preface

Papers

Invited lectures

Posters

Journal of Physics: Conference Series **701** (2016) 011001 doi:10.1088/1742-6596/701/1/011001

Preface

These proceedings comprise the invited lectures of the third international symposium on Emergent Quantum Mechanics (EmQM15), which was held at the Vienna University of Technology in Vienna, Austria, 23–25 October 2015.

The symposium convened at the Festsaal and the adjacent Boeckl-Saal of the Technical University, and was devoted to the open exploration of the quantum state as a reality. The resurgence of interest in ontological quantum theory, including both deterministic and indeterministic approaches, challenges long held assumptions and focuses on the following questions: Is the world local or nonlocal? What is the nature of quantum nonlocality? If nonlocal, i.e., superluminal, influences exist then why can't they be used for superluminal signaling and communication? How is the role of the scientific observer/agent to be accounted for in realistic approaches to quantum theory? How could recent developments in the field of space-time as an emergent phenomenon advance new insight at this research frontier? What new experiments might contribute to new understanding? These and related questions were addressed in the context also of a possible deeper level theory for quantum mechanics that interconnects three fields of knowledge: emergence, the quantum, and information. Could there appear a revised image of physical reality from recognizing new links between emergence, the quantum, and information? The symposium provided a forum for considering (i) current theoretical and conceptual obstacles which need to be overcome as well as (ii) promising developments and research opportunities on the way towards realistic quantum mechanics. Contributions were invited that present current advances in both standard as well as unconventional approaches.

The EmQM15 symposium was co-organized by Gerhard Grössing (Austrian Institute for Nonlinear Studies (AINS), Vienna), and by Jan Walleczek (Fetzer Franklin Fund, USA, and Phenoscience Laboratories, Berlin). After two successful conferences on the same topic in 2011 and 2013, the partnership since EmQM13 between AINS and the Fetzer Franklin Fund in producing the EmQM15 symposium was able to further expand interest in the promise of emergent quantum mechanics.

The articles contained in these proceedings represent the talks of the invited speakers as written immediately after the symposium. The volume starts with an introductory statement by organizers Jan Walleczek and Gerhard Grössing, explaining why emergent quantum mechanics is a promising approach in quantum foundations. The conference proceedings then continue with the presentations as given in their chronological order, as far as they were submitted. The papers by F. Scardigli and T. Nieuwenhuizen were not presented, but are added here as additional contributions to the group of papers on gravity and emergent spacetime. Note that the talks of all speakers are freely available on the conference website (**http://www.emqm15.org/**) and on **http://www.fetzer-franklin-fund.org/** as video presentations.

The organizers wish to express their gratitude to Siegfried Fussy and Herbert Schwabl from AINS for the organizational support. The organizers also wish to thank Bruce Fetzer, President and CEO, John E. Fetzer Memorial Trust, and the Members of the Board of Trustees, for their strong support and for funding this symposium.

Journal of Physics: Conference Series **701** (2016) 011001 doi:10.1088/1742-6596/701/1/011001

The expertise of the Members of the Scientific Advisory Board of the EmQM15 symposium, Ana María Cetto (Mexico), Lajos Diósi (Budapest), Maurice de Gosson (Vienna), Theo Nieuwenhuizen (Amsterdam) and Helmut Rauch (Vienna), is also gratefully acknowledged.

Finally, it is a pleasure to once more thank Sarah Toms and her team at IOP Publishing (Bristol) for their continued support during the preparation of these proceedings..

Vienna, Pisa, Berlin, March 2016

Gerhard Grössing,
Hans-Thomas Elze,
Johannes Mesa Pascasio,
Jan Walleczek

The front cover image shows three bouncing oil droplets on an oscillating oil surface, as they are employed by Couder, Bush, Batelaan and others to show macroscopic analogues of wave-particle complementarity (courtesy of Dan Harris and John Bush, MIT).

Journal of Physics: Conference Series **701** (2016) 011002 doi:10.1088/1742-6596/701/1/011002

Peer review statement

All papers published in this volume of *Journal of Physics: Conference Series* have been peer reviewed through processes administered by the proceedings Editors. Reviews were conducted by expert referees to the professional and scientific standards expected of a proceedings journal published by IOP Publishing.

Is the World Local or Nonlocal? Towards an Emergent Quantum Mechanics in the 21st Century

Jan Walleczek[1] **and Gerhard Grössing**[2]

[1] Phenoscience Laboratories, Novalisstrasse 11, 10115 Berlin, Germany
[2]Austrian Institute for Nonlinear Studies, Akademiehof, Friedrichstr. 10, 1010 Vienna, Austria

E-mail: walleczek@phenoscience.com, ains@chello.at

Abstract. What defines an emergent quantum mechanics (EmQM)? Can new insight be advanced into the nature of quantum *nonlocality* by seeking new links between *quantum* and *emergent* phenomena as described by self-organization, complexity, or emergence theory? Could the development of a *future* EmQM lead to a unified, *relational* image of the cosmos? One key motivation for adopting the concept of emergence in relation to quantum theory concerns the persistent failure in standard physics to unify the two pillars in the foundations of physics: quantum theory and general relativity theory (GRT). The total contradiction in the foundational, metaphysical assumptions that define orthodox quantum theory *versus* GRT might render *inter*-theoretic unification impossible. On the one hand, *indeterminism* and *non*-causality define orthodox quantum mechanics, and, on the other hand, GRT is governed by causality and *determinism*. How could these two metaphysically-contradictory theories ever be reconciled? The present work argues that metaphysical contradiction necessarily implies physical contradiction. The contradictions are essentially responsible also for *the measurement problem* in quantum mechanics. A common foundation may be needed for overcoming the contradictions between the two foundational theories. The concept of emergence, and the development of an EmQM, might help advance a *common foundation* – physical *and* metaphysical – as required for successfull *inter-theory unification*.

1. Introduction

The question "Is the world local or *non*-local?" has long guided work in quantum foundations. At the latest, this started with the introduction, by Einstein, Podolsky, and Rosen (EPR), of the first precise *metaphysical definitions* in relation to nonlocality as a concept (Einstein *et al.* [1]). 80 years on, that question – rather surprisingly – remains unanswered still. On the one hand, there is no doubt any longer that EPR-type nonlocal correlations can be observed in quantum experiments by observers who are separated at *space-like* distances. On the other hand, the *ontological* question remains wholly undecided of whether these nonlocal *observations* might imply the actual existence of a "*nonlocal reality*" – *not* merely in terms of an *operational metaphor* as in orthodox quantum theory. The prospect of fundamentally "real nonlocality" was proposed, for example, by de Broglie–Bohm theory (Bohm [2, 3]). Inspired by both Bohm's proposal (Bohm [2, 3]) and the EPR argument (Einstein *et al.* [1]), John Bell succeeded in proving that no quantum theory based on the joint assumptions of *reality* and *locality* could successfully reproduce the predictions that are yielded by *orthodox*, i.e., *operationalist* quantum mechanics (Bell [4]).

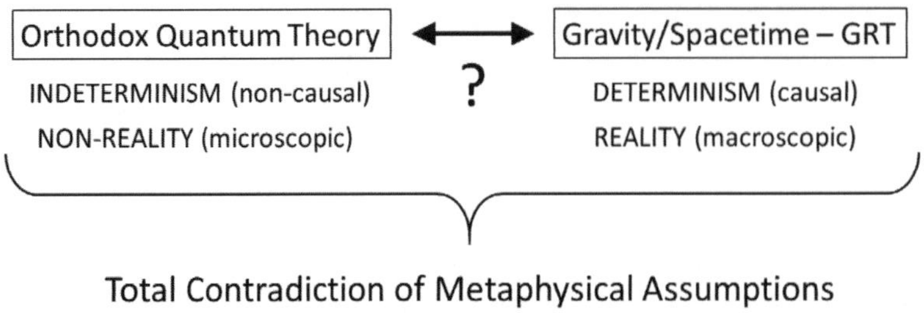

Figure 1. Total contradiction of metaphysical assumptions between orthodox quantum theory and general relativity theory (GRT). *Metaphysical* contradiction implies *physical* contradiction (see Sect. 2.1). How is the reconciliation of metaphysical assumptions possible?

The seminal proof of Bell's theorem left open, however, the extraordinary possibility that reality might be – ontologically-speaking – *nonlocal in nature.* That possibility, which necessarily reaches beyond *operationalist* quantum theory, is pursued by what has become known as the *ontological, realist, approach* to quantum mechanics (e.g., Bohm and Hiley [5]). The project of developing an 'emergent quantum mechanics' (EmQM) is usually placed in the context of *realist* approaches to quantum mechanics.

The implications of an EmQM are startling, however, when viewed through the lens of the orthodox perspective: instead of finding – at reality's deepest levels – absolute "*quantum randomness*", a future EmQM, including also de Broglie–Bohm theory, would find "*quantum interconnectedness*", e.g., possibly in the form of instantaneous *nonlocal influences* across the universe. For example, when John Bell was asked what the meaning was of nonlocality, he answered that *nonlocality* "*... means that what you do here has immediate consequences in remote places*" (Mann and Crease [6]). What might the phenomenon of '*emergence*' offer towards a new understanding of nonlocality in the deeper sense of Bell's "*immediate consequences*" – beyond the standard *operationalism* of orthodox theory?

2. Why 'emergence' in quantum mechanics?

One key motivation for adopting the concept of (irreducible) emergence in relation to quantum theory concerns the much-debated failure to unify the two pillars in the foundations of physics: quantum theory and general relativity theory (GRT). Therefore, the long-term project of inter-theory unification might be injected with fresh thinking via the introduction into quantum mechanics of the concept of emergence. Why might that be so?

On the one hand, *orthodox* quantum theory, as we understand it today, is an entirely *indeterministic* and *non-causal* theory, which presumes the *complete absence* of any fundamental, ontological reality at the level of the quantum. "There is no quantum world." Niels Bohr's explained, "There is only abstract quantum-mechanical description" (Petersen [7]). On the other hand, relativity theory (GRT) represents an ontological theory of *space-time reality*, in a decidely *causal* and *deterministic* manner. Fig. 1 illustrates the fact that the metaphysical assumptions associated with the theories contradict each other: "indeterminism" *versus* "determinism", and "non-reality" *versus* "reality". These contradictions are responsible also, of course, for the so-called *measurement problem* in quantum mechanics.

2.1. Why 'metaphysics' in quantum physics?

Why is there this emphasis on *metaphysical assumptions*? It is helpful to remember that in the EPR argument already, which called for the *incompleteness* of orthodox quantum mechanics (Einstein *et al.* [1]), it was the exact derivation and definition of metaphysical assumptions which

Journal of Physics: Conference Series **701** (2016) 012001 doi:10.1088/1742-6596/701/1/012001

allowed the EPR argument to have relevance to concrete problems facing quantum physics: Is the world local or nonlocal? It was only through the consideration of metaphysical notions like "locality", "nonlocality", "causality", and "reality", that the breakthrough of Bell's theorem was possible (Bell [4]). What is often lost in this picture is the following: *metaphysical* assumptions essentially constrain the application of any *mathematical theory* to concrete *physical* situations. Importantly, "metaphysical" is neither "mystical" nor "irrational". A metaphysical analysis refers to the first principles and the foundational physical assumptions which inevitably underpin *any scientific* or *mathematical analysis* of nature. Often, foundational assumptions represent the *preferred world view* of the working scientist, including *preconceived notions* of what may, or may not, be possible in reality. Thus, by adopting a new metaphysical position, a new vista might open up towards the solution of a previously intractable scientific problem.

It appears likely that *not* any amount of mathematical or technical sophistication will reconcile the two theories – quantum and relativity, *unless the problem of their immediate metaphysical opposition could be resolved also* (compare Fig. 1). Similarly, any resolution of the measurement problem is likely to depend on the "metaphysical reconcilition" – at the *macroscopic* and *microscopic* levels – of any future *physical explanations.* Not suprisingly, it was John Bell [8] again who suggested "... that a real synthesis of quantum and relativity theories requires not just technical developments but radical conceptual renewal."

3. Towards an emergent quantum mechanics

The research project of an EmQM follows the spirit of John Bell's call for "radical conceptual renewal" (Bell [8]), a call consistent with his well-documented *realist* expectations about the future of quantum mechanics (e.g., Bell [9]). EmQM research seeks a common foundation upon which might rest both quantum theory *and* GRT. Presently, the availability of a common foundation is disputed or, at least, entirely unconfirmed. However, the concept of 'emergence' from self-organization, chaos, or complexity, theory – once properly adapted – might offer a *universal framework*, both physically *and* metaphysically, for finally promoting "... a real synthesis of quantum and relativity theories..." (Bell [8]).

For some time now, the concept of emergence has found use already in gravitational theory and in understanding the nature of space and time. Both the puzzles and the possibilities of notions such as 'emergent gravity' and 'emergent space-time' have been well summarized, for example, by David Gross [10]: "Many of us are convinced that space is an emergent, not fundamental concept. We have many examples of interesting quantum mechanical states, for which we can think of some (or all) of the spatial dimensions as emergent. Together with emergent space, we have the emergent dynamics of space and thus emergent gravity. But it is hard to imagine how time could be emergent? How would we formulate quantum mechanics without time as a primary concept? Were time to be emergent, our understanding of quantum mechanics would have to change." See Fig. 2 for a sketch illustrating the proposal that new understanding of quantum mechanics, based on *emergence*, could lessen, or even lift, the inter-theoretic contradiction shown before in Fig. 1.

The key point is the following: once space-time and gravity are recast in terms of fundamentally *emergent* states or dynamics, this invites the new view of the *quantum* nature of reality in terms of *emergent dynamics* as well. Thus, a common conceptual foundation might be developed – based on emergence as a guiding principle – capable of bridging the vast chasm between quantum and relativity theories. Maybe, then, there could be a new way to look at the problem of inter-theory unification. In the future, there might be theories decribing some kind of "emergent quantum gravity" as a result. For example, pioneering work based upon a locally-deterministic form of an "emergent quantum mechanics" was carried out by 't Hooft [11] (2007).

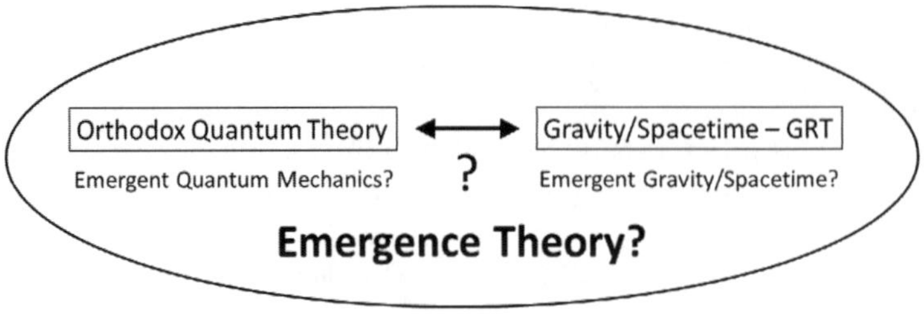

Figure 2. The concept of *emergence* may provide a common *physical* and *metaphysical foundation* in efforts to unify quantum and relativity theories (GRT). A common foundation will be needed for overcoming the deep metaphysical contradictions – in the *orthodox* approach – which have thus far prohibited success in inter-theory unification (compare Fig. 1).

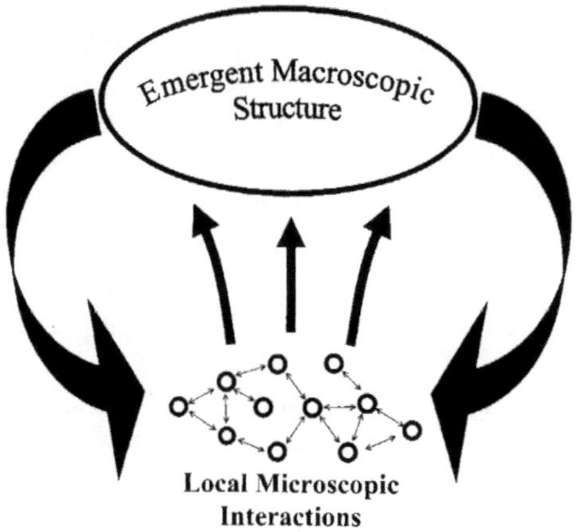

Figure 3. Illustration of self-referential, dynamical interactions across levels of organization – *microscopic* and *macroscopic* (from Walleczek [12]). Both top-down and bottom-up causal flows are indicated in the formation of an *emergent macroscopic structure.* Emergence describes the *spontaneous synchronization* of individual random motions into a unified collective motion. Emergence accounts for the rise of *global* macroscopic order from *local* microscopic randomness. An example is the emergent formation of spatio-temporal, *long-range coherence.*

4. What is emergence?

In a more general context, what is *emergence*? The concept of emergence is present under the guise of many different names and theories: complexity theory, chaos theory, self-organization theory, non-linear dynamical systems theory, synergetics, cybernetics, fractal sets, cellular automata, and so on. Emergent events are characterized by sensitive dependencies on *initial* conditions in combination with *evolving boundary* conditions. Generally, emergence accounts for the rise of global macroscopic *order* from local microscopic *randomness.* Both, top-down *and* bottom-up causal flows are implicated in the formation of an *emergent macroscopic structure* (Fig. 3). These causal flows are considered to be *relational* because vastly different levels in the hierarchy of organization are *actively interconnecting* without exclusive priority of one level over another (see legend to Fig. 3).

4.1. Determination without pre-determination: "effective indeterminism"
An important dimension in the development of an EmQM, i.e., for any theory which connects (classical) emergence theory with quantum mechanics, is the question of the inherently *probabilistic* nature of quantum phenomena. Crucially, in-principle *unpredictability*, as well as *uncontrollability*, of individual microscopic (quantum) events must be ensured by any kind of *non-orthodox* theory which claims success in reproducing the predictions that are yielded by *orthodox* quantum mechanics. Otherwise, for example, the *non-signalling theorem* of quantum mechanics would be instantly violated as we have discussed before at great length (Walleczek and Grössing [13, 14]). Critical in this context is that emergent phenomena are subject to unpredictability as a consequence of the *intrinsically self-referential* nature of the governing dynamics as illustrated in Fig. 3 (e.g., compare also the halting problem in computational theory). A well-known example is the phenomenon of *deterministic chaos*, which provides a vivid image of determination without predetermination, i.e., "effective indeterminism". Future work in EmQM foundations needs to clearly establish the limits and conditions under which such scenarios apply in alternative models for quantum phenomena, including for quantum nonlocality.

5. Outlook: new approaches in realist quantum mechanics
What are the prospects for an 'Emergent Quantum Mechanics'? It is possible – in principle – that the universe is deterministic, e.g., *nonlocally causal* in light of EPR-type nonlocal correlations. Yet – at the same time – even a deterministic universe can have an *open future* in the context of emergence theory, i.e., a future where both the free-choice performances of an observer/agent, and other physical processes in the cosmos, are *not pre-determined* by past events. As was explained in Sect. 4, emergent dynamical processes are well-known for being governed by entirely deterministic relations, and yet these very same processes can be *without* pre-determined outcomes in the future. As a consequence of the *intrinsically self-referential* nature of emergent phenomena, the *in-principle unpredictability* of *individual* microscopic events is granted. Whether such concepts might apply productively in a future quantum mechanics remains for now a promising *vision*. However, the resurgence of interest in *ontological* approaches to quantum mechanics, including those pioneered and envisioned by David Bohm [2, 3] and John Bell [8, 9, 15, 16] may further increase interest in the project of an EmQM (e.g., see also Bohm and Hiley [5]; Holland [17]).

In conclusion, a new wave of work has drawn attention to ontological, realist questions in quantum mechanics: Does the concept of 'nonlocality' reflect the true nature of reality? Is the quantum state real? Is the wave function ψ a reality? On the *theoretical* side, especially work by Harrigan and Spekkens [18] has renewed interest in ontological theory, including de Broglie–Bohm theory, by presenting the productive distinction between ψ-ontic and ψ-epistemic approaches to quantum mechanics. In that context, our own recent work showed that *nonlocal quantum information transfers*, which are *inevitably* associated with any ψ-ontic quantum theory, including Bohm's theory, need *not* violate the *non-signalling theorem* (Walleczek and Grössing [14]). On the *experimental* side, the important work by Kocsis *et al.* [19], Ringbauer *et al.* [20], and Mahler *et al. [21]*, has advanced fresh insight into the *non-orthodox* option of nonlocality as a *reality*, e.g., the reality of the wave function ψ. Finally, the most recent available findings provide a "compelling visualization" – as the authors put it – "of the nonlocality inherent in any realistic interpretation of quantum mechanics" (Mahler *et al. [21]*).

Acknowledgments
Work by Jan Walleczek at Phenoscience Laboratories (Berlin) is partially funded by the Fetzer Franklin Fund of the John E. Fetzer Memorial Trust. Work by Gerhard Grössing at the Austrian Institute for Nonlinear Studies (Vienna) is also partially funded by the Fetzer Franklin Fund of

the John E. Fetzer Memorial Trust. The authors wish to thank Siegfried Fussy, Johannes Mesa Pascasio, Herbert Schwabl and Nikolaus von Stillfried for their excellent contributions.

References

[1] Einstein A, Podolsky B and Rosen N 1935 Can quantum-mechanical description of physical reality be considered complete? *Phys. Rev.* **47**(10) 777–780
[2] Bohm D 1952 A suggested interpretation of the quantum theory in terms of "hidden" variables. I *Phys. Rev.* **85**(2) 166–179
[3] Bohm D 1952 A suggested interpretation of the quantum theory in terms of "hidden" variables. II *Phys. Rev.* **85**(2) 180–193
[4] Bell J S 1964 On the Einstein Podolsky Rosen paradox *Physics* **1**(3) 195–200
[5] Bohm D and Hiley B J 1993 *The Undivided Universe: An Ontological Interpretation of Quantum Theory* (London, UK: Routledge) ISBN 0-415-06588-7
[6] Mann C and Crease R 1988 Interview: John Bell, particle physicist *Omni* **10**(8) 84–92
[7] Petersen A 1963 The philosophy of Niels Bohr *Bulletin of the Atomic Scientists* **19**(7) 8–14
[8] Bell J S 1987 Speakable and unspeakable in quantum mechanics, *Speakable and Unspeakable in Quantum Mechanics* (Cambridge, UK: Cambridge University Press) pp 169–172
[9] Bell J S 2004 La nouvelle cuisine, *Speakable and Unspeakable in Quantum Mechanics* (Cambridge, UK: Cambridge University Press) pp 232–248 revised ed ISBN 978-0-511-81567-6
[10] Gross D 2014 A century of quantum mechanics, ed Struppa D C and Tollaksen J M *Quantum Theory: A Two-Time Success Story* (Milano: Springer) pp 3–8 ISBN 978-88-470-5216-1
[11] 't Hooft G 2007 Emergent quantum mechanics and emergent symmetries *AIP Conf. Proc.* **957**(1) 154–163 (*Preprint* arXiv:0707.4568)
[12] Walleczek J (ed) 2000 *Self-Organized Biological Dynamics and Nonlinear Control* (Cambridge, UK: Cambridge University Press) ISBN 0-521-62436-3
[13] Walleczek J and Grössing G 2014 The non-signalling theorem in generalizations of Bell's theorem *J. Phys.: Conf. Ser.* **504** 012001 (*Preprint* arXiv:1403.3588 [quant-ph])
[14] Walleczek J and Grössing G 2016 Nonlocal quantum information transfer without superluminal signalling and communication *Found. Phys.* In press. (*Preprint* arXiv:1501.07177 [quant-ph])
[15] Bell J S 1976 The theory of local beables *Epistemol. Lett.* **9** (3)(March) repr. in *Dialectica* **39** (1985) 85-96.
[16] Bell J S 1977 Free variables and local causality *Epistemol. Lett.* **15** (2)(February) repr. in *Dialectica* **39** (1985) 103-106.
[17] Holland P R 1993 *The Quantum Theory of Motion* (Cambridge, UK: Cambridge University Press) ISBN 0-521-35404-8
[18] Harrigan N and Spekkens R W 2010 Einstein, incompleteness, and the epistemic view of quantum states *Found. Phys.* **40**(2) 125–157
[19] Kocsis S, Braverman B, Ravets S, Stevens M J, Mirin R P, Shalm L K and Steinberg A M 2011 Observing the average trajectories of single photons in a two-slit interferometer *Science* **332**(6034) 1170–1173
[20] Ringbauer M, Duffus B, Branciard C, Cavalcanti E G, White A G and Fedrizzi A 2015 Measurements on the reality of the wavefunction *Nature Phys.* **11**(3) 249–254 (*Preprint* arXiv:1412.6213 [quant-ph])
[21] Mahler D H, Rozema L, Fisher K, Vermeyden L, Resch K J, Wiseman H M and Steinberg A 2016 Experimental nonlocal and surreal Bohmian trajectories *Science Advances* **2**(2) e1501466

Journal of Physics: Conference Series **701** (2016) 012002 doi:10.1088/1742-6596/701/1/012002

Bell's theorem and the measurement problem: reducing two mysteries to one?

Eric G. Cavalcanti

Centre for Quantum Dynamics, Gold Coast campus, Griffith University, Southport, Queensland 4215, Australia

E-mail: e.cavalcanti@griffith.edu.au

Abstract. In light of a recent reformulation of Bell's theorem from causal principles by Wiseman and the author, I argue that the conflict between quantum theory and relativity brought up by Bell's work can be softened by a revision of our classical notions of causation. I review some recent proposals for a quantum theory of causation that make great strides towards that end, but highlight a property that is shared by all those theories that would not have satisfied Bell's realist inclinations. They require (implicitly or explicitly) agent-centric notions such as "controllables" and "uncontrollables", or "observed" and "unobserved". Thus they relieve the tensions around Bell's theorem by highlighting an issue more often associated with another deep conceptual issue in quantum theory: the measurement problem. Rather than rejecting those terms, however, I argue that we should understand why they seem to be, at least at face-value, needed in order to reach compatibility between quantum theory and relativity. This seems to suggest that causation, and thus causal structure, are emergent phenomena, and lends support to the idea that a resolution of the conflict between quantum theory and relativity necessitates a solution to the measurement problem.

1. Introduction

> "For me then this is the real problem with quantum theory: the apparently essential conflict between any sharp formulation [of quantum theory] and fundamental relativity. That is to say, we have an apparent incompatibility, at the deepest level, between the two fundamental pillars of contemporary theory (...) It may be that a real synthesis of quantum and relativity theories requires not just technical developments but radical conceptual renewal."- J.S. Bell, 1986 [1]

Bell's famous 1964 theorem [2] has challenged our understanding of quantum theory and ignited much debate on its foundations as well as technological spin-offs such as device-independent quantum cryptography and randomness amplification. More than 50 years later, the debate still hasn't settled. As the quote above shows, Bell saw his work as demonstrating a fundamental incompatibility between quantum theory and relativity, and believed that a solution would require a "radical conceptual renewal". Here we take the point of view that at least part of that renewal regards our understanding of causality. In fact, Bell's theorem is best understood as being about the constraints from causal structure to possible correlations among events, where relativity comes in merely to supply the causal structure.

But who cares about causality? Some philosophers, like Bertrand Russell, have argued that it has no role in fundamental physics. In his famous 1913 essay "On the Notion of Cause", Russell that "The law of causality, I believe, like much that passes muster among philosophers, is a relic of a bygone age, surviving, like the monarchy, only because it is erroneously supposed to do no harm" [4]. For Russell, the time-asymmetric notions of cause and effect are not compatible with the deterministic and time-symmetric laws of physics, in which the future determines the past just as the past determines the future. Another point of view is defended by Nancy Cartwright [5]: "Bertrand Russell argued that laws of association are all the laws there are, and that causal principles cannot be derived from the causally symmetric laws of association. (...) Causal principles cannot be reduced to laws of association; but they cannot be done away with". For Cartwright, causal principles are required to distinguish between *effective* and *ineffective* strategies.

For Cartwright, effective action requires that agents represent their world in causal terms. This view is shared by philosophers such as Huw Price, although he argues [6] that nevertheless this position does not require realism about causation, that is, it doesn't require a commitment to causation and its associated agent-centric concepts as a fundamental aspect of the ontology. Rather, Price defends the view that causation emerges from an agent's perspective in the world, while proposing that the world itself may be describable in a view from nowhere, a block universe.

Despite the influence of Russell's causal eliminativist view in the philosophical literature, I believe that it is a minority view within physicists. The word "causality" certainly appears in a wide range of physics literature, often associated with the light-cone structure of relativity theory and the impossibility of *something* (information, physical systems, influences...) being transmitted between space-like separated regions. Specifically within discussions on Bell's theorem, the debate is usually about *what notion of causality* to use, rather than whether causality plays a role in Physics. The views in this realm can be broadly classified according to the holder's philosophical inclination as an *operationalist* or a *realist*. Here the conflict becomes a terminological dispute about what is the physically correct notion of *locality*, with different camps disagreeing about whether or not it is violated by quantum theory. Roughly speaking, the realist (like Bell) focuses on the importance of causal notions as the basis for the explanation of correlations, and believes that quantum correlations violate the causal structure imposed by relativity and is thus nonlocal. The operationalist, on the other hand, focuses on the fact that quantum correlations cannot be used to send signals faster than light and is thus local, and believes that it is Bell's classical notion of *local causality* that is to blame for the apparent conflict.

In a recent article [7], Howard Wiseman and the author have presented a reformulation of Bell's theorem in terms of causal notions, and proposed a way forward for the debate between operationalists and realists. In that formulation, Bell's theorem was shown to be derived from four Axioms and four Principles the definitions of which can be agreed upon by operationalists and realists alike, with the disagreement being merely about which of those to drop in light of the contradiction with observed phenomena. For the realist, it is relativistic causality that must go, to be supplemented perhaps by a preferred foliation of space-time such as that required by Bohmian mechanics. For the operationalist, it is the notion that a cause explains the correlation between two variables by rendering them uncorrelated. In our reformulation, this was accomplished by following an earlier proposal by the author and Ray Lal to break *Reichenbach's principle of common cause* into two logically separate principles [8].

Reichenbach's principle [9] is a fundamental tenet in the classical theory of causation. It states that if two events are correlated, then either one causes the other, or there exists a common cause such that conditioned on it, the events become uncorrelated. In [7] this principle was separated as a principle of *Common Causes*, requiring that two sets of events $\mathcal{A}$ and $\mathcal{B}$ are correlated only if one causes the other or there exists a set of common causes $\mathcal{C}$ that *explains* the correlation (and where the term *explains* is left to be defined by further principles); and a principle of *Decorrelating Explanation*, by which a set of causes $\mathcal{C}$ explains the correlation between $\mathcal{A}$ and $\mathcal{B}$

only if those variables are left uncorrelated by conditioning on C. By dropping the requirement of Decorrelating Explanation, the operationalist can thus keep relativistic causal structure.

But Reichenbach's principle is at the basis of the application of causal reasoning in the sciences. It is for example, a consequence of the basic assumptions behind the theory of causal networks (see e.g. the book of Pearl [10]). That theory has a wide range of applicability, and simply dropping Reichenbach's principle, even if only partly, would be "throwing out the baby with the bathwater". The challenge then becomes to generalise the classical theory of causal networks to accommodate quantum mechanics, in a way that reduces to the classical case in some appropriate limit. Recent work in quantum causal structures has followed that line of reasoning and proposed generalisations of the classical theory of causal networks into a quantum theory [11, 12, 13].

These important developments go a long way towards alleviating the "essential conflict" raised by Bell, and at this stage the operationalist would likely be satisfied that a resolution has been achieved. Here I will argue however, that there is still cause for concern. The reformulations so far seem to fall short of another of Bell's concerns, that the operationalist would do well not to ignore. That is, they seem to require an apparent agent-centric fundamental distinction (at least implicitly) between "controllable" and "uncontrollable" events, and indeed causal concepts are only applicable at the level of "measurements". This suggests that the resolution of the mystery of Bell's theorem so far has been a reduction to another foundational puzzle in quantum theory – the measurement problem. This raises the question of whether this is a necessary condition for the application of causal concepts that are compatible with quantum physics, and if so, does it point to causal concepts having an emergent nature?

This paper is structured as follows: In Sec. 2 I review the classical theory of causal networks and reformulate Bell's theorem within that language. In Sec. 3 I look at recent proposals for quantum theories of causal networks and argue that although they provide a reformulation of the concept of causal explanation that allows for compatibility with relativistic causal structure and reduces to the classical case in the appropriate limit, they all require agent-centric causal concepts. I end with a discussion about how this reduces the puzzle of Bell's theorem to the measurement problem, and points to the possibility that causal concepts are emergent.

2. Classical causal networks

> "Do we then have to fall back on "no signalling faster than light" as the expression of the fundamental causal structure of contemporary theoretical physics? That is hard for me to accept. For one thing we have lost the idea that correlations can be explained, or at least this idea awaits reformulation". J.S. Bell, "La Nouvelle Cuisine", in [3].

What does it mean to explain correlations? Suppose a rare virus infection spreads almost simultaneously in Australia and Brazil. Possible explanations epidemiologists would entertain include: a) The virus was carried from Australia to Brazil; b) The virus was carried from Brazil to Australia or c) The virus was carried to both places from a third country. This idea is encapsulated by Reichenbach's Principle of Common Cause (RPCC) [9]. Informally, it says that if two events are correlated, either one is a direct cause of the other, or they share a common cause. This principle is fundamental in many areas of science and underlies the important idea that correlation does not necessarily imply causation.

A more modern and general theory of causation and its role in explaining correlations can be found in the theory of causal networks [10]. It has been developed as a tool to connect causal inferences and probabilistic observations. For example, in developing a drug, a pharmaceutical company may want to distinguish, from some observed statistics, the effects of the drug from correlations that do not imply causation (e.g. because a subset of the population that is more likely to be cured without the drug is also more likely to use the drug in the first place). Here

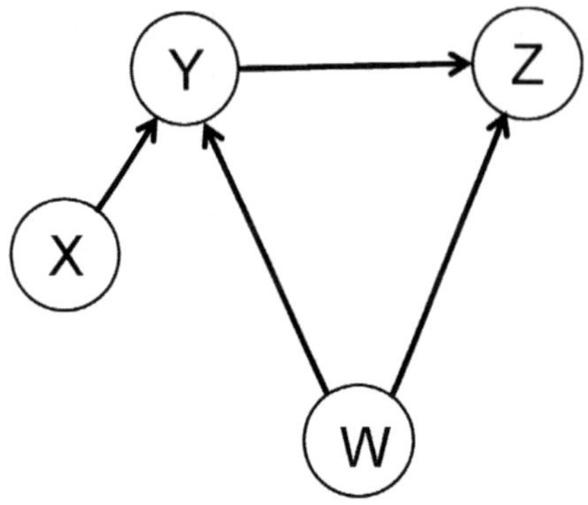

Figure 1. A directed acyclic graph (DAG). Nodes represent random variables associated with possible events and directed edges represent causal connections between events.

causal structure is encoded as a graph, with nodes representing random variables of interest and directed edges between nodes representing causal links. Since we want to exclude causal loops, the resulting structure is that of a *directed acyclic graph* (DAG). An example of a DAG is given in Fig. 1.

In a DAG G, the nodes that have arrows pointing to a given node X are denoted the *parents* of X. The set of all parents of parents (of parents, etc) of X is the set of *ancestors* of X. The *descendants* of X are all variables for which X is an ancestor.

A DAG can be seen as representing the conditional independences associated with any probability distribution $P(\cdot)$ over the variables in the graph that is compatible with that causal structure. We will use lower case letters to denote particular values of a random variable, e.g. $X = x$. Formally, we say that two sets of variables X and Y are *conditionally independent* in a probability distribution $P(\cdot)$ given a set of variables Z if and only if knowledge of Y provides no new information about X given that we already know Z, that is $P(x|y, z) = P(x|z)$. This is represented as $(X \perp\!\!\!\perp Y|Z)_P$.

Conditional independences satisfy the *graphoid axioms*:

Symmetry: $(X \perp\!\!\!\perp Y|Z)_P \iff (Y \perp\!\!\!\perp X|Z)_P$;
Decomposition: $(X \perp\!\!\!\perp YW|Z)_P \implies (X \perp\!\!\!\perp Y|Z)_P$;
Weak union: $(X \perp\!\!\!\perp YW|Z)_P \implies (X \perp\!\!\!\perp Y|ZW)_P$;
Contraction: $(X \perp\!\!\!\perp Y|ZW)_P \,\&\, (X \perp\!\!\!\perp W|Z)_P \implies (X \perp\!\!\!\perp YW|Z)_P$;
Intersection: $(X \perp\!\!\!\perp W|ZY)_P \,\&\, (X \perp\!\!\!\perp Y|ZW)_P \implies (X \perp\!\!\!\perp YW|Z)_P$

A DAG represents the conditional independences of a probability distribution in the following way. Every probability distribution P over n variables $X_1, X_2, ...X_n$ can be decomposed as a product of conditional distributions:

$$P(x_1, ..., x_n) = \prod_j P(x_j|x_1, ...x_{j-1}) \,. \tag{1}$$

Some of those conditional distributions will display independences, say for example $P(x_3|x_1, x_2) = P(x_3|x_1)$. The conditional distribution for a variable X_j will depend only on a subset of variables, the parents of X_j, denoted by PA_j. A DAG G represents P, or is compatible with P if and only if P admits a factorisation of the following form, where the parents PA_j are defined by the variables in G with directed edges terminating at X_j:

$$P(x_1, ..., x_n) = \prod_j P(x_j|pa_j)\,. \tag{2}$$

In classical causal models, one obtains the conditional independencies from the graph using a rule called *d-separation.* Two sets of variables X and Y are d-separated given a set of variables Z (denoted by $(X \perp\!\!\!\perp Y|Z)_d$) if and only if Z "blocks" all paths (sequences of consecutive edges) p from X to Y. A path p is said to be d-separated by Z if and only if (i) the path p contains a chain $i \to m \to j$ or a fork $i \leftarrow m \to j$ such that the middle node m is in Z, or (ii) the path p contains an inverted fork (head-to-head) $i \to m \leftarrow j$ such that the node m is not in Z, and there is no directed path from m to any member of Z.

The implication of d-separation for conditional independences is that if X and Y are d-separated by Z in a DAG G, then for all distributions P that are compatible with G, X is independent of Y given Z. That is, $(X \perp\!\!\!\perp Y|Z)_d \implies (X \perp\!\!\!\perp Y|Z)_P$ for all P compatible with G. In other words, d-separation is a sound criterion for conditional independence. Furthermore, if for all distributions P compatible with G, $(X \perp\!\!\!\perp Y|Z)_P$, then $(X \perp\!\!\!\perp Y|Z)_d$. In other words, d-separation is a complete criterion for conditional independence.

The conditional independencies obtained from d-separation satisfy the following property:

Causal Markov Condition: In any probability distribution P that is compatible with a graph G, A variable X is independent of all its non-descendants, conditional on its parents.

The causal Markov Condition implies a weaker property that we call *Causal Completeness*:

Causal Completeness: A variable X is independent of its non-descendants, conditional on its ancestors.

The reason for considering this weaker notion is that in theories of causal networks beyond the classical case, it may not always be convenient to assume that the parent nodes screen off causal influences from more distant ancestors (see e.g. [13]). With the interpretation that the parents of a variable X are its direct causes, and the ancestors are all of its possible causes, this principle states that an event is independent of its non-effects given its causes. Causal Completeness implies Reichenbach's Principle of Common Cause, with the common causal past of two events X and Y interpreted as the intersection of their respective sets of ancestors.

2.1. Causal networks and Bell's theorem

We are now ready to formulate Bell's theorem in this language. Consider the usual Bell scenario, with Alice and Bob in two space-like separated regions, performing experiments on two entangled quantum systems. Their respective choices of experiment are given by variables X and Y, with the associated outcomes represented by A and B. Assuming relativistic causal structure, there cannot be any direct causal connection between events at Alice's lab and events at Bob's lab. There could, however, be some common causes Λ in their common past light cone (for example, variables associated with the preparation of the entangled quantum state). We assume that X and Y are free variables, meaning that there are no arrows going into the causal graph representing the situation. Given these assumptions, the scenario is represented by the graph in Fig. 2.

The above considerations allow us to determine the set of probability distributions P that are compatible with this graph. We can always decompose the distribution over the random variables in question as

$$P(a, b, x, y, \lambda) = P(a|b, x, y, \lambda)P(b|x, y, \lambda)P(\lambda|x, y)P(x|y)P(y)\,. \tag{3}$$

Causal Completeness implies that $P(a|b, x, y, \lambda) = P(a|x, \lambda)$, $P(b|x, y, \lambda) = P(b|y, \lambda)$, $P(\lambda|x, y) = P(\lambda)$ and $P(x|y) = P(x)$, and so

$$P(a, b, x, y, \lambda) = P(a|x, \lambda)P(b|y, \lambda)P(\lambda)P(x)P(y)\,. \tag{4}$$

Journal of Physics: Conference Series **701** (2016) 012002 doi:10.1088/1742-6596/701/1/012002

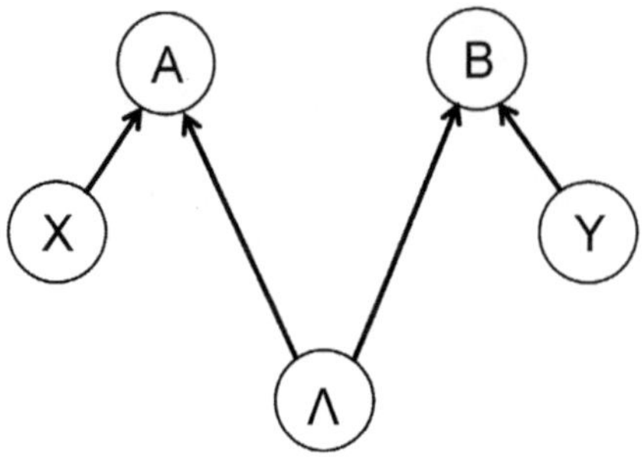

Figure 2. A directed acyclic graph (DAG) representing the causal structure of a Bell correlation scenario, under the assumption of relativistic causality.

Averaging the conditional probabilities $P(a, b, \lambda|x, y) = P(a, b, x, y, \lambda)/P(x, y)$ over λ we thus obtain the usual factorisability condition of a Local Hidden Variable model:

$$P(a, b|x, y) = \sum_{\lambda} P(\lambda)P(a|x, \lambda)P(b|y, \lambda)\,, \tag{5}$$

and as is well known, this leads to the Bell inequalities, which are violated by certain quantum correlations. Therefore, assuming relativistic causal structure and free choices, quantum correlations cannot be reproduced by the classical theory of causation, which assumes Causal Completeness.

3. Quantum causal networks

An alternative to resolve the conflict while maintaining relativistic causality is to reject some of the principles underlying the classical theory of causal networks, including Causal Completeness. Several recent works have proposed generalisations of the theory of causal networks to accommodate quantum correlations [11, 12, 13]. In the following I will argue that they raise an interesting puzzle – the apparent need for agent-centric notions such as controllable/uncontrollable or observed/unobserved in quantum theories of causal structure.

Pienaar and Brukner (PB) [13] developed a theory based on a DAG with two different types of nodes, corresponding to events that are controllable "settings" and those that are observable but not directly controllable "outcomes". Based on this they proposed a theory of quantum causal models that satisfies a:

Quantum Causality Condition: An outcome is independent (conditional on the empty set) of all settings that are not its causes and all outcomes that do not share a common cause.

Note that unlike the case of Causal Completeness, here a fundamental distinction is made between settings and outcomes. Underlying this condition there is an associated notion of 'q-separation' that generalises the notion of d-separation:

> Given a DAG representing a quantum network, two disjoint sets of variables X and Y are said to be *q-separated* by a third disjoint set Z, denoted $(X \perp\!\!\!\perp Y|Z)_q$, if and only if every undirected path between X and Y is rendered inactive by a member of Z. A path connecting two variables is rendered inactive by Z if and only if at least one of the following conditions is met:
>
> (i) both variables are settings, and at least one of the settings has no directed path to any outcome in Z;
>
> (ii) one variable is a setting and the other is an outcome, and there is no directed path from the setting to the outcome, or to any outcome in Z;

(iii) the path contains a collider $i \to m \leftarrow j$ where m is not an outcome in Z, and there is no directed path from m to any outcome in Z.

The authors then proceed to show that q-separation is a sound and complete criterion for conditional independences within their notion of quantum causal models.

Another framework is given by Henson, Lal and Pusey (HLP) [12], who keep d-separation, but at the expense of introducing another form of agent-centric concepts: "observed" and "unobserved" nodes. Classical data, such as outcomes of measurements, are associated to observed nodes. Unobserved nodes output only systems, and have no outcomes associated to them. HLP define a notion of a distribution P being *generalized Markov* with respect to a graph G, with reference to this distinction. They also show that their theory reduces to the classical theory if all nodes are observed, in which case a generalized Bayesian network is a classical Bayesian network.

Here we note that the work of Fritz [11] provides a framework that does not make a distinction between different types of nodes. However, no discussion is given there for a substitute for d-separation or the Causal Markov Condition, and it is not clear therefore whether that framework can accommodate a notion of quantum causality that is sound and complete for the conditional independences implied by quantum phenomena while not making agent-centric distinctions such as the two frameworks discussed above.

Another discussion of causal concepts underlying Bell's theorem and further motivation for the idea that agent-centric notions are required to maintain compatibility with relativistic causality is given in a recent paper of Wiseman and the author [7].

4. Who do we think "we" are?

> "More importantly, the 'no signalling...' notion rests on concepts which are desperately vague, or vaguely applicable. The assertion that 'we cannot signal faster than light' immediately provokes the question: Who do we think *we* are? *We* who can make measurements, *we* who can manipulate external fields, *we* who can signal at all, even if not faster than light? Do *we* include chemists, or only physicists, plants, or only animals, pocket calculators, or only mainframe computers?" - J.S. Bell, "La Nouvelle Cuisine", in [3].

The theories of quantum causal networks discussed in the previous section are important developments in the elucidation of Bell's theorem, and arguably provide at least a partial answer to the question of what counts as a causal explanation of correlations in the quantum case, an idea that "awaited reformulation" according to Bell, as quoted in the beginning of Sec. 2.

However, the frameworks above seem to indicate that we can recover causal principles that conciliate quantum mechanics and relativistic causality, but at the expense of introducing agent-centric notions of causation. For an operationalist, this doesn't seem to be a very high price to pay. However, I would like to suggest that there's cause for further consideration.

As the quote above shows, Bell had two concerns regarding the usage of "no signalling" as "the expression of fundamental causal structure of contemporary theoretical physics". One concern, addressed by quantum causal networks, is the reformulation of the idea of causal explanation, generalising Reichenbach's principle of common cause, which was implicit in Bell's thought. The other concern, as the passage above shows, is that the concept of "no signalling" relies on agent-centric notions, and therefore should not form part of a fundamental physical theory, no more than the concept of "measurement".

Again, this may be an arcane concern for an operationalist. But remember, that if the operationalist had never paid attention to the realist point of view, they would never have realised that there was a whole theory of causation that needed to be modified to accommodate quantum theory.

Now I don't entirely agree with Bell either, mind you. Clearly we *can* talk about "settings" and "outcomes" rather unambiguously, and it's useful to do so. This of course, does not require anthropocentrism – computers and measurements devices are good enough "agents" for defining what we mean by those terms, without the need to involve human beings. But this might be hinting, at least to my mind, that a complete conciliation between relativity and quantum theory will require a deeper understanding of the role these agent-centric notions play in the theory. And here again, I would urge the operationalist to heed the lessons learned from the realist. Realists are not willing to accept agent-centric notions as fundamental concepts in a physical theory. If these causal concepts are stubbornly agent-centric, for a realist that means that they ought to be emergent properties. If that's the case, we must understand how and why they emerge. It's a speculation, but I think it's one worth considering.

Maybe one way of seeing all this is that these new developments have gone to great lengths towards solving the mystery of Bell's theorem, towards the conceptual renewal that Bell sought in order to reconcile relativistic causal structure and quantum theory. However, the efforts so far have done so only at the expense of forcing us to face up to the other major conceptual puzzle of quantum theory: the measurement problem. Which is a great advance: perhaps we are reducing two mysteries to one!

Going further, it would be interesting to determine whether Causal Completeness can be maintained together with relativistic causal structure, perhaps with a modification of the concept of independence, or relaxing some of the implicit or explicit assumptions of the formalism, e.g. that the events considered are part of a single relativistic space-time manifold – would a multiverse theory allow for fundamental causal explanation free of agent-centric terms? Or perhaps it could provide a story about how those notions emerge from a more fundamental background?

Acknowledgments

The author would like to thank Howard Wiseman, Sally Shrapnel and Jacques Pienaar for useful discussions. This publication was made possible through the support of a grant (FQXi-RFP-1504) from the Foundational Questions Institute Fund (fqxi.org) at the Silicon Valley Community Foundation, and Australian Research Council projects DE120100559 and DP140100648.

References

[1] Bell J S 1986 *Phys. Rep.* **137** 7; reprinted in [3], ch. 18
[2] Bell J S 1964 *Physics* **1** 195–200
[3] Bell J S 1987 *Speakable and unspeakable in quantum mechanics* (Cambridge University Press)
[4] Russell B 1912 *Proc. Aristotelian Soc.* **13** 1–26
[5] Cartwright N 1979 *Noûs* **13** 419–437
[6] Price H 2007 *Preprint* `http://philsci-archive.pitt.edu/4487/`
[7] Wiseman H M and Cavalcanti E G 2015 *Preprint* `arXiv:1503.06413 [quant-ph]`
[8] Cavalcanti E G and Lal R 2014 *J. Phys. A: Math. Theor.* **47** 424018
[9] Reichenbach H 1956 *The direction of time* (Berkeley: University of Los Angeles Press)
[10] Pearl J 2000 *Causality: Models, Reasoning and Inference* (Cambridge University Press)
[11] Fritz T 2014 *Preprint* `arXiv:1404.4812 [quant-ph]`
[12] Henson J, Lal R and Pusey M F 2014 *New J. Phys.* **16**, 113043
[13] Pienaar J and Brukner C 2015 *New J. Phys.* **17** 073020

Bohmian Conditional Wave Functions (and the status of the quantum state)

Travis Norsen
Physics Dept., Smith College, Northampton, MA 01063, USA
E-mail: tnorsen@smith.edu

Abstract. The de Broglie - Bohm pilot-wave theory – uniquely among realistic candidate quantum theories – allows a straightforward and simple definition of the wave function of a sub-system of some larger system (such as the entire universe). Such sub-system wave functions are called "Conditional Wave Functions" (CWFs). Here we explain this concept and indicate the CWF's role in the Bohmian explanation of the usual quantum formalism, and then develop (and motivate) the more speculative idea that something like single-particle wave functions could replace the (ontologically problematical) universal wave function in some future, empirically adequate, pilot-wave-type theory. Throughout the presentation is pedagogical and points are illustrated with simple toy models.

1. Introduction

One of the central themes of the workshop was assessing the status (ontological or otherwise) of the quantum state. And, as is always the case in quantum foundations, questions about Bell's theorem and nonlocality were not far below the surface.

The pilot-wave theory of de Broglie and Bohm (a.k.a. "Bohmian Mechanics") is perhaps best known as a realistic model in which quantum non-locality is manifest. This fact is often cited, by critics, as a justification for their negative assessment of the theory. But in fact such an attitude rests on a misunderstanding of Bell's theorem. Dynamical non-locality is not the price one pays for attempting to restore determinism or "realism" to quantum theory, but is instead a necessary feature of any genuine physical explanation of the empirically-observed correlations in EPR-Bell-type experiments. That is what Bell's theorem shows, and in that sense non-locality is simply a fact of nature that we must accept. [1]

But there are actually two very different notions of "non-locality" that we should dis-entangle. One sense is the *dynamical* sense (in which "non-locality" means, roughly, "faster-than-light causal influences"). This is what I meant in the previous paragraph. But the de Broglie - Bohm pilot-wave theory also exhibits the second, more *ontological*, sense of "non-locality." In particular, the (universal) wave function – which is certainly *real* according to the pilot-wave theory – is not a local object (such as something that exists at, or assigns properties to, points or regions of three-dimensional physical space or 4-dimensional space-time). It is, instead, perhaps something like a dynamical field (obeying a kind of wave equation) – but a field that lives on an abstract high-dimensional configuration space. To use Bell's terminology, the wave function may be (for the pilot-wave theory) a "beable", but it is not a "local beable". It is, rather, a "non-local beable."

Figure 1. The familiar world of ordinary perception is, according to Bohmian Mechanics, made of the particles (whose motions are choreographed by, but which are not the same thing as, the wave function Ψ). It is a happy and comfortable world in which one can find things like butterflies and unambiguously-alive cats.

These two senses of "non-locality" are not entirely unrelated. For example, some authors have suggested that a field on configuration space is natural and appropriate for a dynamically-nonlocal theory, and/or that in some sense the theory is (dynamically) nonlocal *because* the wave function lives on configuration space. [2] On the other hand, it is certainly possible to have a theory that contains exclusively local beables but which is nonlocal in the dynamical sense (think Newtonian gravity), or vice versa. So, conceptually, it is possible (and, I think, clarifying) to distinguish the two notions.

In any case, it is the specifically ontological notion of "non-locality" that we will focus on here. After briefly reviewing, in Section 2, the standard formulation of the pilot-wave theory (in which the universal wave function seems to have the status of a "non-local beable"), we introduce, in Section 3, the Bohmian Conditional Wave Function (CWF) and highlight some of its intriguing properties, especially *vis-a-vis* measurement. Section 4 briefly explains how the CWF can be used as part of a rigorously-derivable semi-classical approximation scheme, and then Section 5 introduces a related (so-called "Bohmian Double Semi-Quantum", BDSQ) approximation scheme. Finally, in Section 6, we discuss how something like the BDSQ could perhaps be a useful jumping-off point for producing a (dynamically non-local) theory of exclusively local beables.

2. Bohmian Mechanics

According to Bohmian Mechanics, a complete description of the physical state of the universe involves the (universal) wave function $\Psi = \Psi(\vec{x}_1, \vec{x}_2, ..., \vec{x}_N, t)$ as well as the configuration $Q(t) = \left\{\vec{X}_1(t), \vec{X}_2(t), ..., \vec{X}_N(t)\right\}$ of N particles. In the simple case of non-relativistic, spinless particles the wave function evolves in accordance with Schrödinger's equation

$$i\hbar\frac{\partial\Psi}{\partial t} = \sum_{i=1}^{N} \frac{-\hbar^2}{2m_i}\vec{\nabla}_i^2\Psi + V(\vec{x}_1, ..., \vec{x}_N)\Psi. \tag{1}$$

The particle positions evolve under the influence of the wave function as follows:

$$\frac{d\vec{X}_i}{dt} = \frac{\hbar}{m_i}\mathrm{Im}\left(\frac{\vec{\nabla}_i\Psi}{\Psi}\right)\bigg|_{\vec{x}_j=\vec{X}_j\,\forall\, j}. \tag{2}$$

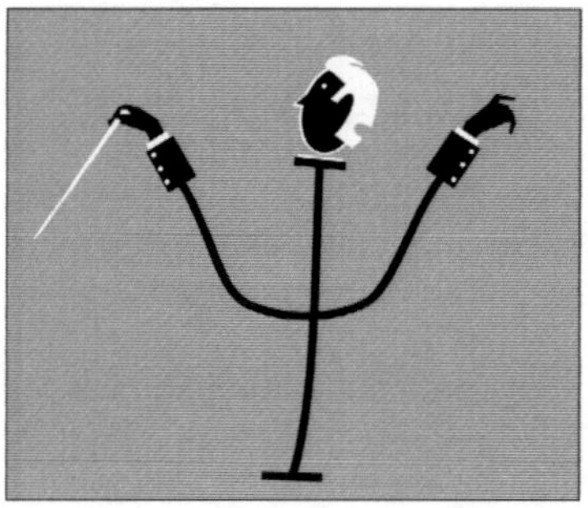

Figure 2. The wave function of the universe (artist's conception).

Note in particular that those two equations completely define the dynamics of the theory, and they apply exceptionlessly. Unlike in textbook/orthodox quantum mechanics, for example, there are no special provisions for violations of the usual dynamical evolution during "measurements."

How should one think about "physical reality" according to the pilot-wave theory? To begin with, it is crucial to appreciate that ordinary material objects (like tables, trees, cats, planets, and pieces experimental measuring equipment) are made of the particles. The particles are (in Bell's terminology) "local beables" and function (in the terminology of Goldstein *et al.*) as the "primitive ontology" of the theory. [3] The wave function is also real, according to the theory, but is by contrast not a "local beable" and not part of the theory's "primitive ontology". Unlike the particles, Ψ is a spooky, invisible, ethereal object which, so to speak, lurks offstage and orchestrates the motion of the particles according to Equation (2). These two components of the theory's posited ontology are illustrated in Figures 1 and 2, respectively.

The greatest virtue of Bohmian Mechanics is the simplicity with which one can understand, in terms of the posited local beables (namely, particles with definite positions), the emergence of a familiar macroscopic world of three dimensional objects. This is what's really illustrated in Figure 1. And in so far as one regards the theory's *dynamical* non-locality as a necessary feature (rather than a bug), the theory's only vice would seem to be that it is very difficult to understand what kind of thing, exactly, the universal wave function Ψ is supposed to be. As a "non-local beable", it does not in any straightforward sense exist in ordinary physical space, and yet it influences the particles in physical space. How does that work? And what does it even mean for a physically real thing to not exist in physical space? Is there some other space where it does exist? Physically? Questions of this sort have led some to suggest that Ψ is not a "thing" at all, but is instead more like a "law". [4]

Our ultimate goal here is to propose an alternative possibility. But first we must put some technical groundwork in place.

3. Bohmian Conditional Wave Functions

It is important to appreciate that the Ψ which appears in Equation (1) is the wave function of the universe. This is not the wave function one normally deals with in quantum mechanics, which is always the wave function for some small system which is (say) later to be *measured* by means of an appropriate interaction with some other object (a measuring device, say) outside the small system of interest. So it is not immediately obvious how to relate the axioms of Bohmian Mechanics to textbook quantum mechanics.

As it turns out, though, Bohmian Mechanics does allow one to fully understand – to derive – the textbook QM formalism, including even the infamous collapse postulate. To see this, notice that Bohmian mechanics provides a natural (and, among candidate quantum theories, unique)

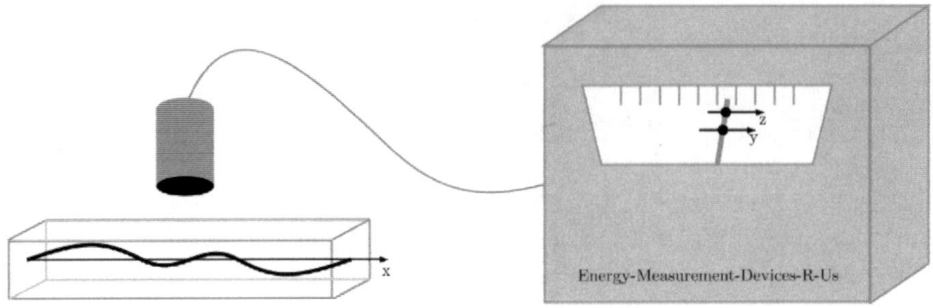

Figure 3. The energy of a "particle-in-a-box" is measured, and the outcome is registered in the final position of the particles composing the pointer.

way of defining the wave function of a sub-system. Consider, for example, a particular particle with degree of freedom x. We can divide the generic configuration point $q = \{x, y\}$, where y now denotes the coordinates of *other* particles, i.e., degrees of freedom from outside our sub-system of interest. We then define the conditional wave function (CWF) for our particle as follows:

$$\phi(x,t) = \Psi(q,t)|_{y=Y(t)} = \Psi\left[x, Y(t), t\right]. \tag{3}$$

That is, the CWF for one particle is simply the universal wave function, evaluated at the actual positions $Y(t)$ of the *other* particles.

Note that the "guidance equation" (according to which the wave function "orchestrates" the motion of the particles) for each particle can be re-expressed in terms of that particle's CWF

$$\frac{d\vec{X}}{dt} = \frac{\hbar}{m}\mathrm{Im}\left(\frac{\vec{\nabla}\phi(x,t)}{\phi(x,t)}\right)\Bigg|_{x=X_j(t)} \tag{4}$$

so that the CWF can be understood as having precisely the dynamical significance, for the associated particle, that one perhaps normally imagines to be involved in a "pilot-wave theory".

It is fairly straightforward to see that, in the case where the particle in question is suitably decoupled from its environment, the associated CWF will obey the obvious single-particle Schrödinger equation that one would have expected from the point of view of textbook quantum mechanics.

The more interesting and surprising thing is that, in the kind of case that we would normally describe as a *measurement*, the CWF *collapses* according to the usual textbook rule, even though nothing like a "collapse postulate" appears in the basic axioms of the theory. Let us illustrate with a simple toy model, in which the quantum system is a one-dimensional "particle-in-a-box" and the energy-measurement apparatus is treated very schematically. In particular, we explicitly track just the horizontal positions of two particles in the apparatus pointer – see Figure 3 – which are taken to interact with the particle-in-a-box according to the interaction Hamiltonian

$$\hat{H}_{int} = \lambda \hat{H}_x(\hat{p}_y + \hat{p}_z) \tag{5}$$

where $\hat{H}_x$ is just the energy of the PIB, and $\hat{p}_x$ and $\hat{p}_y$ are the momentum operators for the pointer particles. (The total Hamiltonian includes not only $\hat{H}_{int}$ and $\hat{H}_x$ but also kinetic energy terms for the two pointer particles.) The interaction is such that, if the initial quantum state is

$$\Psi(x,y,z,0) = \psi_n(x)G_0(y)G_0(z) \tag{6}$$

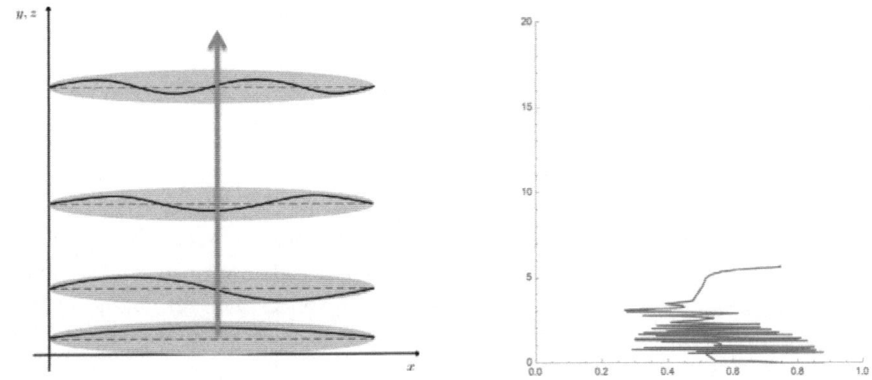

Figure 4. The left panel indicates the evolution of the wave function $\Psi(x, y)$ in configuration space – the z coordinate is suppressed here – while the right panel indicates the evolution of the actual particle configuration $(X(t), Y(t))$ for a random initial condition $(X(0), Y(0))$. Note that the final pointer position $Y(t) \approx 6$ is in the support of the $n = 2$ "branch" of Ψ and so we would say, in this case, that the energy measurement had outcome E_2.

(where ψ_n is an eigenstate of $\hat{H}_x$ with eigenvalue E_n and G_0 is a Gaussian wave packet centered on zero) the final state involes (pseudo-) Gaussian wave packets that have moved, to the right, by a distance proportional to E_n:

$$\Psi(x, y, z, t) = \psi_n(x) e^{-iE_n t/\hbar} G_t(y - \lambda E_n t) G_t(z - \lambda E_n t). \tag{7}$$

It then follows from the linearity of the Schrödinger equation that, if instead the PIB begins in a superposition of energy eigenstates

$$\Psi(x, y, z, 0) = \left(\sum_n c_n \psi_n(x) \right) G_0(y) G_0(z) \tag{8}$$

the final state is the entangled superposition

$$\Psi(x, y, z, t) = \sum_n c_n \psi_n(x) e^{-iE_n t/\hbar} G_t(y - \lambda E_n t) G_t(z - \lambda E_n t). \tag{9}$$

From the orthodox perspective (in which one regards the wave function itself as somehow describing or corresponding to physical reality) this last equation reflects the so-called measurement problem. But from the Bohmian point of view (in which it is the particles, not the wave function, which are supposed to correspond to the observable material world) there is no such problem. The particles move in some particular way and, in particular, the particles composing the pointer end up in some particular location which we interpret as indicating the (totally unambiguous and unproblematic and actual) outcome of the experiment. See Figure 4.

But then consider the time-evolution of the CWF of the particle-in-the-box. At $t = 0$, when the quantum state factorizes, the CWF is just (proportional to) the energy-superposed initial state of that particle: $\phi(x, 0) \sim \sum_n c_n \psi_n(x)$. Then, as the quantum state begins to separate into branches, the CWF evolves in a very complicated non-linear way. But once the branches have ceased to overlap appreciably in configuration space – see the left panel of Figure 4 – the CWF settles down into one of the energy eigenstates $\psi_n(x)$. For example, for the particular initial particle positions illustrated in the Figure (with the final pointer position $Y(T)$ in the support of the $n = 2$ branch of Ψ), it is clear that

$$\phi_1(x,T) = \Psi\left[x, Y(T)\right] \sim \psi_n(x). \tag{10}$$

So the conditional wave function has "collapsed," as a result of the interaction with the energy measuring device, from a superposition of many energy eigenstates, to the one particular energy eigenstate that corresponds to the actual outcome of the experiment. (Some video animations of the particle-in-a-box's CWFs evolving – i.e., "collapsing", but in a perfectly continuous way – can be found in the supplementary materials.) This is the sense in which the pilot-wave theory *explains* the collapse postulate, and in general the rules that are instead postulates for ordinary QM. Although, according to the pilot-wave theory, the "big" (i.e., universal) wave function always evolves in accordance with the Schrödginer equation (and hence never collapses) the "small" wave functions associated with sub-systems obey a sub-system Schrödinger equation when (according to ordinary QM) they should, but also collapse when (according to ordinary QM) they should.

4. Bohmian Semi-Classical Techniques

In addition to helping us understand how, from the Bohmian pilot-wave point of view, the usual textbook quantum formalism can be understood to *emerge*, naturally, from a set of simple and unambiguous axioms, CWFs are also relevant to understanding and justifying semi-classical approaches. Let's discuss this in terms of a simple toy-model system: two interacting one-dimensional simple harmonic oscillators. The total Hamiltonian will be

$$\hat{H} = \left(-\frac{\hbar}{2m}\frac{\partial^2}{\partial x^2} + \frac{1}{2}m\omega^2 x^2\right) + \left(-\frac{\hbar}{2M}\frac{\partial^2}{\partial y^2} + \frac{1}{2}m\omega^2 y^2\right) + \lambda xy. \tag{11}$$

If (say) $M \gg m$ we might treat the system semi-classically, with the wave function $\psi(x,t)$ of the ("light") quantum system obeying

$$i\hbar\frac{\partial\psi}{\partial t} = -\frac{\hbar}{2m}\frac{\partial^2\psi}{\partial x^2} + \frac{1}{2}m\omega^2 x^2\psi + \lambda x Y\psi \tag{12}$$

with a contribution from the ("heavy") classical system whose position $Y(t)$ obeys

$$\ddot{Y} + \omega^2 Y + \frac{\lambda}{M}\mathcal{X} = 0. \tag{13}$$

In the usual ("orthodox") semi-classical treatment

$$\mathcal{X} = \langle\psi|\hat{x}|\psi\rangle \tag{14}$$

is the "quantum average position" of the x-system.

In the pilot-wave theory, though, we have the resources to let the x-system influence $Y(t)$ via the *actual*, Bohmian position, that is

$$\mathcal{X} = X(t) \quad \text{satisfying} \quad \frac{dX}{dt} = \frac{\hbar}{m}\text{Im}\left(\frac{\nabla\psi}{\psi}\right)\Big|_{x=X}. \tag{15}$$

One anticipates that, in situations where the "quantum average position" is very different from any of the individual quantum possibilities (e.g., there is a scattering event, with a transmitted packet and a reflected packet, and so the average position is nowhere near any of the possible positions) the uniquely Bohmian semi-classical technique should yield more accurate results. This is indeed borne out by some preliminary studies, although this issue has not yet received the attention it probably deserves. [6, 7, 8, 9]

Note also that, whereas the orthodox semi-classical approach can only really be motivated in a somewhat hand-wavy way, it is possible to give a more rigorous derivation of the uniquely Bohmian semi-classical (BSC) approximation. To begin with, the full Bohmian description of the two-particle system involves the Schrödinger equation

$$i\hbar\frac{\partial\Psi(x,y,t)}{\partial t} = \left[\frac{-\hbar^2}{2m}\nabla_x^2 + \frac{-\hbar^2}{2M} + V(x,y)\right]\Psi(x,y,t) \tag{16}$$

as well as the guidance equations for the positions of the two particles:

$$\frac{dX}{dt} = \frac{\hbar}{m}\mathrm{Im}\left(\frac{\nabla_x\phi}{\phi}\right)\Big|_{x=X(t)} \tag{17}$$

(where ϕ is the CWF of the x-system) and

$$\frac{dY}{dt} = \frac{\hbar}{M}\mathrm{Im}\left(\frac{\nabla_y\Psi}{\Psi}\right)\Big|_{x=X(t),y=Y(t)}. \tag{18}$$

As is reasonably well-known, if we take a time derivative of this last guidance equation, it can be re-expressed in 2nd order form:

$$M\ddot{Y} = -\nabla_y\left(V+Q\right)|_{x=X(t),y=Y(t)} \tag{19}$$

where Q is the so-called "quantum potential". When (the gradient of) Q is negligible compared (the gradient of) V, we arrive at the semi-classical equation with the Bohmian-type feedback, i.e., Equation (13) with $\mathcal{X} = X(t)$.

It is then also possible to show that, as a consequence of the "big" wave function Ψ obeying Equation (16), the CWF ϕ satisfies the pseudo-Schrödinger equation

$$i\hbar\frac{\partial\phi(x,t)}{\partial t} = \frac{-\hbar^2}{2m}\nabla_x^2\phi(x,t) + V[x,Y(t)]\phi(x,t) + i\hbar\frac{dY}{dt}\frac{\partial\Psi}{\partial y}\Big|_{y=Y(t)} + \frac{-\hbar^2}{2M}\frac{\partial^2\Psi}{\partial y^2}\Big|_{y=Y(t)}. \tag{20}$$

This makes it clear that, when the last two terms on the right hand side can be ignored (i.e., roughly, if M is large and dY/dt is small) then we have exactly the semi-classical dynamical equation for the quantum half – a Schrödinger equation in which the particle is influenced by the "classical" position Y of the other particle through the interaction potential.

Thus, not only does the pilot-wave theory afford a somewhat novel version of the semi-classical treatment (which should and seems to work better than standard approaches in at least some cases), it also allows a more rigorous derivation and hence a better and more explicit understanding of when and why a semi-classical treatment might be appropriate. And it does this, in part, because the pilot-wave theory allows this very natural definition of "the wave function of a sub-system", the CWF.

5. The Bohmian Double Semi-Quantum (BDSQ) Approximation

The last section described the Bohmian approach to breaking a compound system apart into a quantum part and a classical part. It is interesting to note, though, that one could also break a compound system apart and treat both parts in a quantum way, with influences on each sub-system occuring via the Bohmian position of the particle in the other system, as in the

Journal of Physics: Conference Series **701** (2016) 012003 doi:10.1088/1742-6596/701/1/012003

BSC treatment. I call this the "Bohmian Double Semi-Quantum" (or BDSQ) approximation scheme, and the idea centers around the Bohmian CWF. As noted in the previous section, for a two-particle system whose ("big") wave function Ψ obeys the Schrödinger equation, the CWF for particle 1 will obey

$$i\hbar\frac{\partial\phi_1(x,t)}{\partial t} = -\frac{\hbar}{2m_1}\frac{\partial^2\phi_1}{\partial x^2} + V[x,Y(t)]\phi_1 + i\hbar\frac{dY}{dt}\frac{\partial\Psi}{\partial y}\bigg|_{y=Y(t)} - \frac{\hbar}{2m_2}\frac{\partial^2\Psi}{\partial y^2}\bigg|_{y=Y(t)} \tag{21}$$

and similarly the CWF for particle 2 will obey

$$i\hbar\frac{\partial\phi_2(y,t)}{\partial t} = -\frac{\hbar}{2m_2}\frac{\partial^2\phi_2}{\partial y^2} + V[X(t),y]\phi_2 + i\hbar\frac{dX}{dt}\frac{\partial\Psi}{\partial x}\bigg|_{x=X(t)} - \frac{\hbar}{2m_1}\frac{\partial^2\Psi}{\partial x^2}\bigg|_{x=X(t)}. \tag{22}$$

And of course we also have the particle positions $X(t)$ and $Y(t)$, evolving under the influence of their respective CWFs. Dropping the last two terms in both of the previous two equations gives a simple, closed system of two one-particle Bohmian systems (i.e., particles moving under the influence of a pilot-wave which in turn evolves according to a Schrödinger-type equation) which interact: the position of each particle affects the evolution of the other particle's pilot-wave via the interaction potential.

Note that this scheme will exactly reproduce the ordinary Bohmian particle trajectories (and hence the empirical predictions of ordinary quantum mechanics) if the two particles are (and remain!) unentangled. (In this case, the "big" wave function Ψ is a product, and so each particle's Bohmian CWF is simply proportional to the state it would be assigned in ordinary QM.) So this is in some sense a "small entanglement approximation" (SEA). [10]

I illustrate this here with the simple example of the two interacting 1-dimensional SHOs mentioned before. This system is nice because it can be solved exactly. For example, if $\Psi(x,y,0) = \psi_1(x)\psi_0(y)$ (i.e., particle 1 is in the first excited state, $n = 1$, while particle 2 is in the ground state, $n = 0$), the exact solution of Schrödinger's equation is

$$\Psi(x,y,t) = \cos\left(\frac{\lambda t}{\hbar\omega}\right)\psi_1(x,t)\psi_0(y,t) - i\sin\left(\frac{\lambda t}{\hbar\omega}\right)\psi_0(x,t)\psi_1(y,t). \tag{23}$$

That is, the system oscillates back and forth between the initial state and the state in which the one quantum of energy is instead held by particle 2.

If we treat the system using the BDSQ approximation scheme, the wave function of each particle can be written as in, for example,

$$\phi_1(x,t) = a_0(t)\psi_0(x,t) + a_1(t)\psi_1(x,t). \tag{24}$$

To solve the system numerically, we must track, for particle 1, $a_0(t)$, $a_1(t)$ and of course $X(t)$, and then three similar degrees of freedom for particle 2. See Figure 5 for graphs showing the results from this simple example.

One sees that the BDSQ approach reproduces the qualitative, oscillatory behavior of the exact fully-quantum solution, but that the details (such as the period of the oscillations) are not exactly correct. It is of course not surprising that the calculations do not precisely reproduce the exact dynamics; recall that the last two terms in Equations (21) and (22) have been simply ignored. But, I want to suggest, it is interesting that we get oscillations here at all, because these arise from the specifically Bohmian character of the approximation scheme. In particular, at $t = 0$, $\phi_1 = \psi_1$ (the first excited state) and $\phi_2 = \psi_0$ (the ground state). The interaction, $\hat{H} = \lambda\hat{x}\hat{y}$, would seem to be impotent if one replaces $\hat{y} \to \langle\psi_0|\hat{y}|\psi_0\rangle = 0$. But of course, in the Bohmian semi-quantum approximation, one instead has $\hat{y} \to Y$ which can be (and typically is) different from zero even when, for example, ϕ_2 is the perfectly symmetric ground state.

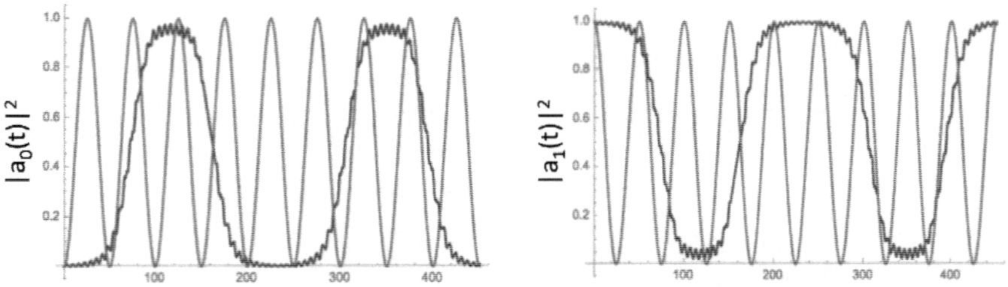

Figure 5. The blue curves show the results of a numerical calculation of the time-dependent weights of the $n = 0$ and $n = 1$ states in ϕ_1. The red curves show the corresponding quantum probabilities from the exact solution. One sees that while the overall qualitative behavior (namely, the back-and-forth oscillations) is basically correct, the details are not quite right. This is not too surprising since the BDSQ is a rather uncontrolled approximation scheme. Still, the result is interesting in comparison to a corresponding non-Bohmian approach, in which each sub-system is influenced only by the quantum average position of the other system. In this case one would never see even the qualitatively appropriate oscillatory behavior seen here. The oscillation, that is, arises specifically from the fact that each quantum half-system feels a force (so to speak) from the Bohmian position of the other half system, and this can be different from zero even when the quantum average would be zero.

Note that we can, in principle, go beyond the BDSQ approximation by including the neglected terms by introducing new fields, for example

$$\phi_1'(x,t) = \left.\frac{\partial \Psi(x,y,t)}{\partial y}\right|_{y=Y(t)} \tag{25}$$

and

$$\phi_1''(x,t) = \left.\frac{\partial^2 \Psi(x,y,t)}{\partial y^2}\right|_{y=Y(t)} \tag{26}$$

which will obey their own dynamical Schrödinger-type equations (e.g.)

$$i\hbar\frac{\partial \phi_1'(x,t)}{\partial t} = -\frac{\hbar}{2m_1}\frac{\partial^2 \phi_1'}{\partial x^2} + V[x,Y(t)]\phi_1' + \frac{\partial V}{\partial y}[x,Y(t)]\phi_1 + i\hbar\frac{dY}{dt}\phi_1'' - \frac{\hbar}{2m_2}\phi_1''' \tag{27}$$

involving even higher-order fields (ϕ_1''', etc.). [11]

Or perhaps this particular re-packaging of the structure in the "big" wave function – i.e., this particular way of sorting the structure that is left out of the BDSQ – is not ideal, and some other method will work better. Sorting this out remains a work in progress. But there should certainly exist *some* sensible, converging expansion which moves the BDSQ approximation toward the exact results by introducing additional beables which capture (what is described in ordinary QM as) entanglement.

Stepping back, though, the BDSQ (and the gestured-at program of pushing beyond the BDSQ to more accurate pilot-wave approximation techniques) can be seen as a uniquely Bohmian alternative to various existing approximation techniques (for example, density functional theory and Hartree-type methods) which break apart intractable N-particle quantum systems into (something like) N single-particle systems, each responding to some appropriately-averaged version of the other $N-1$ particles. In the same way that the specifically Bohmian semi-classical approximation holds some promise of computational benefits compared to more orthodox treatments, so I think the BDSQ deserves further thought and attention.

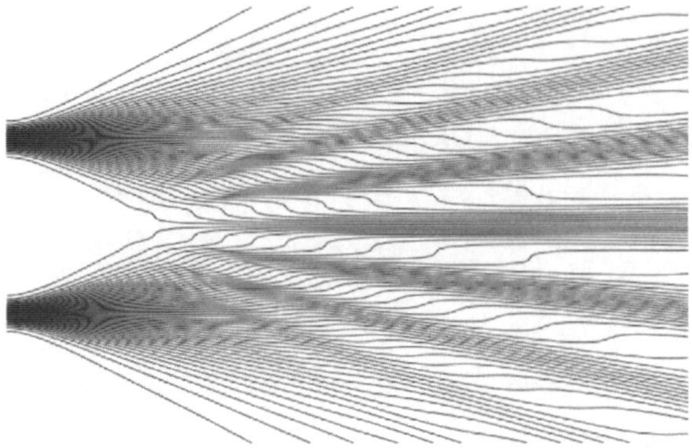

Figure 6. An ensemble of Bohmian trajectories for a double-slit experiment. Each (say) electron, in the pilot-wave picture, comprises a wave – which goes through both slits and produces an interference pattern – as well as a literal particle whose motion is "so influenced [by the interference pattern in the wave] that the particle does not go where the waves cancel out, but is attracted to where they cooperate." [12]

6. Discussion

Although I am quite interested in the potential practical applications (to computationally difficult many-particle systems) of these Bohmian approximation techniques, I also find the program outlined here interesting for the light it may shed on the questions, about the nature of the quantum state, raised at the beginning. Recall, for example, that the only really troubling feature of the de Broglie - Bohm pilot-wave theory is the involvement of the universal wave function. This is, to me, a very strange kind of thing (living as it does in the abstract configuration space) and (since its dynamics seem so comparable to the fields of classical physics) I never really got the hang of interpreting it as a 'law'.

The program outlined above, though, suggests the possibility of reformulating the pilot-wave theory in terms of single-particle pilot-wave fields (like the CWFs of ordinary Bohmian mechanics) which can be understood as normal, physical fields on 3-space. Indeed, what I described in the last section as the BDSQ – but generalized in the obvious way to a system of N particles, perhaps the entire universe – constitutes a full-fledged, well-formulated, consistent theory. Of course, we know the theory isn't *right*: its empirical predictions will only be correct for a universe without (what ordinary QM describes as) "entanglement". But still, in so far as one's goal is to think about how one might formulate an empirically viable quantum theory, in which strange ("non-local beable") objects like the universal wave function play no role, it seems like a useful jumping off point.

Let me close by quoting, perhaps somewhat out of context and in a way he might not endorse, a principle enunciated by Matt Leifer at the workshop:

> **Explanatory Conservatism:** "If there is a natural explanation for a quantum phenomenon then we should adopt an interpretation that incorporates it."

For me, this brings to mind Bell's memorable discussion of the two-slit experiment (See Figure 6):

> "Is it not clear from the smallness of the scintillation on the screen that we have to do with a particle? And is it not clear, from the diffraction and interference patterns, that the motion of the particle is directed by a wave? De Broglie showed in detail how the motion of a particle, passing through just one of the two holes in [the] screen, could be influenced by waves propagating through both holes. And so influenced that

> the particle does not go where the waves cancel out, but is attracted to where they cooperate. This idea seems to me so natural and simple, to resolve the wave-particle dilemma in such a clear and ordinary way, that it is a great mystery to me that it was so generally ignored." [12]

Surely Bell is correct that this idea – of particles being guided by waves – constitutes an exceptionally simple and natural explanation of the phenomenon of one-particle-at-a-time two-slit interference. And so, according to Leifer's principle, we should adopt an interpretation that incorporates this idea.

At present, however, it is impossible to do so, because no candidate theory fully incorporating this idea exists! The pilot-wave theory of de Broglie and Bohm, of course, is in the right ballpark, but in that theory the wave is the universal wave function – not a "local beable" at all and in particular not the kind of thing that can "propagat[e] through both holes" in the two-slit experiment. Unlike the universal wave function, though, the Bohmian conditional wave function of a single particle *is* the right kind of thing to play the role of the pilot-wave in this most-natural explanation of this most-paradigmatic quantum phenomenon.

In closing, I hope I have convinced you that the following is an interesting question: how could we formulate a realistic, empirically viable quantum theory which posited *exclusively local beables*? ... and that the pilot-wave theory and the CWF in particular seem like promising jumping-off points to try to answer it.

References

[1] Goldstein, Norsen, Tausk and Zanghi 2011 Bell's theorem *Scholarpedia* **6**(10) 8378
[2] Dürr and Teufel 2009 *Bohmian Mechanics* (Springer) pp 131 and 202
[3] Allori V, Goldstein S, Tumulka R and Zanghi N 2008 On the common structure of Bohmian mechanics and the Ghirardi–Rimini–Weber theory *Br. J. Philos. Sci.* **59** 353–389
[4] Goldstein and Zanghi 2011 Reality and the role of the wavefunction in quantum theory (*Preprint* arXiv:1101.4575)
[5] Dürr, Goldstein and Zanghi 1992 Quantum equilibrium and the origin of absolute uncertainty *J. Stat. Phys.* **67** 843–907; see also: Quantum equilibrium and the role of operators as observables in quantum theory *J. Stat. Phys.* **116** 959–1055
[6] Gindensperger, Meier and Beswich 2000 Mixing quantum and classical dynamics using Bohmian trajectories *J. Chem. Phys.* **113** 9369–9372
[7] Prezhdo and Brooksby 2001 Quantum backreaction through the Bohmian particle *Phys. Rev. Lett.* **86**(15) 3215–3219
[8] Oriols 2007 Quantum trajectory approach to time-dependent transport in mesoscopic systems with electron-electron interactions *Phys. Rev. Lett.* **98** 066803
[9] Struyve 2015 Semi-classical approximations based on Bohmian mechanics (*Preprint* arXiv:1509.04771)
[10] Norsen, Marian and Oriols 2015 Can the wave function in configuration space be replaced by single-particle wave functions in physical space *Synthese* **192**(10) 3125–3151
[11] Norsen 2010 The theory of (exclusively) local beables *Found. Phys.* **40** 1858–1884
[12] Bell 2004 Six possible worlds of quantum mechanics; reprinted in *Speakable and Unspeakable in Quantum Mechanics*, 2nd ed (Cambridge)

Can decoherence make quantum theories unfalsifiable? Understanding the quantum-to-classical transition without it

X. Oriols
Universitat Autònoma de Barcelona, 08193, Bellatera (Spain)
E-mail: xavier.oriols@uab.es

Abstract. Exact predictions for most quantum systems are computationally inaccessible. This is the so-called many body problem, which is present in most common interpretations of quantum mechanics. Therefore, predictions of natural quantum phenomena have to rely on some approximations (assumptions or simplifications). In the literature, there are different types of approximations, ranging from those whose justification is basically based on theoretical developments to those whose justification lies on the agreement with experiments. This last type of approximations can convert a quantum theory into an "unfalsifiable" quantum theory, true by construction. On the practical side, converting some part of a quantum theory into an "unfalsifiable" one ensures a successful modeling (i.e. compatible with experiments) for quantum engineering applications. An example of including irreversibility and dissipation in the Bohmian modeling of open systems is presented. On the ontological level, however, the present-day foundational problems related to controversial quantum phenomena have to avoid (if possible) being contaminated by the unfalsifiability originated from the many body problem. An original attempt to show how the Bohmian theory itself (minimizing the role of many body approximations) explains the transitions from a microscopic quantum system towards a macroscopic classical one is presented.

1. Introduction: the many body problem

The role of all physical theories (and the quantum theory among them) is to employ mathematical models implementing physical laws to predict natural phenomena. The correctness of a theory is judged by the extent to which its predictions agree with empirical observations. In this sense, it seems to me that the deep consequences of the *many body problem* [1, 2, 3] on ontological issues of quantum mechanics have not been properly recognized.

The many body problem is present in almost all interpretations of quantum mechanics (Copenhagen [4, 5], many-worlds [6, 7], Bohmian mechanics [8, 9, 10, 12, 11, 13], etc). This problem is due to the fact that the Schrödinger equation (or any other equation of motion) is analytically unsolvable most of the times, and its numerical integration is out of today's present computer capabilities, even for quantum systems with very few degrees of freedom. This computational difficulty is the reason why our predicting capability of quantum phenomena is, indeed, very limited. In 1929, Dirac wrote the following [1]:

> The general theory of quantum mechanics is now almost complete. The underlying physical laws necessary for the mathematical theory of a large part of physics and the

Journal of Physics: Conference Series **701** (2016) 012004 doi:10.1088/1742-6596/701/1/012004

entire chemistry are thus completely known, and the difficulty is only that the exact application of these laws leads to equations much too complicated to be soluble.

Three decades later, Born rephrased the issue [2]:

It would indeed be remarkable if Nature fortified herself against further advances in knowledge behind the analytical difficulties of the many body problem.

I really believe that, as indicated by Born, many *advances* in our ability to understand the quantum world are limited by this computational difficulty [3]. We can successfully solve the Schrödinger equation for a single-particle quantum system in an isolated environment (for example, an hydrogen atom whose electronic wave function is solved with an external potential generated by fixed nuclei). However, isolated quantum systems that appear in all textbooks of quantum mechanics are so simple that they have to be understood as a kind of toy-model of the quantum systems of interest. No real quantum system is completely isolated from its surroundings [14, 15, 16].

One of the most common strategies to study quantum systems is to differentiate between the "active" particles (or degrees of freedom) and the rest of particles, which are considered as "external" or "passive". The set of *active* particles are collected under the name *open quantum system*[1] and all the rest, the *passive* part, are named the *environment* or the *bath* [17]. For the type of discussion in the paper, the measuring apparatus can also be considered as part of he environment. In general, a quantum system has few degrees of freedom, while the environment typically has a large number. As emphasized above, the *many body problem* makes the exact solution of the system plus environment impossible. The environment is not explicitly simulated, but its effect is introduced through some type of external operator (or potential energy) acting on the degrees of freedom of the quantum system. *How do we specify the external operators or potentials to get concrete predictions of a natural phenomenon?*

This paper revolves around the implications of the different types of answers to the question posed above. There are (approximated) external operators that are basically based on theoretical developments, while others are justified by the agreement with experiments. This last type of (approximated) external operators can convert any formal quantum theory into an "unfalsifiable" theory, true by construction. In section 2, the modeling of open quantum systems is explained. I discuss how easy it is to develop approximations that traverse the subtle line that differentiates falsifiable and unfalsifiable theories. As an example, in section 3, I discuss a Bohmian modeling of open systems where irreversibility and dissipation in electronic devices is introduced [18]. I also emphasize how useful this type of unfalsifiable model can be (not for ontological discussion but) for practical simulations of quantum phenomena and their applications. On the contrary, in section 4 I will show how the conditions for the classicality for the center of mass of a macroscopic system, not for each degree of freedom, at each individual experiment, not for an ensemble of experiments, can be obtained within the Bohmian theory (minimizing the mentioned many body problem) [19]. Finally, I provide some conclusion of the paper in section 5 comparing the intrinsic tendency towards unfalsifiability in different interpretations of quantum mechanics.

2. Can decoherence make quantum theories unfalsifiable?

Over the course of the past decade, decoherence has become an ubiquitous scientific term popular in all kinds of research, from fundamental theories of quantum physics to applications in nanoengineering and quantum computing [15]. While decoherence has been directly observed in various experiments, its meaning, scope and consequences are still diffusely and ambiguously

[1] For commodity, whenever possible, I will omit the word "open".

understood by the scientific community. For example, decoherence has been invoked as the solution to long-standing foundational problems dating back to the beginnings of quantum mechanics. Perhaps, the key example of the "magical" explanatory power of decoherence is the attempt to explain the quantum-to-classical transition as something simple due to the loss of coherence of quantum particles (but, *why do the particles lose the coherence?*). Certainly, this transition remains one of the most intriguing features of the quantum world [15, 20].

The mathematical definition of decoherence is more simple and concrete than its physical consequences: Decoherence is the result of the entanglement between the quantum system and its environment. [15, 20] If the mathematical definition of decoherence is so simple and clear, *why are its physical implications so ambiguous?* Because of the many body problem. We cannot know which is the exact physical effect of this quantum entanglement by directly solving the many body Schrödinger equation, and we have to relay on (more or less imprecise and vague) approximations.

2.1. Modeling an open quantum system in terms of the reduced density matrix

In order to explain better how the many body problem forces us to explain decoherence with approximations, let us consider a simple example for a non-relativistic quantum world with spinless particles "living" in a 1D physical space. The generalization to a 3D physical space with particles with spin does not change any of the conclusions presented here, but only makes the many body problem more dramatic. We consider N particles of mass m governed by the wave function $\Psi(\vec{x},t)$ solution of the many-particle Schrödinger equation,

$$i\hbar\frac{\partial\Psi(\vec{x},t)}{\partial t} = \hat{H}\Psi(\vec{x},t) = \left(-\frac{\hbar}{2m}\sum_{j=1}^{N}\frac{\partial^2}{\partial x_j^2} + U(\vec{x},t)\right)\Psi(\vec{x},t), \tag{1}$$

where $\vec{x} = \{x_1, x_2, ..., x_N\}$ are the positions of all the particles and the Hamiltonian operator $\hat{H}$ contains the dynamical laws of the particles of the system, of the environment and the interaction among particles. The potential $U(\vec{x},t)$ is responsible for the non-separability of Eq. (1). I emphasize that Eq. (1) can only be solved for $N = 1$, $N = 2$, or perhaps with powerful computational abilities, for $N = 3$ degrees of freedom. This is the many body problem [3, 17].

Alternatively to the (universal) wave function $\Psi(\vec{x},t)$ in Eq. (1), the whole quantum system of N particles can be tackled with the (universal) density matrix $\hat{\rho}_{tot}(t) = |\Psi\rangle\langle\Psi|$. Its (unitary) time evolution is determined by the *Lioville–von Neumann equation* [4]:

$$i\hbar\frac{d\hat{\rho}_{tot}(t)}{dt} = [\hat{H}, \hat{\rho}_{tot}(t)]. \tag{2}$$

Because of the many body problem, obviously, the exact solution of Eq. (2) is inaccessible too. A quite common strategy [2] for treating Eq. (2) is based on defining the density matrix of the open system $\hat{\rho}_{sys}(t)$ by integrating or tracing out the degrees of freedom of the environment of the full (universal) density matrix $\hat{\rho}_{sys}(t) = Tr_{env}\{\hat{\rho}_{tot}(t)\}$. A standard approximation (under the consideration that the environment "lose" memory of their internal correlations [25, 26, 27]) is that the density matrix of the system is governed by the following master equation:

$$\frac{d\hat{\rho}_{sys}(t)}{dt} = -\frac{i}{\hbar}[\hat{H}_{sys}, \hat{\rho}_{sys}(t)] + \hat{\mathbf{D}}\left[\hat{\rho}_{sys}(t)\right], \tag{3}$$

[2] There are other approximations for overcoming the many body problem of time-independent (closed system) Hamiltonian eigenstates with several degrees of freedom. For example, the Hartree - Fock approximation [21, 22], Density Functional Theory [23, 24], etc.

where $\hat{H}_{sys}$ is the part of the Hamiltonian that acts on the degrees of freedom of the system and the superoperator $\hat{\mathbf{D}}$ converts the values of the density matrix operator at one time into new values at another time. This superoperator $\hat{\mathbf{D}}$ is usually defined as the dissipation operator. Perhaps, the name decoherence operator would also be adequate. According to the work of Lindblad, Gorini, Kossakoneski and Sudarshan [25, 26, 27], such operator $\hat{\mathbf{D}}$ can be defined by,

$$\hat{\mathbf{D}}\left[\hat{\rho}_{sys}(t)\right] = -\frac{1}{2}\hat{A}^{\dagger}\hat{A}\hat{\rho}_{sys}(t) - \frac{1}{2}\hat{\rho}_{sys}(t)\hat{A}^{\dagger}\hat{A} + \hat{A}\hat{\rho}_{sys}(t)\hat{A}^{\dagger} \tag{4}$$

where $\hat{A}$ is an unknown operator (not necessarily hermitian). The ultimate reason why $\hat{A}$ is unknown (which means that $\hat{D}$ is also unknown) is the many body problem [3]. To exactly identify $\hat{D}$ from Eq. (3) we require the knowledge of $\hat{\rho}(t)_{tot}$, that implies the full knowledge of $\Psi(\vec{x}, t)$, which is numerically inaccessible [28].

To further elaborate $\hat{D}$ some approximation is mandatory. The second term of the right hand side of Eq. (3), if needed, can include dissipative (inelastic) scattering not achievable from a unitary operator. For example, the electron-phonon interaction. The typical semi-classical Boltzmann-like scattering term (based on the Fermi Golden rule) deals with scattering mechanics in the phase space. In order to be able to use the same tools to tackle electron-phonon interaction we need a type of density matrix in the phase-space. This can be obtained through a Wigner–Weyl transformation of $\hat{\rho}_{tot}$ that is named Wigner distribution function [29]. This Wigner distribution function is a really powerful tool for the simulation of quantum systems of interest, because it mixes the pure quantum unitary dynamic present in the first term of the right hand side of Eq. (3) together with dissipative and decoherence dynamics introduced by the second term. However, *What is the justification for this type of external (approximated) operators that includes electron-phonon interactions?* Part of the justification comes from the theory, but another part from the experimental evidences.

2.2. Modeling an open quantum system in terms of the (Bohmian) conditional wave function

In order to enlarge the previous discussion about different models and approximations for open systems, in this subsection I explain an original algorithm to deal with open systems within Bohmian mechanics. The wave function of an open quantum (sub) system does not exist. This is the reason why the (reduced) density matrix is invoked in section 2.1. Fortunately, the concept of wave function of an open system can be rigorously formulated within the Bohmian explanation of quantum phenomena, by means of the *conditional wave function* [10, 11, 12]. In Bohmian mechanics, a system of N particles is composed by the (universal) wave function $\Psi(\vec{x}, t)$ described by Eq. (1) plus the position of all the particles that I denote here by capital letters $\vec{Y}[t] = \{X_1[t], .., X_a[t], ..., x_N[t]\}$. The time evolution of the wave function $\Psi(\vec{x}, t)$ is described by Eq. (1). Clearly, the many body problem in Bohmian mechanics is hidden here. The time evolution of the particles position is determined by the following guiding equation:

$$v_k(\vec{x}, t) = \frac{dX_k[t]}{dt} = \frac{J_k(\vec{x}, t)}{|\Psi(\vec{x}, t)|^2}, \tag{5}$$

where $J_k(\vec{x}, t)$ is the standard quantum current density. The formal presentation of the theory requires a discussion on the selection of the initial positions according to the quantum equilibrium hypothesis, plus an explanation of the symmetry of the wave function of identical particles. See many reviews, for example [8, 9, 10, 11, 12, 13].

What I have explained about Bohmian mechanics is enough to define the conditional wave function. Let us consider the mentioned many-particle wave function $\Psi(\vec{x}, t)$ solution of the unitary Schrödinger equation (1). If we focus only on the trajectory $X_a[t]$ of one specific particle

a. Bohmian mechanics defines the so-called *conditional wave function* [10, 11, 12, 13, 30] that determines the dynamics of such particle as:

$$\psi_a(x,t) = \Psi(x_a, \vec{Y}[t], t), \tag{6}$$

where $\psi_a(x,t)$ constitutes a slice of the whole multi-dimensional wave function $\Psi(\vec{x},t)$ solution of Eq. (1). [3] I have defined $\vec{Y}[t] = \{x_1[t], ..x_{a-1}[t], x_{a+1}[t], ..., x_N[t]\}$ as the Bohmian trajectories of all particles except a. It has been demonstrated by the author [30] that $\Psi_a(x_a,t)$ obeys the following wave equation:

$$i\hbar\frac{\partial\psi_a(x_a,t)}{\partial t} = \Big\{ -\frac{\hbar}{2m}\frac{\partial}{\partial x_a^2} + U(x_a,\vec{Y}[t],t) + G_a(x_a,\vec{Y}[t],t) + iJ_a(x_a,\vec{Y}[t],t)\Big\}\psi_a(x_a,t), \tag{7}$$

where $U(x_a,\vec{Y}[t],t)$ is the potential that appears in Eq. (1) evaluated at $\vec{x} = \{x_a,\vec{Y}[t],t\}$. The additional potentials $G_a(x_a,\vec{Y}[t],t)$ and $J_a(x_a,\vec{Y}[t],t)$ are defined through the (universal) wave function [30, 31]. By using Eq. (7) for each particle, $x_a[t]$ for $a = 1,\ldots,N$, we obtain a system of N coupled single-particle pseudo-Schrödinger equations. In principle, such system of equations is able to compute many-particle Bohmian trajectories without knowing the many-particle wave function [11]. The great merit of Eq. (7) is to demonstrate that such a single-particle solution of a many-particle problem exists, although we do not know exactly the values of the terms $G_a(x_a,\vec{Y}[t],t)$ and $J_a(x_a,\vec{Y}[t],t)$. The ultimate reason why it is not possible to know such terms is because they depend on the knowledge of the (universal) wave function $\Psi(\vec{x},t)$ that remains inaccessible because of the many body problem [31]. In summary, the situation is quite similar to that found in Eq. (3). We have a quantum equation of motion to tackle a many body system, but there are some operators in Eq. (3) or external potentials in Eq. (7) that require approximations to arrive to some specific predictions.

The formalism of the conditional (single-particle) wave function just described is the base of the simulator BITLLES [32], which already includes electron-electron interaction beyond mean field [33], open system boundaries [34], exchange (fermion) interactions [35, 36], noise computation [37, 38] and high-frequency displacement currents [39]. In section 3, I will use the conditional wave equation (7) to explain an *ad hoc* approximation for including the electron-phonon interactions mentioned in section 2.1. After, in section 4, I will use the same equation (7) to develop a quantum-to-classical transition minimizing the approximations needed to overcome the many body problem.

2.3. Falsifiable and unfalsifiable theories

After two examples on how approximations are mandatory in the quantum treatment of open systems, let us discuss the role of such approximations in the falsifiability of quantum theories. In 1963, Karl R. Popper [40] wrote his idea about the importance of developing theories that can be falsified. He argued that irrefutability is not a virtue of a theory (as people often think) but just the contrary. He wrote the following example explaining how theories that predict things too vaguely can imply that the predictions can hardly fail. He compared astronomers and astrologers in the following sentence [40]:

> It is a typical soothsayer's trick to predict things so vaguely that the predictions can hardly fail: that they become irrefutable.

A similar point of view was shared by Einstein [41] when he emphasized that physical theories have to be risky (avoid ambiguities that minimize the possibility of being refuted):

[3] The generalization of the conditional wave function for a set of particles, not only for one, can be trivially obtained.

> No amount of experimentation can ever prove me right; a single experiment can prove me wrong.

Because of the many body problem, as I repeatedly stressed in this paper, approximations are mandatory to make predictions of most quantum systems. Unfortunately, an approximation introduces some degree of ambiguity into the predictions. There are many types of approximations in the literature. Most of them are basically hanged on the theory. Although the many body problem does not allow us to exhaustively define the external operators or potentials, we can take advantage of some mathematical properties provided by the quantum theory at hand (for example the hermiticity of the observable results or the unitary evolution of the whole system) to get some (partial and) useful information about the shape of such operators.

On the other side, there are approximations to fix the external operators in Eq. (4) (or the potentials in Eq. (7)) whose justification is, in some part, the agreement with experiments. Initially, one can make a hand made proposal for the operators (or potentials). If such operators do not provide good agreement with the experiment, we slightly "tune" them to improve the agreement. In principle, this second route to fix the external potentials seems very satisfactory. However, we have to realize about the ontological implications of this second route. If we select the external operators (or the potentials $J_a(x,t)$ and $G_a(x,t)$ in Eq. (7)) to ensure that we correctly reproduce the experimental results, then, our theory will reproduce experiments by construction.

When dealing with practical applications of quantum mechanics, this route seems less dangerous. However, when dealing with foundational questions much more care is required. Such external operators can explain by themselves the appearance of the arrow of time, why the entropy grows, why quantum systems become classical, etc. All these intriguing quantum phenomena are, somehow, behind the umbrella term of decoherence. One can erroneously conclude that these foundational problems are, already solved, by the "magic" explanatory power of decoherence. Any disagreement between predictions and experiments can always be attributed to the lack of a "proper" modeling of decoherence (a proper computation of the quantum entanglement between the system and the environment), not to a limitation of the theory itself. This flexible understanding of the role of the external operators can convert a quantum theory into a unfalsifiable (true by construction) and universal one (valid for any system by definition).

As far as we cannot avoid the many body problem, all approximations for open quantum systems have a mix of theoretical and experimental justification, with different proportions. Therefore, I am not saying that some approximations are scientific and others non-scientific. I am just saying here that, when discussing foundational questions of quantum mechanics, one has to be worried about the possible contamination of the final conclusions by the unfalsifiability, i.e. because of the unsurmountable many body problem.

3. An example of the practical utility of developing unfalsifiable quantum theories

The quantum theory is today an old theory with more than a century of history. Many researchers work on quantum mechanics applications without worrying at all about foundational issues. I am not criticizing this attitude at all since I sincerely believe that it has provided an unquestionable success in our ability to develop quantum applications. For example, in the next subsection, I explain the recent work of our research group to include irreversibility and dissipation in the Bohmian modeling of open systems [18], and how interesting these capabilities can be in many applied disciplines.

One common approximation to simplify the many body problem in the simulation of electronic devices is the Born-Oppenheimer assumption, which states that the motion of atomic nuclei and electrons in a (semiconductor) crystal can be separated because of the very different masses of the electrons and atomic nuclei [42]. The (approximated) electronic wave-function can be solved alone. However, since it is well-known from experiments that electrons do effectively interact with the nuclei, we can reintroduce the collision of the electron with a phonon (a collective excitation in a periodic arrangement of atoms) through the external potentials discussed in section 2.2. The electron loses/gains energy during the emission/absorption of a phonon. In order to include the effect of the interaction of an electron with phonons we need an effective Schrödinger type evolution for the conditional wave function like Eq. (7), not a single-particle Schrödinger equation. According to Eq. (7), for a system with $U(x,t) = 0$, we suggest the following equation for the electron [18]:

$$i\hbar\frac{\partial\psi(x,t)}{\partial t} = \Big[\frac{-\hbar}{2m}\frac{\partial^2}{\partial x^2} - \frac{i\hbar\lambda f(t)}{m\psi(x,t)}\frac{\partial\psi(x,t)}{\partial x} + \frac{1}{2m}\lambda^2 f^2(t)\Big]\psi(x,t), \tag{8}$$

with

$$J_a(x,t) = -\frac{\hbar\lambda f(t)}{m\psi(x,t)}\frac{\partial\psi(x,t)}{\partial x}, \tag{9}$$

and

$$G(x,t) = \frac{1}{2m}\lambda^2 f^2(t), \tag{10}$$

where $f(t)$ is a time dependent function which mimics the electron-phonon interaction.[4] On the other hand, the parameter λ is the momentum of the phonon emitted ($\lambda > 0$) or absorbed ($\lambda < 0$). The motivation for defining Eq. (8) can be understood by rewriting it as :

$$i\hbar\frac{\partial\psi(x,t)}{\partial t} = \frac{1}{2m}\left(-i\hbar\frac{\partial}{\partial x} + \lambda f(t)\right)^2\psi(x,t). \tag{11}$$

It is quite simple to realize that Eq. (11) satisfies the following desired features. First, the conservation of the total momentum between the electron and the phonon. And second, the electron (i.e. the conditional wave functions assigned to the electron) gains/loses an amount of energy proportional to the energy of the phonon absorbed/emitted . I notice that the momentum operator for Eq. (11) is given by $P = p + \lambda f(t) = -i\hbar\partial/\partial x + \lambda f(t)$.

The trajectories of the particle can be straightforwardly obtained inserting the conditional wave function in the polar form, $\psi(x,t) = R(x,t)e^{iS(x,t)/\hbar}$, into Eq. (11). Taking the imaginary part, one obtains a modified continuity equation

$$\frac{\partial R^2}{\partial t} + \frac{\partial}{\partial x}\left(R^2(\frac{\partial S/\partial x}{m} + \frac{\lambda f(t)}{m})\right) = 0, \tag{12}$$

from which it is easy to realize that the velocity field of the particles is:

$$m\frac{dX[t]}{dt} = mv(x,t)\Big|_{x=X[t]} = \frac{\partial S(x,t)}{\partial x} + \lambda f(t) = \hbar\,\mathrm{Im}\left[\frac{\partial\Psi(x,t)/\partial x}{\Psi(x,t)}\right] + \lambda f(t). \tag{13}$$

[4] In particular, we have tested in [18] two specific examples: an instantaneous interaction $f(t) \equiv \Theta(t) = 0$ for $t < t_1$, $\Theta(t) = 1$ for $t > t_1$ and a gradual interaction $f(t) \equiv \Gamma(t) = 0$ for $t < t_0$, $\Gamma(t) = 1$ for $t > t_1$ and $\Gamma(t) = (t - t_0)/(t_1 - t_0)$ for $t_0 < t < t_1$.

One can see that the velocity field has two terms, the usual one of Bohmian mechanics ($\frac{\partial S(x,t)}{\partial x}$) [11] and an additional term proportional to λ, i.e. the momentum of the phonon. Because of the function $f(t)$ the velocity of the electron changes after the interaction with the phonon. The new velocity of the electron (i.e. the conditional wave function assigned to the electron) is the old one plus a quantity proportional to the momentum of the absorbed/emitted phonon itself. It can increase or decrease depending on whether the phonon is emitted ($\lambda < 0$) or absorbed ($\lambda > 0$). I also notice that the present development includes electron-phonon scattering without violating the continuity equation, a problem that could appear in different formalisms [29]. Before the interaction with the phonon the momentum and the energy of the electron are:

$$\langle P \rangle = \langle p \rangle \tag{14}$$

$$\langle E \rangle = \frac{\langle p \rangle^2}{2m} + \frac{\Delta p^2}{2m}, \tag{15}$$

after the interaction with the phonon they are:

$$\langle P \rangle = \langle p \rangle + \lambda \tag{16}$$

$$\langle E \rangle = \frac{(\langle p \rangle + \lambda)^2}{2m} + \frac{\Delta p^2}{2m}, \tag{17}$$

where $\langle p \rangle = -i\hbar \int \psi^*(\partial\psi/\partial x)dx$. Here I underline that the interaction among electrons and phonons briefly introduced here and explained in detail in Ref. [43] is flexible enough to guarantee an agreement with experiments. The scattering rates (i.e. the number of collisions per second) in our time-dependent scenario are defined according to the standard Fermi golden rule. Therefore, a proper selection of the absorption and emission rates (more emission than absorption) provides dissipation of energy (heat) in electronic devices. Obviously, these rates can be tuned to better reproduce experimental scenarios. I emphasize that the hand made scattering rates in our approach are introduced directly into the (conditional) wave function (not into transitions between phase space points as in the Wigner formalism [29]).

4. Understanding the quantum-to-classical transition with Bohmian mechanics

Since the beginning of quantum theory a century ago, the study of the frontier between classical and quantum mechanics has been a constant topic of debate. Let us now briefly explain an attempt developed by the author and co-workers [19] to discuss the quantum-to-classical transition without being (much) contaminated by the approximations needed to overcome the many body problem. Among the attempts from different interpretations of quantum mechanics to reach the classical limits from quantum theories, Bohmian mechanics [8, 9, 11, 13] has some attractive features. Like in classical mechanics, it describes a unique quantum experiment in terms of a trajectory and, remarkably, the quantum measurement is unproblematically treated as another type of interaction [12]. In section 4.1, I discuss how a single-particle quantum system is an ill-defined scenario to reach classical mechanics. Then, in section 4.2, I briefly explain our recent result [19] on how the quantum-to-classical transition can be successfully tackled with the center of mass of a many body system.

4.1. The unsuccessful single-particle scenario

Let us start by discussing the difficulties of studying the quantum-to-classical transition in single particle scenarios. The Newton law that fully determines a classical trajectory $X[t]$ is :

$$m\frac{d}{dt}v(X[t],t) = \left[-\frac{\partial}{\partial x}\left(U(x,t)\right)\right]_{x=X[t]}, \tag{18}$$

where $v(X[t], t)$ is the velocity of such particle at the position $X[t]$ and $U(x, t)$ is a scalar energy (classical) potential. It is well-known that Newtonian mechanics is compatible with the following Hamilton–Jacobi equation [11]:

$$H\left(x, \frac{\partial S(x,t)}{\partial x}, t\right) + \frac{\partial S(x,t)}{\partial t} = \frac{1}{2m}\left(\frac{\partial S(x,t)}{\partial x}\right)^2 + U(x,t) + \frac{\partial S(x,t)}{\partial t} = 0, \tag{19}$$

where H is the Hamiltonian of the single-particle system that contains a kinetic energy plus a potential energy $U(x, t)$, and $S(x, t)$ can be identified with the Hamilton's principal function. In general, the direct solution of Eq. (19) is more difficult than using the Newton formulation in Eq. (18). In any case, Eq. (19) can be interesting when dealing with an (infinite) ensemble of classical trajectories, i.e. in classical statistics. For example, for a classical (single-particle) experiment with some practical difficulties in specifying the initial position and velocity of the particle, such that different experimental realizations can have slightly different initial conditions. For such an experiment, we can define some distribution of the initial position of the particles of different experiments $R^2(x, 0) \geq 0$. Such an ensemble of trajectories will evolve in time according to Eq. (18) so that we will obtain a function $R^2(x, t)$ that describes the particular distribution of particles at any time. All these classical particles will move, in a *continuous* way, from one unit of volume of the physical space to another. Therefore, we can ensure that the ensemble of trajectories accomplishes the following *local* conservation law [11] :

$$\frac{\partial R^2(x,t)}{\partial t} + \frac{\partial}{\partial x}\left(\frac{1}{m}\frac{\partial S(x,t)}{\partial x}R^2(x,t)\right) = 0. \tag{20}$$

The two previous (real) equations, Eqs. (19) and (20), for $S(x, t)$ and $R(x, t)$ are equivalent to the following (complex) classical wave equation [11]:

$$i\hbar\frac{\partial \psi_{cl}(x,t)}{\partial t} = \left(-\frac{\hbar}{2m}\frac{\partial^2}{\partial x^2} + U(x,t) - Q(x,t)\right)\psi_{cl}(x,t), \tag{21}$$

where $\psi_{cl}(x, t) = R(x, t)\exp(iS(x, t)/\hbar)$ is defined as a classical[5] (complex) wave function. The new function $Q(x, t)$ in Eq. (21) is the so-called quantum potential in the hydrodynamic or Bohmian interpretation of quantum mechanics:

$$Q(x,t) = -\frac{\hbar}{2m}\frac{\partial^2|\psi_{cl}(x,t)|/\partial x^2}{|\psi_{cl}(x,t)|} = -\frac{\hbar}{2m}\frac{\partial^2 R(x,t)/\partial x^2}{R(x,t)}. \tag{22}$$

The reader can be, perhaps, surprised that $Q(x, t)$ appears in a pure classical context, while it is referred to as a quantum entity. The justification is based on historical arguments [11, 8].

Let us move to the quantum modeling of the same single-particle system. The wave equation that describes the quantum wave function $\psi(x, t)$ of such experiment is just the typical single-particle Schrödinger equation:

$$i\hbar\frac{\partial \psi(x,t)}{\partial t} = \left(-\frac{\hbar}{2m}\frac{\partial^2}{\partial x^2} + U(x,t)\right)\psi(x,t). \tag{23}$$

I notice that the $Q(x, t)$ term is missing when comparing the quantum wave equation (23) to the classical wave equation (21). Again, when putting $\psi(x, t) = R(x, t)\exp(iS(x, t)/\hbar)$ into Eq.

[5] For simplicity, I assume that $R(x, t)$ and $S(x, t)$ are single valued. This condition is equivalent to assigning a unique velocity to each position of space. If this is not the case, different wave functions (i.e. a mixed state) will be needed to describe the different velocity fields of the classical system.

(23) we arrive at a quantum continuity equation identical to the classical continuity equation, Eq. (20), and at a quantum Hamilton–Jacobi equation quite similar to the classical one, Eq. (19), but now with the presence of a quantum potential [8]:

$$\frac{1}{2m}\left(\frac{\partial S(x,t)}{\partial x}\right)^2 + U(x,t) + Q(x,t) + \frac{\partial S(x,t)}{\partial t} = 0. \tag{24}$$

The last step in this description of similitudes and differences between classical and quantum modelings is getting a quantum Newton-like equation. It can be easily demonstrated using Eq. (20) and Eq. (24) that a Newton-like equation can be developed for the quantum (Bohmian) trajectories [8, 11]:

$$m\frac{d}{dt}v(X[t],t) = \left[-\frac{\partial}{\partial x}\left(U(x,t) + Q(x,t)\right)\right]_{x=X[t]}. \tag{25}$$

In summary, the development performed in this section provides a common language for classical and quantum theories, in terms of either wave functions or trajectories. Here I emphasize that one has to compare either classical and quantum wave functions or classical and quantum ensembles of trajectories (not a single classical trajectory with a quantum wave function).

The presence of $Q(x,t)$ in the quantum Hamilton–Jacobi equation, Eq. (24), and in the Newton-like equation, Eq. (25), implies that Bohmian trajectories depend not only on the classical potential $U(x,t)$ but also on the quantum potential $Q(x,t)$, which is a function of $R(x,t)$. In fact, it is the *shape* and not the absolute value of $R(x,t)$ that acts on each individual quantum trajectory. On the contrary, each classical trajectory can be computed from Eq. (18) and Eq. (19) independently of the shape of the classical ensemble. In conclusion, the differences between quantum and classical ensembles of trajectories is not a difference between waves and particles, because both waves and particles can be used to study classical or quantum systems. It is a difference on what is the role of $R(x,t)$ on the trajectory of a unique particle. Classically, it has no role. Quantum mechanically, $R(x,t)$ provides the information about how particles are distributed in an ensemble of experiments, and it also affects each individual trajectory, through the quantum potential.

In 1964, Nathan Rosen [44] used these features to develop the conditions needed to reach a classical regime for a one-particle system. I follow here his single-particle arguments. Classical trajectories in Eq. (25) can be obtained by imposing, as a third equation [44, 45, 46], a negligible spatial derivative of a quantum potential:

$$Q(x,t) = -\frac{\hbar^2}{2m}\frac{\partial^2 R(x,t)/\partial x^2}{R(x,t)} = cst. \tag{26}$$

Most of the time, this route to recover classical dynamics from quantum systems is incompatible with a well-defined wave function solution of Eq. (23). The reason is because we have three quantum equations, Eq. (20), Eq. (24) and Eq. (26), imposed on only two unknowns, the amplitude $R(x,t)$ and the phase of the wave function $S(x,t)$. [6]

4.2. The successful many-particle scenario

In this subsection I explain briefly the recent work of the author and co-authors through a generalization of Rosen's attempt to the center of mass of a many-particle system [19]. I show

[6] In the literature, sometimes we see a way of increasing the number of unknowns by assuming the (classical) potential $U(x,t)$ as an additional unknown that has to be fixed by the three equations. So, there are some exotic potentials where quantum and classical solutions can live together [47] (one simple example is a plane wave $R(x,t) = 1$ and $S(x,t) = kx$ with $p = \hbar k$ the momentum in a free space with $U(x,t) = 0$). However, it is a clearly unphysical solution to say that classical scenarios do only appear in quantum systems for very exotic potentials.

that classicality appears as a *natural* quantum limit for most macroscopic objects, i.e. the effect of the quantum potential becomes negligible while retaining a well-defined wave function. Our work opens the path for a new wave equation for decoherent (sub)systems that are neither purely quantum nor classical (similar to Ref. [48]). The ultimate reason of the unsuccessful attempt to reach classical mechanics from a single-particle quantum system is the fact that we cannot introduce, by hand, any new term on the single particle Schrödinger equation. However, by looking at the (conditional) wave equation for an open quantum system, Eq. (7), we realize that apart from kinetic and (classical) potential energy, there is an additional (complex) term. New possibilities appear that were unaccessible for a single-particle (closed) system.

I consider a quantum system described by Eq. (1) and the center of mass x_{cm} and a set of relative coordinates $\vec{y} = \{y_2, .., y_N\}$ defined as

$$x_{cm} = \frac{1}{N}\sum_{i=1}^{N} x_i \tag{27}$$

$$y_j = x_j - \frac{(\sqrt{N}x_{cm} + x_1)}{\sqrt{N}+1}, \tag{28}$$

which allows rewriting Eq. (1) for N particles of mass m in terms of the wave function with the new variables $\Psi(x_{cm}, \vec{y}, t)$ as [19]

$$i\hbar\frac{\partial\Psi(x_{cm},\vec{y},t)}{\partial t} = \left(-\frac{\hbar}{2M}\frac{\partial^2}{\partial x_{cm}^2} - \frac{\hbar}{2m}\sum_{j=2}^{N}\frac{\partial^2}{\partial y_j^2} + U(x_{cm},\vec{y},t)\right)\Psi(x_{cm},\vec{y},t), \tag{29}$$

with $M = Nm$. The force acting on the (Bohmian) center of mass trajectory is given by:

$$M\frac{d^2X_{cm}[t]}{dt^2} = -\left(\frac{\partial}{\partial x_{cm}}(U(x_{cm},\vec{y},t) + Q_{cm}(x_{cm},\vec{y},t) + \sum_{j=2}^{N} Q_j(x_{cm},\vec{y},t))\right)\Bigg|_{\substack{x_{cm}=X_{cm}[t]\\ \vec{y}=\vec{Y}[t]}} \tag{30}$$

with the quantum potentials

$$Q_{cm}(x_cm,\vec{y},t) = -\frac{\hbar}{2MR(x_{cm},\vec{y},t)}\frac{\partial^2 R(x_{cm},\vec{y},t)}{\partial x_{cm}^2} \tag{31}$$

$$Q_j(x_{cm},\vec{y},t) = -\frac{\hbar}{2mR(x_{cm},\vec{y},t)}\frac{\partial^2 R(x_{cm},\vec{y},t)}{\partial y_j^2}, \tag{32}$$

Hereafter, I develop the wave equation that determines the (conditional) wave function for the center of mass $\psi(x_{cm},t) = \Psi(x_{cm},\vec{Y}[t],t)$ computed from Eq. (29). According to section 2.2, the trajectory $X_{cm}[t]$ can be equivalently computed from ψ or Ψ. Following Refs. [19, 30], we get:

$$i\hbar\frac{\partial\psi}{\partial t} = -\frac{\hbar}{2M}\frac{\partial^2\psi}{\partial x_{cm}^2} - \frac{\hbar}{2m}\sum_{j=2}^{N}\frac{\partial^2\Psi(x_{cm},\vec{y},t)}{\partial y_j^2}\Bigg|_{\vec{y}=\vec{Y}[t]} - i\hbar\sum_{j=2}^{N} v_j^h[t]\frac{\partial\Psi(x_{cm},\vec{y},t)}{\partial y_j}\Bigg|_{\vec{y}=\vec{Y}[t]} + U\psi. \tag{33}$$

Eq. (33) is just our previous equation Eq. (7) where the terms $G(x_{cm},t)$ and $J(x_{cm},t)$ can be identified as follow. If we define $\psi(x_{cm},t) = r(x_{cm},t)\exp(i\,s(x_{cm},t)/\hbar)$ into Eq. (33), one can then develop a continuity-like equation:

$$0 = \frac{\partial r^2}{\partial t} + \frac{\partial}{\partial x_{cm}}\left(r^2\frac{1}{M}\frac{\partial s}{\partial x_{cm}}\right) + J, \tag{34}$$

with

$$J = \hbar \sum_{j=2}^{N} \left[\frac{\partial r^2}{\partial y_j} v_j[t] - \frac{\partial}{\partial y_j} \left(\frac{1}{m} r^2 \frac{\partial s}{\partial y_j} \right) \right], \tag{35}$$

where $v_j[t]$ is the Bohmian velocity of y_j. In addition, we get a quantum Hamilton–Jacobi-like equation:

$$0 = \frac{\partial s}{\partial t} + \frac{1}{2M} \left(\frac{\partial s}{\partial x_{cm}} \right)^2 + U + G, \tag{36}$$

with

$$G = Q_{cm} + \sum_{j=2}^{N} \left(\frac{1}{2m} \left(\frac{\partial s}{\partial y_j} \right)^2 + Q_j - v_j^h[t] \frac{\partial s}{\partial y_j} \right). \tag{37}$$

The relevant point in this many-particle generalization of Rosen's attempt is that the new equation (33) has the additional environment degrees of freedom, i.e. $G(x_{cm}, t) + iJ(x_{cm}, t)$, included. Now, contrarily to the single-particle case, we will see that the number of equations and unknowns in the quantum-to-classical transition is well balanced.

To obtain a classical solution for the (quantum) center of mass (not other degrees of freedom) using Eq. (30), the additional requirement is needed:

$$\frac{\partial U(x_{cm}, \vec{y}, t)}{\partial x_{cm}} \gg \frac{\partial Q_{cm}(x_{cm}, \vec{y}, t)}{\partial x_{cm}} + \sum_{j=2}^{N} \frac{\partial Q_j(x_{cm}, \vec{y}, t)}{\partial x_{cm}}, \tag{38}$$

along the path $X_{cm}[t]$. Therefore, we have again three real equations (that is Eq. (34), Eq. (36) and Eq. (38)) to get a classical solution and, now, we have three unknowns $r(x_{cm}, t)$, $s(x_{cm}, t)$ and $G(x_{cm}, t)$[7] In summary, a classical solution for the center of mass is possible, while still remaining a quantum solution.

In our recent work [19] we have shown that the *natural* conditions required for a quantum system of identical particles to ensure that its center of mass has a classical behavior. In particular, we discus when and why it is reasonable to expect that the spatial derivatives of the classical potential in Eq. (38) are expected to be larger than those of the quantum potentials. We conclude there [19], without invoking ambiguous arguments about the role of decoherence, that a large number of identical particles ($N \to \infty$) under an external potential is enough to ensure classical dynamics for its center of mass in a single experiment. This result can also be interpreted as a single-experiment generalization of the well-known multiple-experiment Ehrenfest theorem.

5. Conclusion

In summary, I have discussed the ontological and practical implications in quantum theories of the many body problem that makes exact predictions for most quantum systems computationally inaccessible. This problem is present in most common interpretations of quantum mechanics. Therefore, most predictions of natural quantum phenomena have to rely on some type of approximations. Such approximations are, in some part, justified by theoretical developments

[7] One can argue that $J(x_{cm}, t)$ is an additional unknown. It is true. However, although it is not mandatory for a conditional wave function, one can usually fix $J(x_{cm}, t)$ using Eq. (34) to ensure a conservation of probability. In any case, for the many-body system, the number of unknowns is equal or higher than the number of equations to satisfy. In the single-particle case, the balance was opposite.

and, in another part, by the agreement of the outputs (of such approximate model) with experiments.

The progress on many different quantum applications, some of them already present in our society, is due to these approximations that allow us to go far beyond the single-particle understanding of quantum mechanics. An example of including irreversibility and dissipation in the Bohmian modeling of open systems is presented in section 3. On the ontological level, however, the present-day foundational problems related to controversial quantum phenomena (such as the arrow of time, the quantum-to-classical frontier or the increment of the entropy) have to avoid being contaminated by the unfalsifiability originated from the many body problem discussed in this paper. An original attempt to show how the Bohmian theory itself (minimizing the role of many body problem) explains the transitions from a microscopic quantum system towards a macroscopic classical one is shown in section 4. In all cases, I emphasize in this paper that, either for practical or foundational issues, there is no exact solution for quantum systems of interest with most common quantum theories. Therefore, we always have some degree of ambiguity and vagueness in our predictions.

Finally, apart from the many body problem discussed in this paper, I want to make a brief comment about the intrinsic ambiguity (and exposure to refutation) of the different interpretations of quantum mechanics. Paraphrasing Karl R. Popper [40], "every test of a theory is an attempt to falsify it, or to refute it". However, there are quantum theories that are more testable, more exposed to refutation, than others. They take a greater risk. I want to compare here Copenhagen and Bohmian interpretations. One of the basic postulates of the Copenhagen interpretation is: *For every observable in classical mechanics, a linear, Hermitian operator in quantum mechanics exists* [5]. But, *Which operator do I have to choose?* Although the Copenhagen theory does not specify which operator, the answer seems obvious. You have to select that operator that exactly reproduces the experiment. This seems not a very concrete answer and not too risky. Somehow, perhaps, this vagueness in the definition of the operator can explain the success of the Copenhagen (or orthodox) interpretation in front of others. [8] There is a general belief that the Copenhagen interpretation perfectly explains all phenomena of the quantum world, for few or many degrees of freedom. You just need to find the proper operator that ensures the agreements of the predictions with the experiments. I summarize my argument here, by rephrasing the famous sentence *"shut-up-and-calculate"* by the new one, *"shut-up-and-find-the-proper-operator"*. These ideas are lined up with the point of view of John Bell. He said that the orthodox theory is "unprofessional" [49] because of this vagueness of operators (either the external ones discussed in this paper or the operator that encapsulates the measurement process). He wrote [50]:

> The concept of "measurement" becomes so fuzzy on reflection that it is quite surprising to have it appearing in physical theory at the most fundamental level. Less surprising perhaps is that mathematicians, who need only simple axioms about otherwise undefined objects, have been able to write extensive works on quantum measurement theory-which experimental physicists do not find it necessary to read.

There are other interpretations of quantum phenomena whose postulates are, in my opinion, more concrete, more risky. For example, Bohmian mechanics is a quantum theory without operators [10, 51, 52]. In simple words, operators are not needed in Bohmian mechanics, although they can be a very helpful mathematical trick in practical computational issues by encapsulating the environment in a simple mathematical function. These ideas are emphasized by Goldstein,

[8] Here, in this last paragraph, I am talking about the operators associated to the measurement processes. In rest of the paper, I was talking about external operators related to environment. In both cases, the orthodox definition of such operators is ambiguous. In the external operators, the reason is the many body problem, while in the measurement operators it is because of the own theory.

Dürr, Teufel, and coworkers when they refer to the "naive realism about operators" [10, 52, 53]. In Bohmian mechanics, the measurement process is explained as any other type of interaction among particles. There is no measuring postulate. In this sense, the Bohmian theory provides a more definitive answer of how to treat a particular measurement. You just have to include the kinetic energy of the particles of the measuring apparatus and their interaction with the particles of the quantum system into the global Hamiltonian. By knowing the trajectories of the particles of the pointer of the apparatus, you will finally get the observable result without ambiguity.

In summary, in my opinion, the Bohmian explanation of the quantum world (and quantum measurement in particular) is more concrete, more risky, with less ambiguity than the orthodox explanation. However, the many body problem makes the exact solution of the system plus apparatus (or system plus environment) computationally inaccessible in both interpretations. Therefore, at the end of the day, both theories are subject to approximations that imply, more or less, a similar ambiguity in (practical and foundational) predictions. In case of disagreement with the experiments, *is the theory incorrect or the approximation too rude?* It seems that Nature (through the many body problem) fortified herself against giving a concrete answer to this question. *Is it possible to develop other quantum theories free from the many body problem?* [54, 55]

Acknowledgments
The author is grateful to A. Benseny and D. Tena for the discussions in the quantum-to-classical transitions, and to D. Marian, E. Colomés and Z. Zhan for their discussion on the modeling of dissipation in open systems with Bohmian mechanics. This work has been partially supported by the "Ministerio de Ciencia e Innovación" through the Spanish Project TEC2015-67462-C2-1-R, the Generalitat de Catalunya (2014 SGR-384), the Grant agreement no: 604391 of the Flagship initiative "Graphene-Based Revolutions in ICT and Beyond".

References
[1] Dirac P A M 1929 Quantum mechanics of many-electron systems *Proc. Royal Soc. London* **A 123** 714
[2] Born M 1960 *The Classical Mechanics of the Atom* (New York: Ungar)
[3] The ultimate origin of the many body problem is that the many-particle wave function "lives" in the N-dimensional configuration space (not in the physical space). Let me roughly estimate the hard disk space that would be required to store a wave function of $N = 10$ particles confined in a 1D region of 10 nm, which we discretize with a spatial step of $\Delta x = 0.1$ nm. We have a grid of 100 points for each dimension of the configuration space. Then, the total number of points in the configuration space for the 10 particles is $100^{10} = 10^{20}$. Using 4 bytes (32 bits) to store the complex value of the wave function at each grid point, the information contained in a 10-particle wave function would require more than 3×10^8 Terabytes (TB) (and more than 3×10^{28} TB for 20 particles) at each time of the simulation. Since one typical hard disk has less than 1 TB, we require 10^8 hard disk! Now, it would be interesting to address the previous problem of storing the state of a 10-particle system from the classical mechanics perspective. Solving the problem means finding the many-particle trajectory that can be computed alone, without knowing the rest of the trajectories of the ensemble (with different initial positions). At a computational level, this feature of classical mechanics implies a dramatic simplification. Basically, at each time step of the simulation, if the analytical expression for the potential is known, we just need to save the position and velocity of each particle. This means 20 real values for the 10-particle systems mentioned above.
[4] von Neumann J 1932 *Mathematische Grundlagen der Quantenmechanik* (Berlin: Springer Verlag); English translation by: Beyer R T 1995 *Mathematical Foundations of Quantum Mechanics* (Princeton: Princeton University Press)
[5] Landau L D and Lifschitz E M 1981 *Quantum Mechanics, Non-Relativistic Theory* (Boston: Butterworth–Heinemann)
[6] Everett H 1957 Relative state formulation of quantum mechanics *Rev. Mod. Phys.* **29** 454
[7] Wallance D 2012 Chaos, decoherence, and branching; in *The Emergent Multiverse: Quantum Theory according to the Everett Interpretation* (Oxford: Oxford University Press)

[8] Bohm D 1952 A Suggested interpretation of the quantum theory in terms of hidden variables *Phys. Rev.* **85** 166
[9] Holland P R 1993 *The Quantum Theory of Motion* (Cambridge: Cambridge University Press)
[10] Dürr D and Teufel S 2009 *Bohmian Mechanics: The Physics and Mathematics of Quantum Theory* (Springer)
[11] Oriols X and Mompart J (eds.) 2012 *Applied Bohmian Mechanics: From Nanoscale Systems to Cosmology* (Singapore: Pan Stanford)
[12] Dürr D, Goldstein S and Zanghì S 2013 *Quantum Mechanics without Quantum Philosophy* (Berlin: Springer-Verlag)
[13] Benseny A, Albareda G, Sanz A S, Mompart J and Oriols X 2014 Applied Bohmian mechanics *Eur. Phys. J. D* **68** 286
[14] Zurek W H 2003 *Rev. Mod. Phys.* **75** 715
[15] Schlosshauer M 2007 *The quantum-to-classical transition and decoherence* (Berlin, Heidelberg: Springer-Verlag)
[16] Diósi L 2015 Testing spontaneous wave-function collapse models on classical mechanical oscillators *Phys. Rev. Lett.* **114** 050403
[17] This separation is more or less arbitrary. Some times the criteria for the separation is based on different spatial locations. For example, in an electronic device, the electrons in the channel are part of the open system, while those electrons in the battery are external or passive. Other times, the active and the passive degrees of freedom share the same locations. For example, it is quite common to invoke the Born-Oppenheimer approximation to decouple the electrons in the channel from the atoms (phonons), which are also in the channel.
[18] Oriols X, Zhen Z, Colomés E and Marian D 2015 Dissipative quantum transport using one-particle time-dependent (conditional) wave functions *Intl. Workshop on Computational Electronics (IWCE), USA*
[19] Oriols X, Tena D and Benseny A 2016 (*Preprint* `arxiv:1602.03988`)
[20] Hagar A 2012 *Phil. Trans. R. Soc. A* **370**
[21] Hartree D R 1928 The wave mechanics of an atom with a non-Coulomb central field. Part I. Theory and methods *Math. Proc. Cambridge Philos. Soc.* **24**(01) 89
[22] Fock V 1930 Näherungsmethode zur Lösung des quantenmechanischen Mehrkörperproblems *Zeitschr. Physik* **61** 126
[23] Hohenberg P and Kohn W 1964 Inhomogeneous electron gas *Phys. Rev.* 136(3B) B864
[24] Kohn W and Sham L J 1965 Self-consistent equations including exchange and correlation effects *Phys. Rev.* **140**(4A) A1133
[25] Lindblad G 1976 *Commun. Math. Phys.* 48(2) 119
[26] Kossakowski A 1972 *Rep. Math. Phys.* **3** 247
[27] Gorini V, Kossakowski A and Sudarshan E C G 1976 *J. Math. Phys.* **17** 821
[28] Since we can easily get $d\hat{\rho}_{sys}(t)/dt = Tr_{env}\{d\hat{\rho}_{tot}(t)/dt\}$, we conclude from Eq. (2) and Eq. (3) that $\hat{\mathbf{D}}\hat{\rho}_{sys}(t) = \frac{i}{\hbar}\left([\hat{H}_{sys}, \hat{\rho}(t)_{sys}] - Tr_{env}\{[\hat{H}, \hat{\rho}_{tot}(t)]\}\right)$. This result clearly emphasizes the dependence of $\hat{D}$ on $\hat{\rho}_{tot}(t)$.
[29] Frensley W R 1990 *Rev. Mod. Phys.* **62**(3) 745–791
[30] Oriols X 2007 *Phys. Rev. Lett.* **98**(6) 066803
[31] These last two terms in Eq. (7) can be defined as $G_a(\vec{x}, t) = \sum_{k=1;k\neq a}^{N}(\frac{1}{2\,m}\left(\frac{\partial S(\vec{x},t)}{\partial x_k}\right)^2 + Q_k(\vec{x}, t) - \frac{\partial S(\vec{x},t)}{\partial x_k} v_k(\vec{X}[t], t))$ where $X[t] = \{X_a[t], \vec{Y}[t]\}$ and $Q_k(\vec{x}, t)$ as the quantum potential energy [13] and $J_a(\vec{x}, t) = \sum_{k=1;k\neq a}^{N} \frac{\hbar}{2R^2(\vec{x},t)}(\frac{\partial R^2(\vec{x},t)}{\partial x_k} v_k(\vec{x}[t], t) - \frac{\partial}{\partial x_k}\left(\frac{R^2(\vec{x},t)}{m}\frac{\partial S(\vec{x},t)}{\partial x_k}\right))$ where $R(\vec{x}, t)$ and $S(\vec{x}, t)$ are defined as the polar form of the complex (universal) wave function $\Psi(\vec{x}, t) = R(\vec{x}, t)exp(iS(\vec{x}, t)/\hbar)$.
[32] *http://europe.uab.es/bitlles*
[33] Albareda G, López H, Cartoixà X, Suñé J and Oriols X 2010 *Phys. Rev. B* **82** 085301
[34] Albareda G, Suñé J and Oriols X 2009 *Phys. Rev. B* **79** 075315
[35] Alarcón A, Yaro S, Cartoixà X and Oriols X 2013 *J. Phys.: Condens. Matt.* **25** 325601
[36] Marian D, Colomés E and Oriols X 2015 *J. Phys.: Condens. Matt.* **27** 245302
[37] Marian D, Colomés E, Zhan Z and Oriols X 2015 *J. Comput. Electron.* **14**(1) 114–128
[38] Traversa F L *et al.* 2011 *IEEE Trans. Electron Devices* **58** 2104
[39] Alarcón A and Oriols X 2009 *J. Stat. Mech: Theory Exp.* **P01051**
[40] Pooper K 1963 *Conjectures and Refutations: The Growth of Scientific Knowledge* (London: Routledge)
[41] Calaprice A 2005 *The New Quotable Einstein* (USA: Princeton University Press and Hebrew University of Jerusalem)
[42] Born M and Oppenheimer J R 1927 *Annalen der Physik* **389**
[43] Colomés E, Marian D, Zhen Z and Oriols X, *in preparation*

[44] Rosen N 1964 Identical motion in quantum and classical mechanics *Am. J. Phys.* **32** 377
[45] Allori V, Dürr D, Goldstein S and Zanghì N 2002 Seven steps towards the classical world *J. Opt. B* **4** 482
[46] Allori V and Zanghì N 2009 On the classical limit of quantum mechanics *Found. Phys.* **39** 20
[47] Makowski A J 2002 Exact classical limit of quantum mechanics: Central potentials and specific states *Phys. Rev. A* **65** 032103
[48] Richardson C D, Schlagheck P, Martin J, Vandewalle N and Bastin T 2014 Nonlinear Schrödinger wave equation with linear quantum behavior *Phys. Rev. A* **89** 032118
[49] Bell J S 1990 Against measurement *Phys. World* **3** 33
[50] Bell J S 1982 On the impossible pilot wave *Found. Phys.* **12** 989
[51] Goldstein S 1998 Quantum mechanics without observers-part one *Phys. Today* **51**(3) 42
[52] Daumer M, Dürr D, Goldstein S and Zanghì N 1997 Naive realism about operators *Probability, Dynamics and Causality* **45**(2–3) 379
[53] Dürr D, Goldstein S and Zanghì N 2004 Quantum equilibrium and the role of operators as observables in quantum theory *J. Stat. Phys.* **116** 9595
[54] Norsen T, Marian D and Oriols X 2014 Can the wave function in configuration space be replaced by single-particle wave functions in physical space? *Synthese* 1–27
[55] Norsen T 2010 The theory of (exclusively) local beables *Found. Phys.* **40** 1858

The super-indeterminism in orthodox quantum mechanics does not implicate the reality of experimenter free will

J Walleczek[1]

[1] Phenoscience Laboratories, Novalisstrasse 11, 10115 Berlin, Germany

E-mail: `walleczek@phenoscience.com`

Abstract. The concept of 'super-indeterminism' captures the notion that the *free choice assumption* of orthodox quantum mechanics necessitates *only* the following requirement: an agent's free-choice performance in the selection of measurement settings must *not* represent an exception to the rule of irreducible *quantum indeterminism* in the physical universe (i.e, "universal indeterminism"). Any *additional* metaphysical speculation, such as to whether quantum indeterminism, i.e., intrinsic randomness, implicates the reality of experimenter "freedom", "free will", or "free choice", is redundant in relation to the predictive success of orthodox quantum mechanics. Accordingly, super-indeterminism views as redundant also, from a technical standpoint, whether an affirmative or a negative answer is claimed in reference to universal indeterminism as a necessary *precondition* for experimenter freedom. Super-indeterminism accounts, for example, for the circular reasoning which is implicit in the free will theorem by Conway and Kochen [1, 2]. The concept of super-indeterminism is of great assistance in clarifying the often misunderstood meaning of the concept of "free variables" as used by John Bell [3]. The present work argues that Bell sought an *operational, effective free will theorem*, one based upon the notion of "determinism *without* predetermination", i.e., one wherein "free variables" represent *universally uncomputable* variables. In conclusion, the standard interpretation of quantum theory does not answer, and does not need to answer in order to ensure the predictive success of *orthodox* theory, the question of whether either *incompatibilism* or *compatibilism* is *valid* in relation to free-will metaphysics and to the *free-will phenomenology* of experimenter agents in quantum mechanics.

1. Introduction

The question concerning the reality of experimenter free will in science in general, and in quantum mechanics in particular, has a long and controversial history. As is well-known, the concept of free will, as such, has eluded universal definition, no less in the context of quantum theory and quantum indeterminism (e.g., Kane [4, 5]). The present analysis will limit itself to the introduction – into the larger discussion on "quantum free will" – of the concept of "super-indeterminism". Its chief purpose is to draw attention to the fact that neither the metaphysical position of 'determinism' nor of 'indeterminism' is capable of offering any decisive answers towards the reality of experimenter free will in quantum mechanics; both, the *compatibilist*, and *incompatibilist*, interpretation of free will finds strong support in the community of free-will philosophers (Kane [4, 5]). For explanation, *compatibilism* is the position that determinism and free will are compatible, whereas *incompatibilism* states that they *cannot* be. Again, the critical

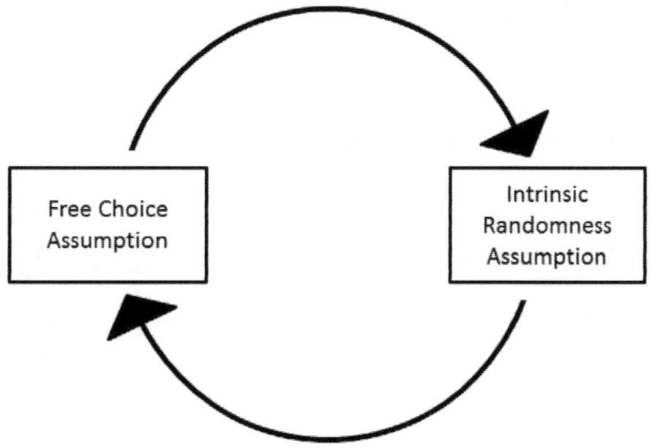

Figure 1. Super-indeterminism in the orthodox interpretation of quantum mechanics. Illustrated is the standard assumption that experimenter *free choice* depends on the existence of a process demonstrating *intrinsic randomness* and *vice versa*. Again, the reality of *universal indeterminism* depends equally – in turn – on experimenter free will being *undetermined*. For example, the concept of 'super-indeterminism' accounts for the much-cited, yet controversial, interpretation of experimenter free will in orthodox quantum mechanics by Conway and Kochen [1] who famously argued that "if indeed we humans have free will, then [so do] elementary particles."

question of determinism *versus* indeterminism in quantum theory, such as, for example, in the evaluation of the possibility of de Broglie–Bohm theory (Bohm [6, 7]) need not be conflated with the *additional* question of whether determinism or indeterminism is compatible with, or even indicative of, the reality of experimenter freedom. Fig. 1 illustrates the metaphysical position of 'super-indeterminism'.

1.1. Physical definitions of experimenter free will: indeterminism versus determinism

Regarding the free choice assumption, *physical definitions* in the literature on quantum foundations typically invoke concepts such as the *complete independence* from, or *complete lack of correlation* with, *any* past physical events in nature. A recent example of a concise definition of free choice in the context of *orthodox* quantum mechanics is the one by Colbeck and Renner [8]. They suggest that a choice called "... A is free if the only variables it is correlated with are those it could have caused." By that definition, it is apparent that a free choice action A is one that (i) must be "uncaused" itself, and (ii) sets in motion an entirely new causal chain, literally from "nothing", or at least starting from an assumption of "intrinsic randomness" (compare Fig. 1). The strict co-dependence of (non-causal) 'intrinsic randomness' and (uncaused) 'experimenter freedom' in the definition offered by Colbeck and Renner [8] is matched, for example, by the metaphysical assumptions also underlying the free will theorem by Conway and Kochen [1, 2]. These latter authors summed up the metaphysical import of their (free will) theorem by claiming that "if indeed we humans have free will, then [so do] elementary particles" (see Fig. 1). It is significant for the present work that Colbeck and Renner [8] conclude their own efforts at defining *physically* a free-will concept by referring to the authority of *John Bell.* They refer to Bell's well-known, but often *misinterpreted*, statement concerning the meaning of "free variables" in the context of quantum-mechanical predictions (Bell [3]): "For me this means that the values of such variables have implications only in their future light cones." In particular, Colbeck and Renner [8] have implicated a correspondence between Bell's position on experimenter freedom and their own incompatibilist (e.g., super-indeterministic) position on free choice. In the following, the incongruence, however, of these separate views will be demonstrated. The subsequent analysis will demonstrate that John Bell did *not* anticipate, and in fact denied, any conflict between experimenter freedom and an entirely *deterministic* world view (Bell [3, 9, 10]).

2. John Bell's own "free will theorem" in quantum mechanics is compatibilist

In contrast to both Conway and Kochen [1] and Colbeck and Renner [8], John Bell did *not* adopt an *incompatibilist* approach to the problem of free will in quantum mechanics, e.g., the super-indeterministic approach illustrated in Fig. 1. Instead, Bell was a *compatibilist* when it came to the free choice assumption and the notion of freedom in general. He argued consistently in the tradition of compatibilism starting in 1976 (Bell [3, 9, 10]). Although Bell never presented his position on free will as a theorem in name, he discussed in explicit terms the conceptual foundations, including key metaphysical assumptions, which lead him to propose a free choice concept on an *operational basis*, a position which might be called "effective indeterminism", or "determinism *without* predetermination". Specifically, Bell proposed that an experimenter agent serves in quantum correlation experiments as a source of *effectively* free variables, i.e., of variables that are, according to Bell, "... effectively free for the purposes at hand" (Bell [3]). "For me this means", Bell explained, "that the values of such variables have implications only in their future light cones. They are in no sense a record of, and do not give information about, what has gone before. In particular they have no implications for the hidden variables v in the overlap of the backward light cones..." (Bell [3]). Contrary to the free choice concept proposed by Colbeck and Renner [8], Bell's own approach to the free choice assumption is *not* founded upon intrinsic randomness or fundamental quantum indeterminism, i.e., his approach does *not* reflect *super-indeterminism* (see Fig. 1).

2.1. John Bell's concept of "free variables" contradicts "super-indeterminism"

The fact that Bell was a compatibilist and that he thought of 'freedom' as a concept that was *not* in opposition to 'determinism' is amply evident from his explanation in 1977 (Bell [3]). Then, Bell admitted making an error in his original presentation of 1976 when he had wrongly proposed that 'free' variables are "... not determined in the overlap of the backward light cones" (Bell [9]). "Here I must concede at once", Bell [3] admitted "that the hypothesis becomes quite inadequate when weakened in this way. The theorem no longer follows. I was mistaken." Following this admission, Bell affirmed that 'free variables' are variables that are neither 'not determined' nor 'uncaused' therefore, contradicting the orthodox concept of super-indeterminism described in Fig. 1. Instead, Bell's free variables are defined *operationally*, as was mentioned already, as those that are "... effectively free for the purposes at hand" (Bell [3]). Bell's crucial insight was this: a concept of *effective* experimenter free will must account for the fact that choices (actions) of the experimenter agent are *not pre-determined by the past history of the universe*. In particular, the variables associated with the choices by the epistemic agent must not be *pre-determined*, although they will be *unpredictably determined*, by "the hidden variables v in the overlap of the backward light cones..." (Bell [3]). For the technical meaning of the notion of "unpredictably determined" variables, in terms of 'universally uncomputable' variables, consult Sect. 4.

2.2. John Bell's concept of "free variables" contradicts "super-determinism"

That Bell was well aware of the crucial distinction between, on the one hand, "*pre*-determinism", such as in the form of "*super*-determinism", and, on the other hand, standard "determinism", is clearly evident from his final statements on the free will problem (Bell [10]). "One can envisage theories", Bell noted, "in which there just *are* no free variables... In such 'super-deterministic' theories the apparent free will of experimenters, and any other apparent randomness, would be illusory." The fact that Bell was a compatibilist and that he argued for the possibility of *effective* experimenter free will in relation to a deterministic world view, is apparent from his clear rejection of the *super-deterministic* approach. "I do not expect to see a serious theory of this kind", Bell [10] explained further. Instead, he proposed that "I would expect a serious theory to permit 'deterministic chaos' or 'pseudorandomness', for complicated subsystems (e.g., computers) which would provide variables sufficiently free for the purpose at hand."

2.3. John Bell's concept of "free variables" implies emergent unpredictability

The chief characteristic of "deterministic chaos", i.e., of the strictly *nonlinear*, *emergent* phenomenon that is accounted for by *self-organization* or *complexity theory*, is the *uncertainty*, or *unpredictability*, of observable outcomes. This is despite the fact that the chaotic behavior is entirely governed by *deterministic* relations. In principle, this essential feature of deterministic chaos, holds true for physical, chemical, biological, and even psycho-physical systems, including in neurophysiological brain states potentially associated with the free-choice performance by an experimenter agent (for an overview, e.g., Walleczek [11]). In short, the image used by Bell of "deterministic chaos" represents well the notion of "effective indeterminism" or "determinism *without* predeterminism". At the same time, Bell's reference to *deterministic chaos* and *pseudorandomness* is, of course, wholly *incompatible* with the intrinsic randomness assumption of orthodox quantum mechanics shown in Fig. 1. Consequently, it is misleading to characterize John Bell's own, well-documented views on free will (Bell [3]) as being in any way consistent with the *incompatibilist* free-choice concept described by Colbeck and Renner [8].

Finally, while the "super-indeterminism" in orthodox quantum mechanics has been recognized before, although not as a general metaphysical position (e.g., Hossenfelder [12]), it was nevertheless argued by many others that a *compatibilist free will concept*, i.e., determinism, must be in violation of the *non-signalling constraint* (e.g., Barrett *et al.* [13]; Kofler *et al.* [14]). The non-signalling constraint mediates the *apparent* conflict between quantum theory and the theory of special relativity (e.g., Eberhard [15]), and its role in deciding the viability of *compatibilist* free-will concepts in quantum mechanics is reviewed next.

3. The non-signalling theorem in the decision whether only an incompatibilist free will concept is viable for quantum mechanics

A standard view holds that *only* quantum theories invoking *universal* or *fundamental indeterminism* could successfully reproduce the predictions that are yielded by orthodox quantum mechanics. Specifically when negating the involvement of "cosmic conspiracy" theories, such as *super-determinism* (e.g., Bell [10]), in the interpretation of measurement outcomes, this view is widely thought to hold true. One major reason is the suggested violation of the standard, i.e., axiomatic, non-signalling constraint in any possible ontological quantum theory (e.g., Colbeck and Renner [16, 17]; Gallego *et al.* [18]). *Axiomatic* non-signalling, which was presumed in Bell's *original* interpretation of his theorem (Bell [19]), is often considered mandatory, therefore, from the mainstream perspective (see Fig. 2). However, as was argued recently, and is illustrated in Fig. 2, the case for, or against, an axiomatic non-signalling assumption is undecidable on logical grounds alone (Walleczek and Grössing [20, 21]). The reason is the non-reducible, *relational interdependence* of the metaphysical assumptions that underlie such considerations (see legend to Fig. 2).

3.1. The two interpretations by John Bell of the non-signalling theorem

As was reviewed recently, John Bell shifted his interest – after 1976 – from an "axiomatic non-signalling" assumption (Figs. 2 and 3) towards an assumption of "effective non-signalling" (Fig. 3), which is the non-signalling assumption that is compatible with de Broglie–Bohm theory (Walleczek and Grössing [21]). In the context of the present analysis, an *effective* non-signalling theorem *only*, but not an *axiomatic* one, is compatible also with Bell's notion of "free variables" as described in Sect. 2. That is, an effective non-signalling theorem allows for nonlocal

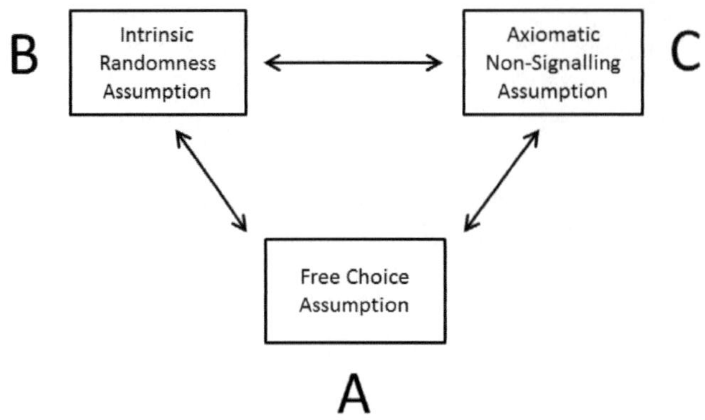

Figure 2. An axiomatic non-signalling assumption does not represent a conclusive argument for or against determinism and nonlocal hidden-variables theories in quantum mechanics (from Walleczek and Grössing [21]). As illustrated in Fig. 1 also, there exists – in orthodox quantum mechanics – an *interdependence* of free-choice (A) and intrinsic randomness (B) assumptions. The validity of an *axiomatic* non-signalling assumption (C), a standard assumption in orthodox quantum mechanics, depends in turn on the *independent* validity of assumptions A and B. Importantly, the concept of *super-indeterminism* (Fig. 1) demonstrates that an *independent* validation is out of reach however. This implies that the validity of the *axiomatic* interpretation of the non-signalling theorem cannot be proven based on present knowledge (for more details see Walleczek and Grössing [21]).

quantum *information transfers* while – at the same time – it prohibits *superluminal signalling* and communication. Fig. 3 provides a summary (i) of the two contrasting interpretations by John Bell of the non-signalling theorem and (ii) of the *communication-theoretic criterion* which was shown to account for the difference between Bell's original position of 1964 (Bell [19]) and his later position which he first introduced in 1976 (Bell [9]).

For the current analysis of the free will concept, the critical reason for introducing an effective, instead of an axiomatic, non-signalling condition is the following: an *effective* non-signalling concept eliminates the danger of contradiction between the free choice assumption and the non-signalling principle, for example, in the context of de Broglie–Bohm theory (Walleczek and Grössing [21]). Next, given the different options for interpreting the notion of experimenter freedom – compatibilism *versus* incompatibilism, Sect. 4 will offer an outlook towards the possible future of the free will concept in quantum mechanics and beyond.

4. The future of free will in quantum mechanics and beyond: How is freedom possible?

What is meant by the reality of experimenter free will? This work defined *the reality* of freedom – until now – simply as the *negation* of the statement that free will is an *illusion.* This preliminary (circular) definition drew on the earlier statement by John Bell [10] who had explained that in "... 'super-deterministic theories' the apparent free will of experimenters, and any other apparent randomness, would be illusory." Thus, the concept of "super-determinism" implicates the complete *absence* of experimenter freedom. However, for those theories that do *not* implicate super-determinism, how could experimenter free will be possible?

An informal, intuitive understanding of free will assumes that the free-willed agent is "free to choose" among different options or alternatives. Also known as the principle of alternative possibilities (e.g., Kane [5]), the intuitive understanding instantly raises the problem of "control" in free-willed choices: Who or what is in charge when is performed the "intentional" selection

Journal of Physics: Conference Series **701** (2016) 012005 doi:10.1088/1742-6596/701/1/012005

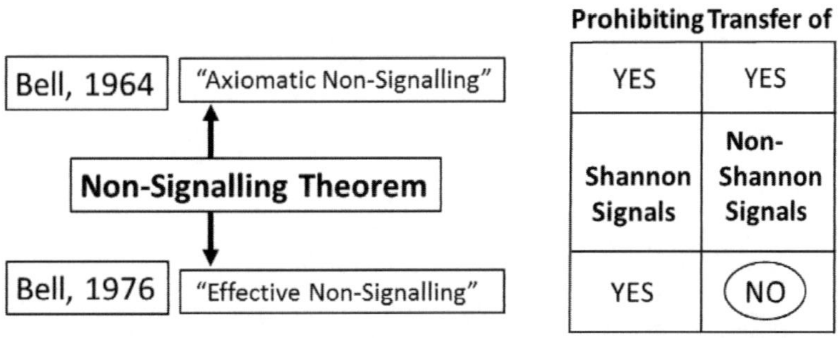

Figure 3. The two different interpretations by John Bell of the non-signalling theorem (from Walleczek and Grössing [21]). Bell's first interpretation (Bell [19]), here called "axiomatic non-signalling", rejects any form of deterministic quantum mechanics, upon assuming experimenter freedom in the incompatibilist, super-indeterministic sense (Fig. 1). The second, subsequent interpretation (Bell [9]), here called "effective non-signalling", is compatible with ontological, deterministic quantum approaches of the *nonlocal* type. Importantly, unlike axiomatic non-signalling, an effective non-signalling theorem does *not* prohibit the nonlocal transfer of *uncontrollable*, i.e., *non*-Shannon, signals. For definitions of Shannon and non-Shannon signals in the context of Shannon's mathematical theory of communication see Walleczek and Grössing [21]. Effective non-signalling ensures the possibility of de Broglie–Bohm theory, under an *effective free choice* assumption, in agreement with Bell's compatibilist notion of "free variables" (Bell [3]).

of *one* option from a set of *many* alternative options? The concept of 'experimenter *freedom*' cannot, therefore, be reasonably discussed without considering also the concept of 'experimenter *control*'. Indeed, generations of free-will philosophers have pointed out the close connection between freedom *and* control (for reviews see Kane [5]). For example, free will would be an illusion if an experimenter agent lacks (operational) control over his or her performance, such as in setting up a *certain* experimental configuration for the purpose of "asking specific questions of nature". An agent who can serve as a "source of operational control" was previously defined as a *type-1 agent*, whereas a "source of random variables" – *without* the element of control – was characterized as a *type-2 source*. For explanation, a '*random event generator*' represents a *type-2 source*, but it does *not* possess *type-1 agency*, according to these definitions (Walleczek and Grössing [21]).

4.1. Quo Vadis quantum mechanics? Dualism or monism?
Again, since "randomness" is usually viewed as the opposite of "control" it is hard to comprehend how 'universal indeterminism' – by itself – could provide a realistic basis for the free-willed actions of (type-1) agents. Similarly, "non-randomness", in the form of (pre)determinism, is equally unable to provide for the capacity of *controlling* physical events by an agent's own free will. Consensus appears to be building towards the view that neither determinism nor indeterminism provides an obvious pathway towards freedom *and* control (e.g., Kane [4, 5]). Regarding the free-willed control by an agent of physical events in nature, only an "*extra-physical factor*" might advance the possibility of genuine freedom in the strong sense of manifesting agent control. It certainly *appears* to be impossible to conceive of a scenario in which this extra-physical "causal power" (e.g., a concept of "free mental causation" by an agent) would not be in violation of *the* central tenet of modern science: the rejection of *dualism* in the form of the *Cartesian separation* of agent "mind" *from* "matter" (e.g., compare Cartesian 'Res Cogitans' *versus* 'Res Extensae'). *Quo Vadis* quantum mechanics? *Un-physical* dualism or *physical* monism? The final Fig. 4 illustrates a comparison of the two competing metaphysical frameworks that underlie the

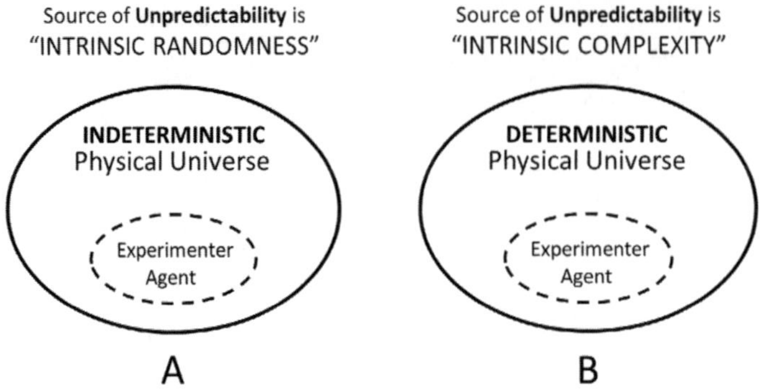

Figure 4. Illustrating (A) *universal indeterminism* and (B) *universal determinism.* (A) In 'universal indeterminism', agents and the physical universe are subject to the same *fundamental indeterminism.* Here, the in-principle *unpredictability* of physical observables is a function of "intrinsic randomness". Note that – by itself – indeterminism neither rejects nor affirms the reality of "free will" or "free choice" as a result of *super*-indeterminism (see Fig. 1). (B) In 'universal determinism', agents and the physical universe are subject to the same *fundamental determinism.* Here, the in-principle *unpredictability* of physical observables is a function of "intrinsic complexity". Note that determinism need *not* be identified with *pre*-determinism. This unnecessary, even false, identification of determinism with pre-determinism would render any notions of "free will" or "free choice" *obsolete* as part of scenario (B).

available options – universal indeterminism and determinism. What do these options mean for the future of quantum mechanics, e.g., regarding the pressing issue of *operationalism versus realism* in quantum theory?

Modern science rejects the Cartesian split, i.e., metaphysical *dualism*, between agents and world. Instead, modern science posits that experimenter agents are *integral elements* (of reality) of the physical universe, i.e., agents and universe – together – constitute a *relational physical continuum.* This *non-dual* continuum principle is indicated by the *open* line in Fig. 4 which encloses the (operational) presence of the experimenter agent as an intrinsic element of the universe. Importantly, natural science posits that the behavior of both experimenter agent and the rest of the physical universe is constrained by the same *Laws of Nature*, e.g., quantum theory and relativity theory. This reflects the (monistic) position of *scientific physicalism.* Must the physicalist position be *radically revised* in light of *orthodox* quantum mechanics and of the concept of *super-indeterminism* in Fig. 1?

4.2. Sources of unpredictability in universal indeterminism and determinism

The above analysis suggests that the relationship between experimenter freedom and experimenter control is neither addressed by universal *indeterminism* (Fig. 4a) nor by universal *determinism* (Fig. 4b). To repeat, the notion of 'control' is usually identified with *determination*, whereas the notion of 'freedom' with *indetermination.* In any case, essential to any minimal conception of experimenter freedom, as defined in the beginning of Sect. 4, appears to be the *in-principle unpredictability* of *individual events* in the universe. Naturally, this must include also an agent's free-choice performance in the selection of measurement settings as an entirely unpredictable event *itself* (compare Fig. 4). On the one hand, orthodox quantum mechanics offers the *traditional* metaphysical position in regards to the source of in-principle unpredictability: "*intrinsic randomness*" (Fig. 4a). On the other hand, concepts such as Bell's notion of "free variables", for example, offer an *alternative* metaphysical position: "*intrinsic complexity*". This alternative position is consistent with a *universally* deterministic universe (see

Fig. 4b). Finally, Bell's free-variables concept will be briefly discussed against the background of self-organization, complexity, or emergence, theory which advances the notion of observables in nature that are *universally uncomputable*.

4.3. John Bell's "free variables" as universally uncomputable variables
Free variables are those that are "... effectively free for the purposes at hand" (Bell [3]). As was reviewed in Sect. 2, in reference to "free variables", John Bell raised the similarity to phenomena from nonlinear dynamics such as 'deterministic chaos', e.g., 'pseudorandomness'. The present work suggests that Bell's concept of free variables can be justifiably defined as variables that *emerge* as a function of the *intrinsic complexity* of universally deterministic processes (compare Fig. 4b). These complex and self-referential processes may generate physical observables whose values are universally uncomputable, i.e., their computation would require an *infinite* amount of computational resources (e.g., compare the halting problem in computational theory). In this scenario, even *Laplace's demon*, and "*the universe itself*", would be incapable of predicting with absolute certainty the future outcome of *individual* microscopic events in nature. The concept of *universal uncomputability*, which is inherent in self-referential systems dynamics, plausibly offers the kind of effective unpredictability which John Bell sought. For example, see also the notions of "effective indeterminism" and "determination without predetermination" in Sect. 2.

5. Conclusions
The here introduced concept of *super-indeterminism* for orthodox quantum mechanics, which was shown to underlie the free will theorem by Conway and Kochen [1, 2], is consistent only with the metaphysical position shown in Fig. 4a: *intrinsic randomness*. By contrast, John Bell's search for an *effective free will* concept was shown to be consistent only with the position shown in Fig. 4b: *intrinsic complexity*. It is likely that significant advances towards a *realist* quantum mechanics will depend on new insights into what might be called also "*quantum complexity*", i.e., complexity at the smallest dimensions of reality. Summarizing, this work has identified two distinct routes towards avoiding the "cosmic pre-determination" of the free-choice performance by an experimenter agent: intrinsic randomness and intrinsic complexity (Fig. 4). Importantly, neither route was found to address the problem of the "free-willed control" – by an agent – of any physical events in nature. Importantly, both routes were argued to satisfy the minimal requirement of *unpredictability*, i.e., lack of pre-determination, of an agent's choices based upon past information about the universe. Finally, a better understanding of the limits and possibilities of the alternative position of *quantum complexity* – as a source of unpredictability – will require the development of a deeper understanding of the relationship between quantum theory and the theory of complexity, self-organization, and emergence. The prospects for the possibility of an *emergent quantum mechanics* critically depends on fashioning a deeper understanding also – in agreement with Bell's *effective free will* concept – of the *relational* nature of an emergent universe, including of emergent space and time.

Acknowledgments
Work by Jan Walleczek at Phenoscience Laboratories (Berlin) is partially funded by the Fetzer Franklin Fund of the John E. Fetzer Memorial Trust. The author wishes to thank Gerhard Grössing and the research group at the Austrian Institute for Nonlinear Studies, as well as Nikolaus von Stillfried, for invaluable insights during many excellent discussions.

References
[1] Conway J and Kochen S 2006 The free will theorem *Found. Phys.* **36**(10) 1441–1473 (*Preprint* `arXiv:quant-ph/0604079`)
[2] Conway J and Kochen S 2009 The strong free will theorem *Notices of the AMS* **56**(2) 226–232

[3] Bell J S 1977 Free variables and local causality *Epistemol. Lett.* **15** (2)(February) repr. in *Dialectica* **39** (1985) 103-106.

[4] Kane R 2005 *A Contemporary Introduction to Free Will* (New York, NY: Oxford University Press) ISBN 978-0-19-514970-8

[5] Kane R (ed) 2011 *The Oxford Handbook of Free Will* 2nd ed (UK: Oxford University Press) ISBN 978-0-19-539969-1

[6] Bohm D 1952 A suggested interpretation of the quantum theory in terms of "hidden" variables. I *Phys. Rev.* **85**(2) 166–179

[7] Bohm D 1952 A suggested interpretation of the quantum theory in terms of "hidden" variables. II *Phys. Rev.* **85**(2) 180–193

[8] Colbeck R and Renner R 2013 A short note on the concept of free choice (*Preprint* arXiv:1302.4446 [quant-ph])

[9] Bell J S 1976 The theory of local beables *Epistemol. Lett.* **9** (3)(March) repr. in *Dialectica* **39** (1985) 85-96.

[10] Bell J S 2004 La nouvelle cuisine, *Speakable and Unspeakable in Quantum Mechanics* (Cambridge, UK: Cambridge University Press) pp 232–248 revised ed ISBN 978-0-511-81567-6

[11] Walleczek J (ed) 2000 *Self-Organized Biological Dynamics and Nonlinear Control* (Cambridge, UK: Cambridge University Press) ISBN 0-521-62436-3

[12] Hossenfelder S 2012 The free will function (*Preprint* arXiv:1202.0720 [physics.hist-ph])

[13] Barrett J, Hardy L and Kent A 2005 No signaling and quantum key distribution *Phys. Rev. Lett.* **95**(1) 010503

[14] Kofler J, Paterek T and Brukner C 2006 Experimenter's freedom in Bell's theorem and quantum cryptography *Phys. Rev.* A **73**(2) 022104 (*Preprint* arXiv:quant-ph/0510167)

[15] Eberhard P H 1978 Bell's theorem and the different concepts of locality *Nuov. Cim.* B **46**(2) 392–419

[16] Colbeck R and Renner R 2011 No extension of quantum theory can have improved predictive power *Nature Commun.* **2** 411–415

[17] Colbeck R and Renner R 2012 Free randomness can be amplified *Nature Phys.* **8**(6) 450–454 (*Preprint* arXiv:1105.3195 [quant-ph])

[18] Gallego R, Masanes L, De La Torre G, Dhara C, Aolita L and Acín A 2013 Full randomness from arbitrarily deterministic events *Nature Commun.* **4** 2654 (*Preprint* arXiv:1210.6514 [quant-ph])

[19] Bell J S 1964 On the Einstein Podolsky Rosen paradox *Physics* **1**(3) 195–200

[20] Walleczek J and Grössing G 2014 The non-signalling theorem in generalizations of Bell's theorem *J. Phys.: Conf. Ser.* **504** 012001 (*Preprint* arXiv:1403.3588 [quant-ph])

[21] Walleczek J and Grössing G 2016 Nonlocal quantum information transfer without superluminal signalling and communication *Found. Phys.* In press. (*Preprint* arXiv:1501.07177 [quant-ph])

Conditions for Lorentz-invariant superluminal information transfer without signaling

G Grössing[1], S Fussy[1], J Mesa Pascasio[1,2], H Schwabl[1],

[1] Austrian Institute for Nonlinear Studies, Akademiehof, Friedrichstr. 10, 1010 Vienna, Austria
[2] Atominstitut, TU Wien, Operng. 9, 1040 Vienna, Austria

E-mail: ains@chello.at

Abstract. We understand emergent quantum mechanics in the sense that quantum mechanics describes processes of physical emergence relating an assumed sub-quantum physics to macroscopic boundary conditions. The latter can be shown to entail top-down causation, in addition to usual bottom-up scenarios. With this example it is demonstrated that definitions of "realism" in the literature are simply too restrictive. A prevailing manner to define realism in quantum mechanics is in terms of pre-determination independent of the measurement. With our counter-example, which actually is ubiquitous in emergent, or self-organizing, systems, we argue for *realism without pre-determination.* We refer to earlier results of our group showing how the guiding equation of the de Broglie–Bohm interpretation can be derived from a theory with classical ingredients only. Essentially, this corresponds to a "quantum mechanics without wave functions" in ordinary 3-space, albeit with nonlocal correlations.

This, then, leads to the central question of how to deal with the nonlocality problem in a relativistic setting. We here show that a basic argument discussing the allegedly paradox time ordering of events in EPR-type two-particle experiments falls short of taking into account the contextuality of the experimental setup. Consequently, we then discuss under which circumstances (i.e. physical premises) superluminal information transfer (but not signaling) may be compatible with a Lorentz-invariant theory. Finally, we argue that the impossibility of superluminal signaling – despite the presence of superluminal information transfer – is not the result of some sort of conspiracy (á la "Nature likes to hide"), but the consequence of the impossibility to exactly reproduce in repeated experimental runs a state's preparation, or of the no-cloning theorem, respectively.

1. Emergent Quantum Mechanics: realism without pre-determination

The term "Emergent Quantum Mechanics" (EmQM) has been used in the literature for several years by now, albeit with different meanings regarding the word "emergence". As a major option, the term refers to the possibility that quantum theory might be a (very good) approximation to some "deeper level theory". To some, though, EmQM just stands for quantum theory as a special case for a particular set of parameters of a more encompassing theory. The term would thus refer to the *emergence of a theory.* However, the meaning of EmQM may also be more specific in that it refers to *physical emergence*, i.e., to the modeling of quantum systems as *emergent systems.* It is the latter option that our group has dealt with throughout the last couple of years.

There is, however, a communication problem in getting the relevant ideas across, mainly because the quantum physics and the self-organization/emergence communities, respectively, hardly communicate with each other. Specifically, the problem of accepting physical emergence as a possibility within the quantum physics community seems to be the rather exotic looking theme of *top-down causation* (next to bottom-up causation). Despite the fact that there are numerous examples in hydrodynamics, self-organizing systems, etc., for top-down causation, most quantum physicists seem unaffected by the possibility of this "unusual" type of causality, although the usual understanding of causality is apparently insufficient for a description and understanding of quantum processes. To give just one example, consider the Rayleigh–Bénard cells of hydrodynamics. There, one witnesses microscopic random movement that spontaneously becomes ordered on a macroscopic level. The top-down causality is manifest in that emergent particle trajectories depend crucially on the boundary conditions of the system.

Certainly, this classical example of top-down causation is neither "weird" (as quantum mechanics is claimed to be), nor "surreal" (as the trajectories' behaviors in the convection cells might be called, did one not have a perfectly rational explanation for it). Rather, this example must by necessity be covered by any definition of realism that one would claim as being generally applicable to any physical system. However, in the recent literature, definitions of realism have been proposed in the context of quantum foundations that would deny the above-mentioned examples of top-down causation to be considered "real". In other words, the phenomena of Rayleigh–Bénard cells in particular, but also all other processes of self-organizing, or emergent, systems, are actually counter-examples to recent definitions of "realism" in the quantum foundations literature, such as the following: "all measurement outcomes are determined by pre-existing properties of particles independent of the measurement (realism)." [1]

This is now to be contrasted with our proposal for a physical EmQM, i.e., *realism without pre-determination*. Instead of pre-determination, in the systems of interest we consider the case of co-evolution, i.e., permanently updated co-determination, with essential influences on the microphysics by changing boundary conditions, or measurement arrangements, respectively. With this perspective, it becomes clear that via importing ill-defined concepts of "realism" into the debate about Bell's theorems, all sorts of wrong conclusions become possible. After all, in this scenario "realism" effectively just becomes a "red herring", with the task of killing all proposals of thus understood "realistic" hidden variable theories – an exercise that in the end is rather fruitless. Looking back on the development of quantum foundations throughout the last decades, one actually may get the impression that this kind of strategy to exclude realism in this way has kept a lot of people busy. A historian of science might be well advised to consider a study on "A Brief History of Red Herrings", or the like – this might turn into quite a voluminous book.

To give just one more example of such a red herring, not of recent times, but of half a century ago, consider the following passage from Richard Feynman's famous description of electrons passing a double slit:

"We now make a few remarks on a suggestion that has sometimes been made to try to avoid the description we have given [i.e., of the double slit experiment with electrons]: 'Perhaps the electron has some kind of internal works – some inner variables – that we do not yet know about. Perhaps that is why we cannot predict what will happen. If we could look more closely at the electron, we could be able to tell where it would end up.' So far as we know, that is impossible. We would still be in difficulty. Suppose we were to assume that inside the electron there is some kind of machinery that determines where it is going to end up. That machine must also determine which hole it is going to go through on its way. But we must not forget that what is inside the electron should not be dependent on what *we* do, and in particular upon whether we open or close one of the holes.

So, if an electron, before it starts, has already made up its mind (a) which hole it is going to use, and (b) where it is going to land, we should find P_1 for those electrons that have chosen hole *1*, P_2 for those that have chosen *2*, and necessarily the sum $P_1 + P_2$ for those that arrive through the two holes. There seems to be no way around this. But we have verified experimentally that this is not the case. And no one has figured a way out of this puzzle. So at the present time we must limit ourselves to computing probabilities. We say 'at the present time,' but we suspect very strongly that it is something that will be with us forever – that it is impossible to beat that puzzle – that this is the way nature really *is*." [2]

It is interesting to see how strongly Feynman insists on his conclusion about "the way nature really *is*", although the logic his argument is based on depends on the unquestioned assumption that the electron's behavior is pre-determined, and independent of the measurement. This assumption, however, is in stark contrast to an alternative scenario which is proposed in our EmQM, i.e., that the electron does not propagate in a completely empty space, but is embedded in a "medium". The latter we identify as the vacuum's zero-point field, which can mediate information about the boundary conditions as given by the source and the measurement apparatus. The zero-point field is also the decisive agent in describing the electron's behavior in stochastic electrodynamics [3], with which our ansatz shares some characteristics, and a prototype to illustrate wave-particle duality via particle-medium interactions is given by the famous experiments with oil droplets "walking" on an oil bath [4–6].

So, there is one essential characteristic that radically contrasts particle behavior in our EmQM to that insinuated by above-quoted definition of "realistic" hidden variable theories: Instead of pre-determined velocities of the latter, the particles in our EmQM exhibit emergent velocities stemming from the constant interplay of the particle forward velocity at a particular instant in time with the wave-like embedding surroundings of the zero-point field. Said constant interplay, or mutual influencing, of particle and wave dynamics (which has its classical analogy in Couder's oil droplets, or "walkers"), is a manifestation of what we call *relational causality*: the interlocking of bottom-up and top-down causalities. Assuming local microscopic interactions in a sub-quantum domain, these form – together with the macroscopic boundary conditions – emergent structures in the quantum domain which may exhibit spontaneous nonlocal order. In turn, the thus created emergent structures affect the local microscopic interactions in a top-down manner, which thus closes the causal circle relating vastly different spatial scales at the same time (Fig. 1.1).

We have applied the concept of relational causality to the situation of double-slit interference (Fig. 1.2). Considering an incoming beam of, say, electrons with wave number $\mathbf{k}$ impinging on a wall with two slits, two beams with wave numbers $\mathbf{k}_A$ and $\mathbf{k}_B$, respectively, are created, which one may denote as "pre-determined" quantities, resulting also in pre-determined velocities $\mathbf{v}_I = \frac{1}{m}\hbar\mathbf{k}_I$, $I{=}A$ or B. The definition of "realism" (but also Feynman's dictum) that we criticized above would now imply that any realistic hidden variable theory just has these pre-determined velocities at its disposal for modeling double-slit interference. This, however, would constitute a very naive form of realism which, to our knowledge, nobody in the quantum physics community supports (... thus making it a classical red herring)!

However, if one considers that the electrons are not moving in empty space, but in an undulatory environment created by the ubiquitous zero-point field "filling" the whole experimental setup, a very different picture emerges. For then one has to combine all the velocities/momenta at a given point in space and time in order to compute the resulting, or "emergent", velocity/momentum field $\mathbf{v}_i = \frac{1}{m}\hbar\boldsymbol{\kappa}_i$, $i{=}1$ or 2 (Fig. 1.2), where i is a bookkeeping index not necessarily related to the particle coming from a particular slit [7]. The relevant contributions other than the particle's forward momentum $m\mathbf{v}$ originate from the so-called osmotic (or diffusive) momentum $m\mathbf{u}$. The latter is well known from Nelson's stochastic theory [8], but its identical form has been derived by one of us from an assumed sub-quantum nonequilibrium thermodynamics [9, 10]. Introducing the osmotic momentum in a sub-quantum hidden variable theory constitutes a decisively concrete step beyond the much older, but only rather general proposal by one of us of "quantum systems as 'order-out-of-chaos'

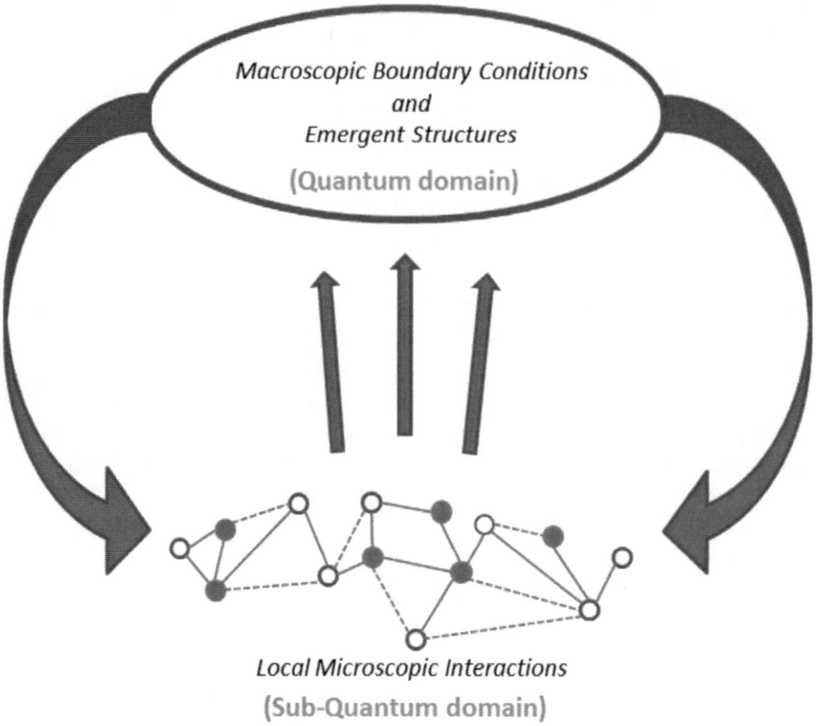

Figure 1.1. Scheme of relational causality: mutual (bottom-up and top-down) processes at the same time.

phenomena" [11]. For now it becomes possible to model double slit interference in more detail, with momentum conservation guaranteed as soon as one takes both the co-evolving forward and the osmotic velocity fields into account [7, 12]. This constitutes, among others, a viable causal model with its implied violation of what is called "causal parameter independence". The latter would state that in EPR-type scenarios Alice's measurement outcomes would not depend on Bob's measurement settings. However, as local changes of boundary conditions such as the settings of an apparatus nonlocally affect the whole system, our relationally causal model does describe said dependence and is therefore not excluded by recent no-go principles for certain causal hidden variable theories [13]. Not only that, our model also provides an understanding and deeper-level explanation of the microphysical, causal processes involved, i.e., of the "guiding law" [14], which also happens to be identical with the central postulate of the de Broglie–Bohm theory.

2. Identity of the emergent kinematics of N bouncers in real 3-dimensional space with the configuration-space version of de Broglie–Bohm theory for N particles

As in our model the "particle" is actually a bouncer in a fluctuating wave-like environment, i.e. analogously to the Couder's bouncers, one does have some (e.g. Gaussian) distribution, with its center following the Ehrenfest trajectory in the free case, but one also has a diffusion to the right and to the left of the mean path which is just due to that stochastic bouncing. Thus the total velocity field of our bouncer in its fluctuating environment is given by the sum of the forward velocity $\mathbf{v}$ and the respective diffusive (or "osmotic") velocities $\mathbf{u}_{\mathrm{L}}$ and $\mathbf{u}_{\mathrm{R}}$ to the left and the right. As for any direction α the diffusion velocity $\mathbf{u}_\alpha = D\frac{\nabla_\alpha P}{P}$, $\alpha = L$ or R, does not necessarily fall off with the distance, one has long effective tails of the distributions which

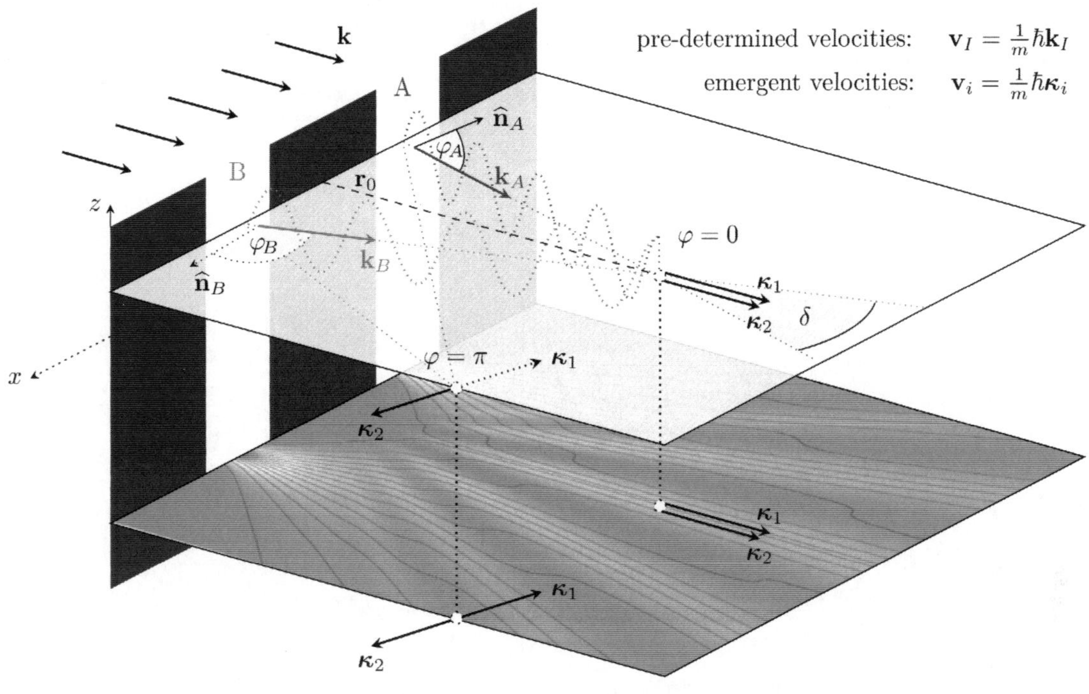

Figure 1.2. Scheme of interference at a double slit. Considering an incoming beam of electrons with wave number $\mathbf{k}$ impinging on a wall with two slits, two beams with wave numbers $\mathbf{k}_A$ and $\mathbf{k}_B$, respectively, are created, which one may denote as "pre-determined" quantities, resulting also in pre-determined velocities $\mathbf{v}_I = \frac{1}{m}\hbar\mathbf{k}_I, I{=}A \,\text{or}\, B$. Recent definitions of "realism" in the quantum foundations literature would now imply that *any* realistic hidden variable theory just has these pre-determined velocities at its disposal for modeling double-slit interference. However, if one considers that the electrons are not moving in empty space, but in an undulatory environment created by the ubiquitous zero-point field "filling" the whole experimental setup, a very different picture emerges. For then one has to combine all the velocities/momenta at a given point in space and time in order to compute the resulting, or "emergent", velocity/momentum field $\mathbf{v}_i = \frac{1}{m}\hbar\boldsymbol{\kappa}_i$, $i = 1 \,\text{or}\, 2$. The relevant contributions differing from the particle's forward momentum $m\mathbf{v}$ originate from the so-called osmotic (or diffusive) momentum field $m\mathbf{u}$. Thus it becomes possible to model double slit interference in microscopic detail, as can be seen from the lower plane where intensity distributions and average trajectories are shown. Local momentum conservation is thereby guaranteed as soon as one takes both the co-evolving forward and the osmotic velocity fields into account.

contribute to the nonlocal nature of the interference phenomena [15]. In sum, one has three, distinct velocity (or current) channels per slit in an n-slit system.

We have previously shown [7, 16] how one can derive the Bohmian guidance formula from our bouncer/walker model. Introducing classical wave amplitudes $R(\mathbf{w}_i)$ and generalized velocity field vectors $\mathbf{w}_i$, which stand for either a forward velocity $\mathbf{v}_i$ or a diffusive velocity $\mathbf{u}_i$ in the direction transversal to $\mathbf{v}_i$, we account for the phase-dependent amplitude contributions of the total system's wave field projected on one channel's amplitude $R(\mathbf{w}_i)$ at the point $(\mathbf{x}, t)$ in the following way: We define a *conditional probability density* $P(\mathbf{w}_i)$ as the local wave intensity $P(\mathbf{w}_i)$ in one channel (i.e. $\mathbf{w}_i$) upon the condition that the totality of the superposing waves is given by the "rest" of the $3n - 1$ channels (recalling that there are 3 velocity channels per slit). The expression for $P(\mathbf{w}_i)$ represents conditions which we describe as "relational causality": any change in the local intensity affects the total field, and *vice versa*, any change in the total field affects the local one. In an n-slit system, we thus obtain for the conditional probability densities and the corresponding currents, respectively, i.e. for each channel component i,

$$P(\mathbf{w}_i) = R(\mathbf{w}_i)\widehat{\mathbf{w}}_i \cdot \sum_{j=1}^{3n} \widehat{\mathbf{w}}_j R(\mathbf{w}_j) \tag{2.1}$$

$$\mathbf{J}(\mathbf{w}_i) = \mathbf{w}_i P(\mathbf{w}_i), \qquad i = 1, \ldots, 3n, \tag{2.2}$$

with

$$\cos\varphi_{i,j} := \widehat{\mathbf{w}}_i \cdot \widehat{\mathbf{w}}_j. \tag{2.3}$$

Consequently, the total intensity and current of our field read as

$$P_{\text{tot}} = \sum_{i=1}^{3n} P(\mathbf{w}_i) = \left(\sum_{i=1}^{3n} \widehat{\mathbf{w}}_i R(\mathbf{w}_i) \right)^2 \tag{2.4}$$

$$\mathbf{J}_{\text{tot}} = \sum_{i=1}^{3n} \mathbf{J}(\mathbf{w}_i) = \sum_{i=1}^{3n} \mathbf{w}_i P(\mathbf{w}_i), \tag{2.5}$$

leading to the *emergent total velocity field*

$$\mathbf{v}_{\text{tot}} = \frac{\mathbf{J}_{\text{tot}}}{P_{\text{tot}}} = \frac{\sum_{i=1}^{3n} \mathbf{w}_i P(\mathbf{w}_i)}{\sum_{i=1}^{3n} P(\mathbf{w}_i)}. \tag{2.6}$$

In [7, 16] we have shown with the example of $n = 2$, i.e. a double slit system, that Eq. (2.6) can equivalently be written in the form

$$\mathbf{v}_{\text{tot}} = \frac{R_1^2 \mathbf{v}_1 + R_2^2 \mathbf{v}_2 + R_1 R_2 (\mathbf{v}_1 + \mathbf{v}_2)\cos\varphi + R_1 R_2 (\mathbf{u}_1 - \mathbf{u}_2)\sin\varphi}{R_1^2 + R_2^2 + 2R_1 R_2 \cos\varphi}. \tag{2.7}$$

The trajectories or streamlines, respectively, are obtained in the usual way by integration. As first shown in [17], by re-inserting the expressions for convective and diffusive velocities, respectively, i.e. $\mathbf{v}_i = \frac{\nabla S_i}{m}$, $\mathbf{u}_i = -\frac{\hbar}{m}\frac{\nabla R_i}{R_i}$, one immediately identifies Eq. (2.7) with the Bohmian guidance formula. Naturally, employing the Madelung transformation for each path j ($j = 1$ or 2),

$$\psi_j = R_j \mathrm{e}^{\mathrm{i}S_j/\hbar}, \tag{2.8}$$

and thus $P_j = R_j^2 = |\psi_j|^2 = \psi_j^* \psi_j$, with $\varphi = (S_1 - S_2)/\hbar$, and recalling the usual trigonometric identities such as $\cos\varphi = \frac{1}{2}\left(\mathrm{e}^{\mathrm{i}\varphi} + \mathrm{e}^{-\mathrm{i}\varphi}\right)$, one can rewrite the total average current immediately in the usual quantum mechanical form as

$$\begin{aligned} \mathbf{J}_{\text{tot}} &= P_{\text{tot}} \mathbf{v}_{\text{tot}} \\ &= (\psi_1 + \psi_2)^*(\psi_1 + \psi_2)\frac{1}{2}\left[\frac{1}{m}\left(-\mathrm{i}\hbar\frac{\nabla(\psi_1 + \psi_2)}{(\psi_1 + \psi_2)}\right) + \frac{1}{m}\left(\mathrm{i}\hbar\frac{\nabla(\psi_1 + \psi_2)^*}{(\psi_1 + \psi_2)^*}\right)\right] \\ &= -\frac{\mathrm{i}\hbar}{2m}\left[\Psi^*\nabla\Psi - \Psi\nabla\Psi^*\right] = \frac{1}{m}\mathrm{Re}\left\{\Psi^*(-\mathrm{i}\hbar\nabla)\Psi\right\}, \end{aligned} \tag{2.9}$$

where $P_{\text{tot}} = |\psi_1 + \psi_2|^2 =: |\Psi|^2$.

Eq. (2.6) has been derived for one particle in an n-slit system. However, it is straightforward to extend this derivation to the many-particle case [14]. *Therefore, what looks like the necessity in the de Broglie–Bohm theory to superpose wave functions in configuration space, can equally be obtained by superpositions of all relational amplitude configurations of waves in real 3-dimensional space. The central ingredient for this to be possible is to consider the emergence of the velocity field from the interplay of the totality of all of the system's velocity channels.*

It can also be shown that the average force acting on a particle in our model is the same as the Bohmian quantum force. For, due to the identity of our emerging velocity field with the guidance formula, and because they essentially differ only via the notations due to different forms of bookkeeping, their respective time derivatives must also be identical. Thus, from Eq. (2.6) one obtains the particle acceleration field (using a one-particle scenario for simplicity) in an n-slit system as

$$\begin{aligned}\mathbf{a}_{\text{tot}}(t) = \frac{\mathrm{d}\mathbf{v}_{\text{tot}}}{\mathrm{d}t} &= \frac{\mathrm{d}}{\mathrm{d}t}\left(\frac{\sum_{i=1}^{3n}\mathbf{w}_i P(\mathbf{w}_i)}{\sum_{i=1}^{3n} P(\mathbf{w}_i)}\right)\\ &= \frac{1}{\left(\sum_{i=1}^{3n} P(\mathbf{w}_i)\right)^2}\left\{\sum_{i=1}^{3n}\left[P(\mathbf{w}_i)\frac{\mathrm{d}\mathbf{w}_i}{\mathrm{d}t} + \mathbf{w}_i\frac{\mathrm{d}P(\mathbf{w}_i)}{\mathrm{d}t}\right]\left(\sum_{i=1}^{3n} P(\mathbf{w}_i)\right)\right.\\ &\qquad\left. - \left(\sum_{i=1}^{3n}\mathbf{w}_i P(\mathbf{w}_i)\right)\left(\sum_{i=1}^{3n}\frac{\mathrm{d}P(\mathbf{w}_i)}{\mathrm{d}t}\right)\right\}.\end{aligned} \tag{2.10}$$

Note in particular that (2.10) typically becomes infinite for regions $(\mathbf{x}, t)$ where $P_{\text{tot}} = \sum_{i=1}^{3n} P(\mathbf{w}_i) \to 0$, in accordance with the Bohmian picture.

From (2.10) we see that even the acceleration of one particle in an n-slit system is a highly complex affair, as it nonlocally depends on all other accelerations and temporal changes in the probability densities across the whole experimental setup! In other words, this force is truly emergent, resulting from a huge amount of bouncer-medium interactions, both locally and nonlocally. This now leads to the central question of how to deal with the nonlocality problem in a relativistic setting. We shall discuss this problem by considering EPR-type experiments.

3. Nonlocality and relativity: A case for Lorentz invariant superluminal information transfer

As is well known, in an EPR-type experiment, where two photons are emitted in opposite directions from one spin-zero state, two different observers would form mutually inconsistent pictures of reality (Fig. 3.1).

However, Maudlin has correctly pointed out that "Relativity also reveals some of the apparent contradictions between frames to be merely matters of equivocation." [18] He explains: "The unprimed (here: l.h.s.) frame says that the right-hand photon is detected before the left while the primed (here: r.h.s.) frame has it the other way around. How could they both be right? In this case the answer is clear: they are simply talking about different things. The unprimed

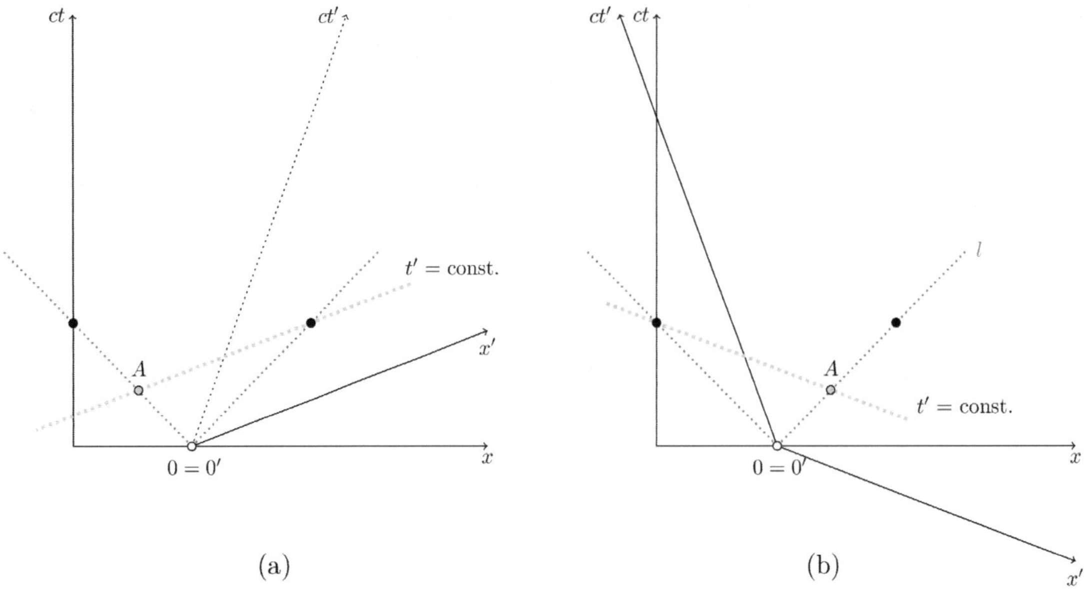

Figure 3.1. (a) An EPR-correlated particle pair is emitted at source *0* and later simultaneously registered by detectors (symbolized by black circles) in the laboratory rest frame. However, an observer in a reference frame moving to the right, would see the right-hand particle registered at a time *t'* at which the left-hand particle is located at spacetime point A. Thus, a "state jump" would occur at A before the particle is registered. (b) Same as in (a), but with observer in a reference frame moving to the left. The latter will see the left-hand particle registered at time *t'* at which the right-hand particle is located at spacetime point A, so that a "state jump" would occur at A before the particle is registered. Taken together, (a) and (b) point at a conflict as to which order of events "really" is happening in a realistic world view.

frame notes precedence in its t-coordinate, which we might call 'time', while the primed frame is concerned with precedence in its own t-coordinate, which we could call 'primetime'. There is no more contradiction between saying that the right detection event precedes the left in time but follows it in primetime than there is in saying that Idaho precedes New Jersey in geographical area but follows it in terms of population." [18]

So, the riddle posed in Fig. 3.1 is solved by realizing that there does not exist the one "true time ordering". All that matters is the Lorentz invariance of the theory. Paraphrasing Maudlin, we add "space" and "primespace" in the scenario, just to obtain frames of "spacetime" and "prime-spacetime". Then, our two different viewpoints are two versions of a Lorentz invariant behavior in space and time, which can be transformed into each other via simple rotation, as required by the Lorentz transformations (Fig. 3.2).

Again, we agree with Maudlin as to the consequences of this insight. He argues that many would agree that "Relativity prohibits something from going faster than light:

Matter or energy cannot be transported faster than light.

Signals cannot be sent faster than light.

Causal processes cannot propagate faster than light.

Information cannot be transmitted faster than light."

But, alternatively, one just needs to require that theories must be Lorentz invariant: "This requirement is compatible with the violation of every one of the prohibitions listed above." [18] However, in order to clearly see the implications of the required Lorentz invariance, one must appreciate that the whole experimental arrangement has to be taken into account, i.e., including

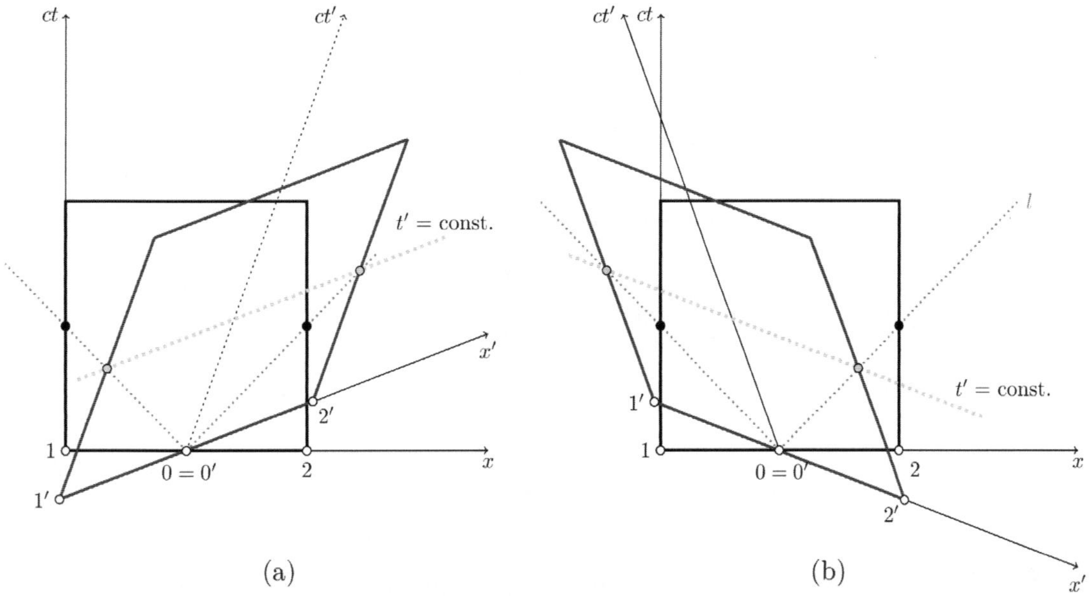

Figure 3.2. As Lorentz transformations correspond to simple rotations in spacetime diagrams, the two situations of Fig. 3.1 just correspond to two different viewpoints. However, what is crucially important to be able to show this, is that one has to take into account the whole experimental arrangement including the source *0* and the detectors *1* and *2*, in each reference frame. Then, even nonlocal correlations (symbolized by the green line) are part of a relativistically invariant description. Moreover, this renders obsolete the question of which of the options (a) and (b) was the "true" time ordering.

the source *0* and the detectors *1* and *2*, in each reference frame. Then, even nonlocal correlations (symbolized by the green line in Fig. 3.2) are part of a relativistically invariant description.

An especially interesting scenario is given for the case that the two frames meet such that the spacetime points at the origin coincide, i.e. *1=1'* (Fig. 3.3). In the resting laboratory frame, the experiment is arranged along an area between points *1* and *2*, with the source *0* emitting photon pairs at time $t_0 \geq 0$. Imagine now that to each element of the experimental setup is attached a traffic light that shows "red" when the preparation is not yet completed, and "green" when it is. In the rest frame, the totality of all traffic lights showing "green" will occur at some first instance, i.e., at some time t_0 = const. Now, what would a moving observer see? The answer is given by Rindler's "wave of simultaneity": whereas the green lights will light up in a "flash" which in the rest frame occurs simultaneously at the time $t_0 = 0$, in the moving observer's frame they will light up simultaneously at some time $t_0^{'} = 0$. (For a more detailed description, with arguments involving a "rigged Hilbert space", see [19].) This has the strict consequence that the photon pairs can be emitted from source *0'* only at times $t_0^{'} \geq 0$. And this is also the source of misleading arguments in the literature. For, although the lightcone is defined uniquely in all reference frames, this is true only as to its spreading in spacetime due to the universality of the vacuum speed of light. However, the *timing* of the emission of the photon pairs must in general be different for two observers moving with relative velocities to each other. In other words, one has to consider the relativity of simultaneity of the whole experimental setup including the source and the extension of the apparatus. Note also that if the moving observer would register at A a photon that was emitted at *0*, we would have the same kind of dilemma as in Figs. 3.1 (a) and (b). For then the "state jump" of the left-hand photon would occur at $\underline{A}$, i.e., before its arrival at the detector $\underline{B}$. However, if one correctly describes the source *0'* of the particle

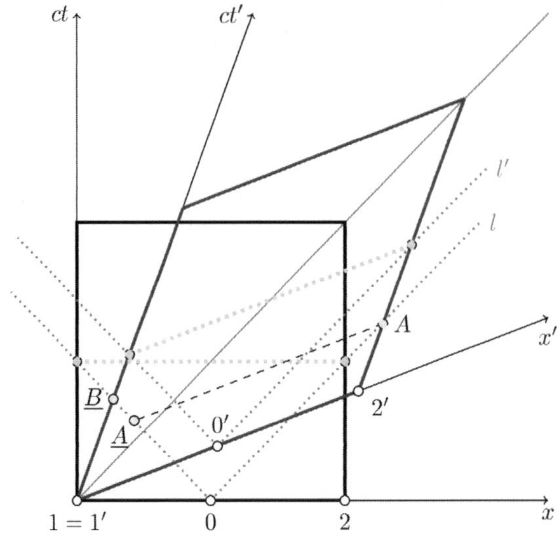

Figure 3.3. In the laboratory rest frame, the EPR-correlated particle pair is emitted at source *0* and later simultaneously registered by detectors (symbolized by green circles). If, under the wrong assumption, the moving observer would register at A a photon that was emitted at *0*, then the "state jump" of the left-hand photon would occur at $\underline{A}$, i.e., before its arrival at the detector at $\underline{B}$. However, since the preparation of an entangled photon pair requires the preparation of the whole experimental setup, and not just of the source, the world lines of all elements of the apparatus must be considered as forming one unseparable whole. Therefore, the photon pairs can be emitted from source 0' only at times $t_0' \geq 0$. Consequently, also in the moving frame will the photons arrive simultaneously at the respective detectors, and Lorentz invariance is again established.

pair in the moving observer's rest frame, then the detectors at points *1'* and *2'*, respectively, would always simultaneously register the corresponding photons. Thus, by relating the axis of equitemporality to the whole extension of an experimental apparatus in spacetime, which is different for each reference frame, one avoids the error of attributing some "idealistic" time to the system (like, e.g., $t' = \text{const}$ in Figs. 3.1 (a) and (b)).

We have thus shown that even nonlocal correlations can be part of a relativistically invariant description of events in spacetime. The conditions for Lorentz-invariant superluminal information transfer are essentially given by the evolution in spacetime of the whole, encompassing system (i.e., quantum system plus macroscopic apparatus, including source and detectors). A crucial question then arises as to interventions in such a system, which lead to the phenomenon of *dynamical nonlocality* [20]. In the above-mentioned paper by one of us [11], a type of experiment called "late choice experiment" was proposed that we then believed would provide an effect with dramatic consequences. The question was clarified for us only fairly recently in the paper by Tollaksen *et al.* [20], with the consequence that although the effect exists its measurable consequences are not dramatic at all. We shall here only refer to the variant that Tollaksen *et al.* discuss, for it is the relevant one. Essentially, the authors ask what happens in a realistic scenario (where one electron goes through just one of two slits present) if at the very moment the particle passes the slit, the other slit is being closed (or opened, in case it was closed before). The authors show within the Heisenberg picture that the opening or closing of a slit results in the nonlocal transfer of (what they call "modular") momentum. (Our group has shown that this effect can also be demonstrated when using the Schrödinger picture [15].)

However, there is an in-principle uncontrollability of that momentum transfer: Tollaksen *et al.* [20] speak of "complete uncertainty" in this regard. This means that here one has the case of a nonlocal transfer of information (i.e., from a slit to the particle), which, however, cannot be used for signaling: due to the necessary uncertainty of the location of the particle before the intervention, the nonlocal momentum transfer just shifts the particle within the wave packet, figuratively speaking, so that its detection at some location cannot in principle give any indication of whether or not that information transfer has happened.

A corollary to this concerns an implication of the no-cloning theorem. Opening or closing of a slit amounts to the preparation of a new state with a certain phase relative to the one associated with the path of the particle through the other slit. Considering many runs with such an intervention highlights an important feature of the system: it cannot be controlled. In other words, the identical phase can *never* be reproduced, which means that the no-cloning theorem is here ultimately responsible for the impossibility of superluminal signaling. (For, if one could control the phase during intervention at a slit, one could do this massively in parallel, and thus on average steer the electron to a desired position.)

Thus, both the consistencies of relativity and of quantum theory are confirmed, despite the superluminal information transfer involved. In other words, complete uncertainty makes it possible to have *nonlocal information transfer without superluminal signaling.* The latter is exactly the option shown in [21] to remain valid for viable nonlocal hidden variable theories.

Acknowledgments
We thank Jan Walleczek for many enlightening discussions, and the Fetzer Franklin Fund of the John E. Fetzer Memorial Trust for partial support of the current work.

References
[1] Gröblacher S, Paterek T, Kaltenbaek R, Brukner C, Zukowski M, Aspelmeyer M and Zeilinger A 2007 An experimental test of non-local realism *Nature* **446** 871–875 (*Preprint* arXiv:0704.2529 [quant-ph])

[2] Feynman R P, Leighton R B and Sands M 1966 *The Feynman Lectures on Physics: Quantum Mechanics* vol 3 (Addison-Wesley)

[3] de la Peña L, Cetto A M and Valdés-Hernándes A 2014 *The Emerging Quantum: The Physics behind Quantum Mechanics* (Berlin: Springer) ISBN 978-3-319-07893-9

[4] Couder Y, Protière S, Fort E and Boudaoud A 2005 Dynamical phenomena: Walking and orbiting droplets *Nature* **437** 208–208

[5] Bush J W 2015 Pilot-wave hydrodynamics *Annu. Rev. Fluid Mech.* **47**(1) 269–292

[6] Batelaan H, Jones E, Huang W C W and Bach R 2016 this volume

[7] Fussy S, Mesa Pascasio J, Schwabl H and Grössing G 2014 Born's rule as signature of a superclassical current algebra *Ann. Phys.* **343** 200–214 (*Preprint* arXiv:1308.5924 [quant-ph])

[8] Nelson E 1966 Derivation of the Schrödinger equation from Newtonian mechanics *Phys. Rev.* **150**(4) 1079–1085

[9] Grössing G 2008 The vacuum fluctuation theorem: Exact Schrödinger equation via nonequilibrium thermodynamics *Phys. Lett.* A **372**(25) 4556–4563 (*Preprint* arXiv:0711.4954 [quant-ph])

[10] Grössing G 2009 On the thermodynamic origin of the quantum potential *Physica* A **388** 811–823 (*Preprint* arXiv:0808.3539 [quant-ph])

[11] Grössing G 1989 Quantum systems as «order out of chaos» phenomena *Nuov. Cim.* B **103**(5) 497–509

[12] Grössing G, Fussy S, Mesa Pascasio J and Schwabl H 2010 Emergence and collapse of quantum mechanical superposition: Orthogonality of reversible dynamics and irreversible diffusion *Physica* A **389**(21) 4473–4484 (*Preprint* arXiv:1004.4596 [quant-ph])

[13] Ringbauer M, Giarmatzi C, Chaves R, Costa F, White A G and Fedrizzi A 2016 Experimental test of nonlocal causality (*Preprint* arXiv:1602.02767 [quant-ph])

[14] Grössing G, Fussy S, Mesa Pascasio J and Schwabl H 2015 Implications of a deeper level explanation of the deBroglie–Bohm version of quantum mechanics *Quantum Stud.: Math. Found.* **2**(1) 133–140 (*Preprint* arXiv:1412.8349 [quant-ph])

[15] Grössing G, Fussy S, Mesa Pascasio J and Schwabl H 2013 'Systemic nonlocality' from changing constraints on sub-quantum kinematics *J. Phys.: Conf. Ser.* **442** 012012 (*Preprint* arXiv:1303.2867 [quant-ph])

Journal of Physics: Conference Series **701** (2016) 012006 doi:10.1088/1742-6596/701/1/012006

[16] Grössing G, Fussy S, Mesa Pascasio J and Schwabl H 2014 Relational causality and classical probability: Grounding quantum phenomenology in a superclassical theory *J. Phys.: Conf. Ser.* **504** 012006 (*Preprint* arXiv:1403.3295 [quant-ph])

[17] Grössing G, Fussy S, Mesa Pascasio J and Schwabl H 2012 An explanation of interference effects in the double slit experiment: Classical trajectories plus ballistic diffusion caused by zero-point fluctuations *Ann. Phys.* **327**(2) 421–437 (*Preprint* arXiv:1106.5994 [quant-ph])

[18] Maudlin T 2011 *Quantum Non-Locality and Relativity: Metaphysical Intimations of Modern Physics* 3rd ed (West Sussex, UK: Wiley-Blackwell) ISBN 978-1-4443-9697-3

[19] Grössing G 2000 *Quantum Cybernetics: Toward a Unification of Relativity and Quantum Theory via Circularly Causal Modeling* (New York, NY: Springer) ISBN 978-0-387-98960-0

[20] Tollaksen J, Aharonov Y, Casher A, Kaufherr T and Nussinov S 2010 Quantum interference experiments, modular variables and weak measurements *New J. Phys.* **12**(1) 013023 (*Preprint* arXiv:0910.4227v1 [quant-ph])

[21] Walleczek J and Grössing G 2016 Nonlocal quantum information transfer without superluminal signalling and communication *Found. Phys.* In press. (*Preprint* arXiv:1501.07177 [quant-ph])

Momentum exchange in the electron double-slit experiment

Herman Batelaan, Eric Jones, Wayne Cheng-Wei Huang, Roger Bach
Department of Physics and Astronomy, University of Nebraska—Lincoln, 208 Jorgensen Hall, Lincoln, Nebraska 68588-0299, USA

Abstract. We provide support for the claim that momentum is conserved for individual events in the electron double slit experiment. The natural consequence is that a physical mechanism is responsible for this momentum exchange, but that even if the fundamental mechanism is known for electron crystal diffraction and the Kapitza–Dirac effect, it is unknown for electron diffraction from nano-fabricated double slits. Work towards a proposed explanation in terms of particle trajectories affected by a vacuum field is discussed. The contentious use of trajectories is discussed within the context of oil droplet analogues of double slit diffraction.

1. Introduction

Recently we performed the electron double slit experiment, and the pattern was recorded one electron at-a-time [1]. The electron detection rate was about one electron per second. This made it possible to manually turn off the electron source after the first electron was recorded. This electron can, by chance, land in a first diffraction order (see Fig.1). This can be considered a completed single-event experiment. Often single events experiments are only considered in a probabilistic way as the best theory available to compare with, that is Quantum Mechanics, is probabilistic. Nevertheless, a quantum description also includes the correct prediction that the individual, in this case position, outcomes are eigenvalues of operators. Even more is known about single events. This becomes clear upon asking the question: "Is momentum conserved for this experiment?" We will provide support for the claim that the generally accepted answer is yes. The natural follow-up question that is the central theme of this paper is: "By what mechanism do the electron and the slit exchange momentum?" We claim that the answer is not known and that the question is a valid one. Some discussion on possible mechanisms is given. In particular, the role of image charge interaction between the electron in double slit walls and the vacuum field is discussed. The proposed explanation that the double slit provides a boundary condition for the vacuum field, which in turn provides a means by which the electron trajectory exchanges momentum [2–6] with the slit is discussed within the context of the theory Stochastic Electrodynamics (SED) [2,7]. The provocative possibility of any trajectory explanation is considered in view of the well-known oil-droplet double slit analogue. The validity range of SED and the relation with the Heisenberg uncertainty relation are discussed for the harmonic oscillator. The intent of this paper is to raise questions and discuss ongoing work that is unfinished and as of yet inconclusive.

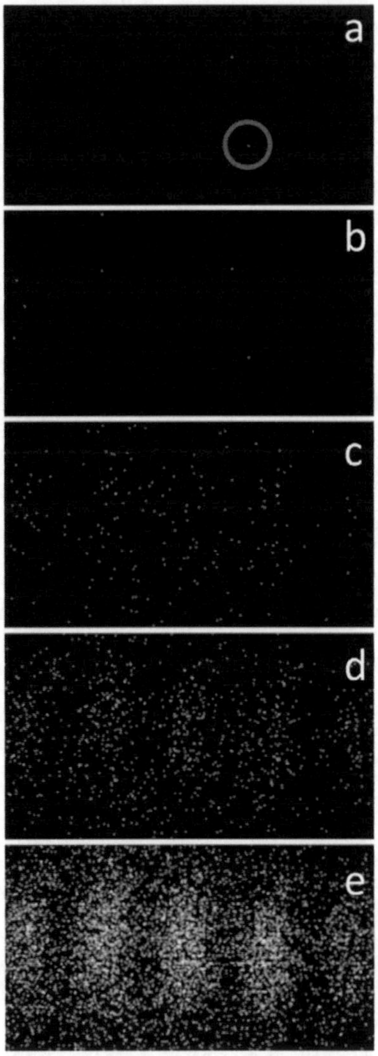

Figure 1. Buildup of Electron Diffraction. "Blobs" indicate the locations of detected electrons. Shown are intermediate build-up patterns from the central five orders of the diffraction pattern, with 2, 7, 209, 1004, and 6235 electrons detected (a-e) [1]. The circle indicates where the first electron landed for this data run.

2. Momentum conservation in double slit diffraction

The famous Einstein-Bohr dialogue sheds light on the question of the momentum conservation [8]. Einstein attempted to prove that quantum mechanics is incomplete. In this famous series of discussions between Einstein and Bohr, several topics in quantum mechanics were debated.

One of these is the double slit interference experiment [9,10]. Quantum mechanics predicts an interference pattern with perfect contrast, but only if we have no knowledge of which slit the particle went through. Einstein devised a thought-experiment that would measure through which slit the particle went *and* show the interference pattern. If such an experiment could be performed he hoped that it would show that quantum mechanics is incomplete. Einstein considered a particle beam illuminating a single slit screen placed in front of a double slit screen. Ingeniously he suggested that the recoil of the single slit screen could be measured to determine

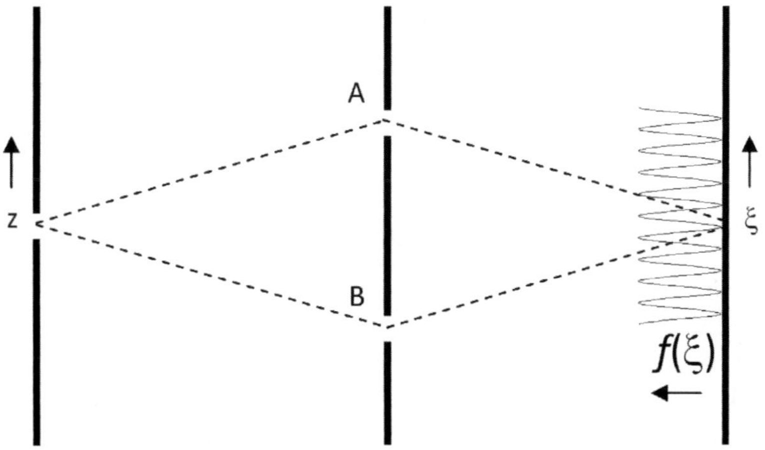

Figure 2. Schematic of Zurek's analysis of the Bohr-Einstein double-slit thought experiment.

through which slit the particle would move. Einstein used momentum conservation to predict the recoil. Because the recoil would be present regardless if it is measured or not, it appears that such a measurement would not affect the experiment in any way, and the interference pattern would remain. Bohr replied that when we measure the recoil, i.e. the momentum of the screen, accurately enough to determine through which slit the particle went, the uncertainty in our knowledge of the position of the slit is so large that the interference pattern is obscured. In other words, quantum mechanics, through Heisenberg's uncertainty relation, protects itself. Also Bohr assumed that momentum conservation holds. In a later treatise Wootters and Zurek analyze this thought experiment quantitatively [11]. This allows setting up a quantitative relation between the probability of going through one slit with the contrast of the interference pattern.

First we will summarize the main results of Wootters and Zurek's approach, using the same notation as used by these authors. The system discussed is schematically presented in figure 2. The position of the single slit plate is given by z, the position of the particle at the detection screen is given by ξ, the positions of each of the double slits of labeled with A and B and the observed position distribution is indicated by $f(\xi)$.

The wavefunction describing both the slit and particle in position representation is given by

$$\psi(z,\xi) = \frac{1}{\sqrt{2a}\pi^{1/4}} \int e^{-x^2/2a^2}[e^{ik_0(\xi+x)} + e^{-ik_0(\xi+x)}]\delta(z-x)dx \tag{1}$$

The momentum of the particle is given by $\hbar k_0$, and the parameter a is a measure of the width of the Gaussian distribution of the slit. If $a = 0$ then we know where the slit is, but we do not know what its momentum is, and the interference pattern has full contrast. Einstein's modification of the double slit experiment means we would know something about the momentum of the slit, which, as Bohr pointed out, would smear out the position of the slit, because of Heisenberg's uncertainty principle, and thus cause the interference contrast to disappear. Wootter and Zurek's equation 1.1 allows that statement to be made quantitative. For a given value of a, the interference pattern can be calculated;

$$f(\xi) = \int |\psi(z,\xi)|^2 \, dz \tag{2}$$

by integrating over the unobserved slit position z. Specifically,

$$\begin{aligned} f(\xi) &= \int |\psi(z,\xi)|^2 \, dz = \\ &= \int \frac{1}{2a\sqrt{\pi}} e^{-z^2/a^2} (e^{ik_0(\xi+z)} + e^{-ik_0(\xi+z)})(e^{-ik_0(\xi+z)} + e^{ik_0(\xi+z)}) = \\ &= \int \frac{1}{a\sqrt{\pi}} e^{-z^2/a^2} (1 + \cos(2k_0(\xi+z)) = \\ &= 1 + e^{-a^2k_0^2} \cos(2k_0\xi). \end{aligned} \tag{3}$$

In the first step the integration over the delta-function, $\delta(z-x)$, has been performed, while in the last step the integral, $\int \frac{1}{a\sqrt{\pi}} e^{-z^2/a^2} \cos(2k_0 z) = e^{-a^2k_0^2}$, has been used. In the limit of $a \to 0$ the position of the slit is exactly known, and the contrast is maximal. We now turn our attention to the wavefunction of the slit and the particle in the momentum representation. Wootters and Zurek state that it can be verified that this is "equal" to the position distribution, and give

$$\psi(k_0,\xi) = \frac{a}{2\pi^{1/4}} \int D^{1/2}(\kappa)[p_A^{1/2}(\kappa)e^{ik_0\xi} + p_B^{1/2}(\kappa)e^{-ik_0\xi}]e^{i\kappa z} d\kappa \tag{4}$$

In this expression, κ is the slit momentum, while the functions D and $p_{A,B}$ are given by

$$\begin{aligned} D(\kappa) &= \frac{a}{2\sqrt{\pi}} \left\{ e^{-a^2(\kappa+k_0)^2} + e^{-a^2(\kappa-k_0)^2} \right\} \\ \frac{p_A}{p_B} &= \frac{e^{-a^2(\kappa+k_0)^2}}{e^{-a^2(\kappa-k_0)^2}}; \qquad p_A + p_B = 1. \end{aligned} \tag{5}$$

The functions $p_{A,B}$ give the probability of going through slit A or B, respectively, and D normalizes the wavefunction. We note that the limit $a \to \infty$ justifies the statement in the introduction that diffraction essentially creates an entangled state. In this limit the probabilities act as delta-functions, $\delta(\kappa + k_0)$, $\delta(\kappa + k_0)$. After integration the wavefunction becomes

$$\begin{aligned} \lim_{a\to\infty} \psi(k_0,\xi) &= \frac{1}{2}\sqrt{2}\left(e^{ik_0\xi}e^{-ik_0\xi} + e^{-ik_0\xi}e^{ik_0\xi}\right) \\ &= |\psi(-k_0)\rangle_{slit} |\psi(k_0)\rangle_{photon} + |\psi(k_0)\rangle_{photon} |\psi(-k_0)\rangle_{slit} . \end{aligned} \tag{6}$$

We can now proceed and calculate the probability to find the photon on position ξ on the detection screen,

$$\begin{aligned} f(\xi) &= \int \lim_{a\to\infty} |\psi(z,\xi)|^2 \, dz = \\ &= \lim_{a\to\infty} \frac{1}{(2a)^2} \int_{-a}^{a} 1 + \cos(2k_0(\xi - z)) \, dz = 1/2a. \end{aligned} \tag{7}$$

In words, this states that, if a is made very large then the momentum of the slit is known, and at the same time the interference pattern is known. Combining this with the earlier statement that for very small a the position of the slit is known, while at the same time the interference pattern has full contrast, it can be recognized that this result can be reached by using Bohr's argumentation using the Heisenberg uncertainty relation. Specifically, if the momentum of the slit is measured exactly the position is completely unknown and incoherently averaging over this position blurs the interference pattern completely. Wootters and Zurek not only justified this reasoning, but also give quantitative expressions when not considering the extreme cases. The

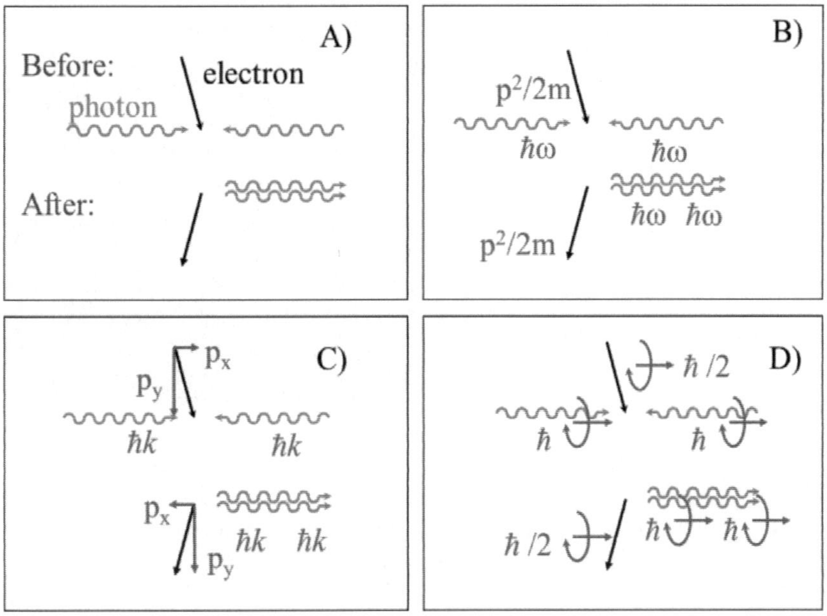

Figure 3. Kapitza–Dirac effect. Electrons can be diffracted from a standing wave of light [13]. A) Stimulated Compton scattering is the basis for the Kapitza–Dirac effect. B) Energy is conserved in this process. C) Momentum is conserved in this process. D) Angular momentum is conserved when the electron spin does not flip.

probability to go through one of the holes and the interference contrast can be calculated for arbitrary values of a.

For this paper, the second line in equation (6) is relevant. It expresses that momentum conservation between the electron and the slit holds. This is used to set up the main question posed in this paper: "By what process does the electron exchange momentum with the double slit?"

3. Momentum exchange mechanisms

Upon asking this question in Physics Colloquia (presented by HB), the answer provided by physics professors is surprisingly varied. Answers range from: "the electron induced an image charge in the double slit which back-acts on the electron," and "the electron excites phonons in the double slit, which back-acts on the electron," to "the vacuum field bounded by the double slit structure acts on the electron." Additional to such answers, the comment is often made that if one does entertain this question, one should not forget that atoms, photons, and neutrons all diffract and thus the mechanism should have some rather ubiquitous elements in it. About half of the comments made support the idea that this is a question that one should not ask or is already answered by the presence of a potential that describes the double slit structure.

To address the comment, whether or not we should ask the question by what process the electron exchanges momentum with the slit, let's consider the same question for the Kapitza–Dirac effect [12].

When electrons pass through a standing wave of light with periodicity $d \approx \lambda/2$, it is possible for the electron wave to diffract from the periodic light structure with a diffraction angle $\theta \approx \lambda_e/d$, where $\lambda_e = 2\pi\hbar/p_e$, and p_e is the electron's momentum. This effect, known as the Kapitza–Dirac effect, was proposed in 1933 [4] and we realized this experiment in 2001

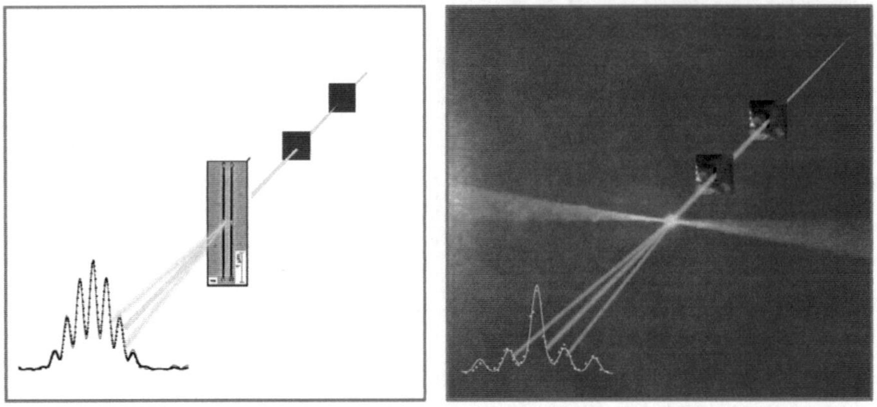

Figure 4. Electron double-slit and Kaptiza-Dirac diffraction. Left) Electrons (blue lines, starting right top) pass through a nanofabricated double slit (distance between the slits is 300 nm, and shown is an electron microscope image). The measured diffraction pattern (dots) agrees with a quantum mechanical analysis (lines). Right) Electrons (blue lines, starting right top) pass through a standing wave of light (periodicity 266 nm, and shown is a photograph of the laser focus made visible with smoke). The measured diffraction pattern (dots) agrees with a quantum mechanical analysis (lines).

[13,14]. The process by which the electron exchanges momentum with light is stimulated Compton scattering. One photon is absorbed, while the emission of another is stimulated (Fig. 2A). Energy and angular momentum are conserved in this process (Fig. 2B, 2D). As the absorption and stimulated emission are due to photons coming from opposite directions, the electron experiences a recoil of $2\hbar k$ momentum (Fig. 2C), where $k = 2\pi/\lambda$. At a basic level it is easy to verify that the scattering angle $\theta \approx 2\hbar k/p_e$and the diffraction angle are identical supporting the explanation of electron diffraction by a "light grating" as stimulated Compton scattering. At a more formal level, perturbation theory and second quantization of the light field can be used to support this claim [15]. The understanding of the mechanism also leads to predictions. When the polarization of the counter propagating light beams is chosen to be perpendicular, no standing wave forms and the electrons do not diffract. Or in the particle picture; angular momentum conservation does not work for photons that carry opposite angular momentum. Before the interaction the two photons carry a total of zero angular momentum, while after the stimulated emission the two photons carry two units of $\hbar$ angular momentum. The electron can at most change its angular momentum by one unit of $\hbar$ in a spin flip process [15].

Inspection of the electron diffraction pattern from light, and from the double-slit reveal, not surprisingly, a very similar phenomenology (Fig. 3). And, a standard quantum mechanical description of the experiments gives good agreement in both these cases. The surprise is that the mechanism can be explained for a light-grating, but not the double-slit case. Let's consider electron diffraction from an ionic crystalline lattice as in the famous Davisson-Germer experiment [16]. Can in this case the mechanism for momentum exchange be explained? An electron experiences the periodic Coulomb potential of the ionic lattice. Coulomb scattering at the particle level is understood as the scattering of virtual photons. Thus the basic mechanism for momentum exchange between the diffracting electron and the ionic lattice is understood. Additionally, predictions can be made. As the lattice heats and the potential shape is modified due to the averaging of the motion of the ionic lattice, the diffraction pattern is modified.

Is the electron diffraction by a nano-fabricated grating [17] a case that is very similar in

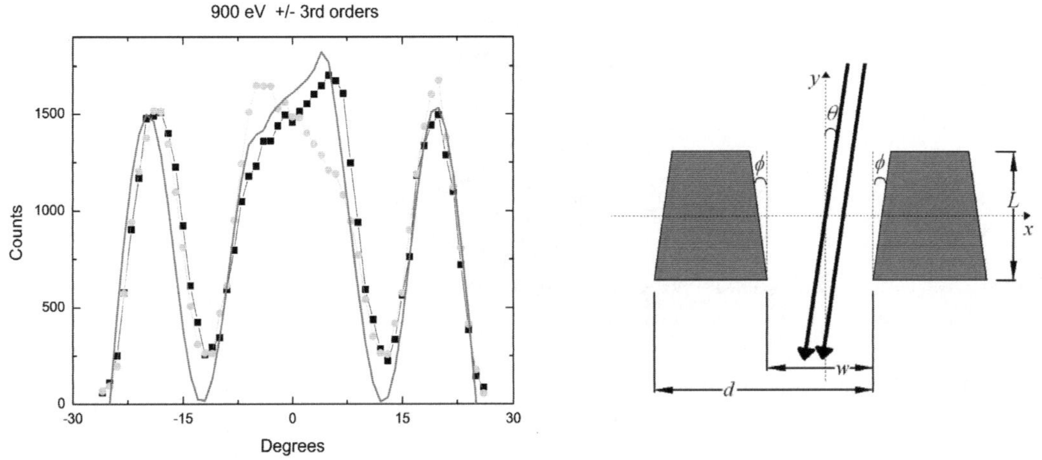

Figure 5. Image charge effect. An electron beam is diffracted from a nanofabricated grating [17,18]. The electron transmission into the third diffraction order is given as a function of grating tilt angle (left). The squares and circles are data for the positive and negative third order. The solid line is the result of a path integral calculation for the positive third order with an image charge potential and the geometry of the grating included. The geometry is given by the cross-sectional cut of the nanofabricated grating (right). The bevel angle of the slit is given by ϕ, the electron beam angle with respect to the grating is θ, the slit width is given as w, and the grating thickness is given by L.

nature to the ionic crystal, in that simply the lattice constant is much larger? The ionic crystal lattice has a priodicity of about 2 Angstrom, while the grating has about 1000 Angstrom. Even though the periodicity is very different, the physics appears similar. The electrons are blocked by the grating bars by scattering of the material that the bars are made of. This is due to the Coulomb interaction with the ionic lattice of the material. However, in another sense the ionic crystal is much more like the light grating. For both the light grating and the ionic lattice, the electron experiences a phase grating. That is, the electron wavefunction accumulates a phase that is dependent on space. For the nanofabricated grating the electron experiences an amplitude grating . That is, the electron wavefunction experiences a modulation of its amplitude as a function of position. Additionally the grating bars have no charge or other field in between the grating bars.

Returning to the double slit, it is analogous to the nanofabricated grating, but instead of many slits it has only two. One of the suggestions for a mechanism for electron diffraction is that the electron would induce an image charge in the double slit material walls, which in turn would back-act on the electron. This would, after all, provide real fields, and thus lead to the exchange of virtual photons with the electron within the slits. This has been investigated experimentally. The image charge interaction has been measured to weakly modify the diffraction pattern [18]. The slope on the diffraction rocking curve around $\theta = 0$ is due to the image charge. In the absence of image charge this slope is zero. The basic idea is that electrons that move with angle where $\theta \approx \phi$ (see Fig.5 right) are closer to one grating wall and thus experience an image charge potential gradient, modifying the diffraction pattern. The removal of the image charge term in the theoretical treatment leaves the diffraction intact, thus the image charge does not explain electron diffraction.

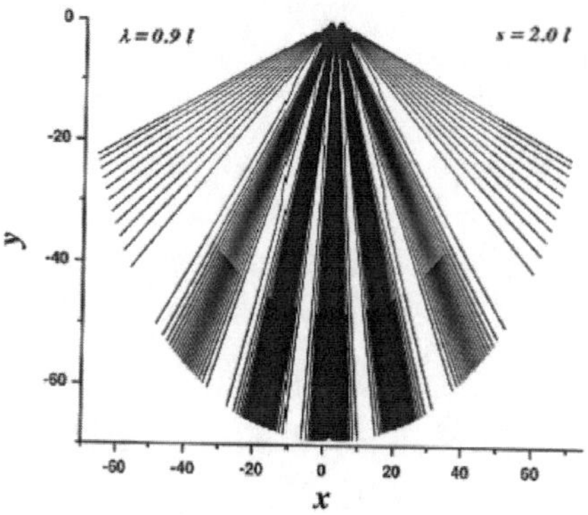

Figure 6. SED trajectory simulation. Trajectories are shown for a simulation of electrons for the double slit experiment. Figure courtesy of J. Avedano .

Another suggestion is that the electron would excite phonons in the double slit material. This does not directly identify the mechanism of diffraction, but would be specific about what two objects interact and exchange momentum. It would be the electron and a phonon. This idea has never been tested, but may lead to such ideas as exciting phonons in the double slit material before the electron diffracts and thereby controlling the electron diffraction pattern.

Another idea is that the double-slit poses a boundary condition on the vacuum field and the modified vacuum field affects the motion of the electron. This may appear akin to a Casimir effect, but this is not what it is. The leading term of the vacuum energy interaction of an electron and a wall is the image charge [19,20]. The QED corrections to that term are very small and do not explain diffraction. In Stochastic Electrodynamics the vacuum field interacts with a charged point particle. This theory has had some successes in for example correctly obtaining Casimir forces [21] and the absorption spectrum of the harmonic oscillator [22]. For an extensive recent review see [23]. On page 323 of this book a bundle of trajectories is shown for the double slit (Fig. 6). No probability distribution has been calculated yet. It is perhaps surprising that a theory that supports real trajectories is entertained to provide insight on a physics phenomenon that is considered to be a hallmark of quantum mechanics. Quantum mechanics is not consistent with non local-real theories as evidenced by experimental tests of Bell's inequalities [24–26]. The notion of trajectories in this context leads inevitably to discussion.

4. On the possibility of trajectory explanations

Recently, there has been a report on oil-droplets that show a behavior that is analogous to electron diffraction from a double slit [27]. The diffraction pattern has been recorded one droplet at-a-time, and the oil droplet trajectories have been recorded. This experiment and its interpretation appear at odds with many claims that have been made on double slit diffraction. For example, Feynman stated: "We choose to examine a phenomenon which is impossible, *absolutely* impossible, to explain in any classical way, and which has in it the heart of quantum mechanics. In reality, it contains the only mystery." [10] As the size of the oil droplet is about 0.8 mm, it is clearly not isolated sufficiently from its environment to require a quantum mechanical description. A classical description is completely sufficient. The oil droplet report should thus be considered to describe a remarkable phenomenon that requires extensive scrutiny. Other research groups are attempting to repeat the oil droplet experiment [28]. In particular, the group of Bohr claims: "that the single-particle statistics in such an experiment will be fundamentally different from the single-particle statistics of quantum mechanics" [29]. Our group is also in the process of repeating the oil droplet experiment. Single- and double slit

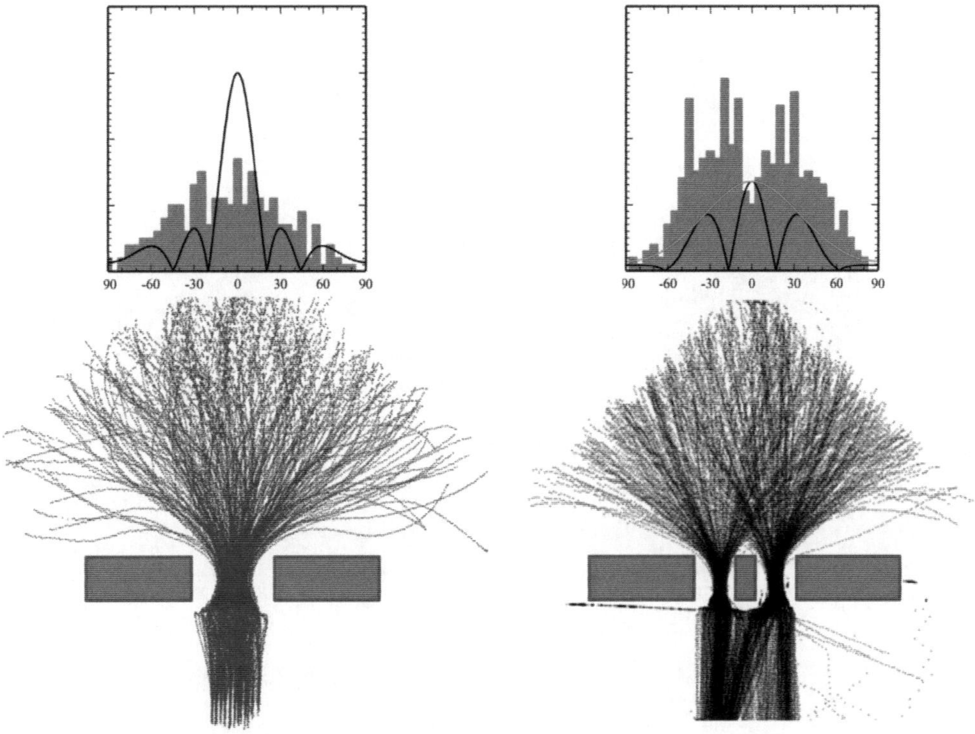

Figure 7. Oil droplet diffraction analogue. Oil droplet trajectories for a single slit (left bottom) and for a double slit (right bottom) measured at UNL are shown. The corresponding angular trajectory distributions are given in the top graphs. The viscosity used for the single and double slit is 20 cSt and 50 cSt, respectively. The solid line is the fit given by Couder to the data reported in [27]. For a description see text.

diffraction was studied. The experimental set-up follows the design of the Couder group and a short description of the experiment is given below for completeness.

A flat, horizontal square dish is filled with approximately 4 mm of oil. The dish oscillates in the vertical direction with a frequency of 50 Hz, which allows a droplet of about 0.8 mm size to bounce on the oil surface for periods up to hours. Just below the Faraday threshold for surface excitation, the oil droplet and the mostly circularly shaped wave that it excites, move together at a constant velocity. This association of a droplet and a wave has been called a "walker." The droplet surfs the wave and the wave guides the droplet motion. As the wave is modified when it moves in the vicinity of a physical boundary, it steers the droplet. The physical boundary used in this experiment is a submerged slit structure. A bundle of trajectories is shown in Fig. 7. The trajectories are recorded with a web camera. Data analysis software finds the position of the droplet at regular intervals and a trajectory is built as a series of timed positions. In Fig. 7 the probability distribution of trajectory angles, after the droplet has passed the slit, has been given. The angles are found at a distance of two slit widths (measured from the center of the slit). The angular distribution does not have distinct peaks. The angles for the single slit case, at which the peaks should occur are approximately given by $\theta \approx \lambda_F/d$, where λ_F is the Faraday wavelength of the oil bath waves and d is the width of the slit. These are the locations of the maxima in the solid curve. The angles observed further from the slit are affected by the walls. The idea is

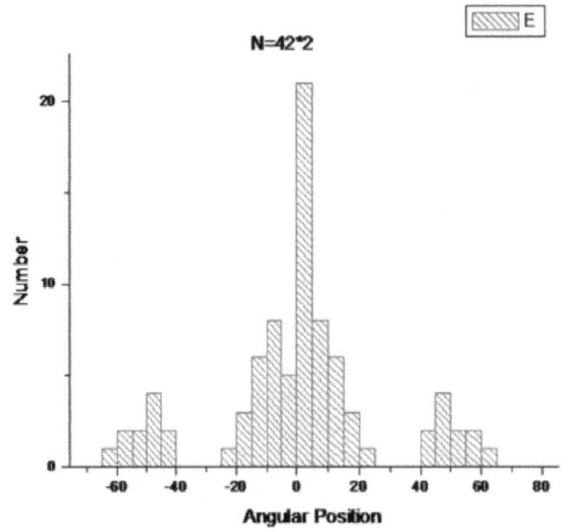

Figure 8. Single-slit oil droplet distribution. The angular distribution shows diffraction-like peaks (oil viscosity of 50 cSt).

that the extent of the oil waves should be sufficient to interact with the entire slit, but smaller than the square walls of the bath, so that the oil droplet's motion is affected by the slit and not the walls. However, after the oil droplet has passed through the slit, the oil waves will interact with the wall and change the droplet motion. For the double slit an example is given where a double lobed distribution is found in contrast with the results reported by Couder. However, this data is for one particular fluid depth and shaking amplitude of the oil bath. Additionally, the initial angular distribution is not post-selected in the double slit example shown. In analogy with an electron diffraction experiment, it is important to limit the divergence of the incoming particle "beam" to less than the diffraction angle. The initial position distribution should be flat to represent a beam with a uniform distribution. The final angular distribution depends strongly on these choices and values. Figure 8 shows another single slit run for slightly different parameters (that is symmetrized following the approach by Couder) that exhibits diffraction like peaks at approximately the correct angle. At this time we cannot draw general conclusions due to the limited amount of data available.

The interpretation of the thought-provoking experiments by Couder, have also been reexamined by Bohr. Bohr has repeated the double slit experiments and does not find diffraction peaks. Bohr's analysis leads to the claim that the oil droplet experiments do not have the capability to show interference-like behavior. It appears that the data provided by Bohr does not use an initial uniform distribution. It is anticipated that more refined results from multiple research groups are forthcoming and may shed light on the veracity of the oil droplet experiment and the meaning of the analogy. Are then explanations of double slit diffraction using trajectories even possible? If the Couder oil droplet experiments find some confirmation this question could be affirmatively answered.

5. The role of the vacuum field

Returning to the proposed vacuum field explanation of electron diffraction, we decided to study the theory of SED for the harmonic oscillator for the purpose of establishing its validity regime. The motivation is that SED yields the ground state of the harmonic oscillator which obeys the Heisenberg uncertainty relation. The Heisenberg relation is central to diffraction. A simple argument can illustrate this. In electron diffraction from a single slit of width d, the extent of diffraction angles θ obeys the diffraction equation $\theta = \lambda_{dB}/d$ for small angles. On the other hand the extent of the diffraction angles are an indication of the momentum uncertainty in the direction of the slit $\theta = \Delta p_x/p$. Using the de Broglie wavelength $\lambda_{dB} = h/p$, equating the slit width with the uncertainty of the position $d = \Delta x$, the uncertainty relation $\Delta x \Delta p_x = h$ is now recovered. As SED gives a model explanation for the ground states of the harmonic oscillator, it appears natural to suspect that it can also explain diffraction.

Boyer has shown that the moments $\langle x^n \rangle$ of a harmonic oscillator immersed in the SED vacuum field are identical to those of the quantum harmonic oscillator in the ground state [8]. As a consequence, the Heisenberg minimum uncertainty relation is satisfied for an SED harmonic oscillator, and the probability distributions of the SED harmonic oscillator is also the same as that of the ground state quantum harmonic oscillator. In [30] we ask the question: "What happens to the dynamics of the classical harmonic oscillator in the presence of the SED vacuum field such that its probability distributions would become Gaussian?" and "Why do the widths of these distributions satisfy Heisenberg's minimum uncertainty relation?" The answer is that the vacuum field drives the electron motion while radiation damps it. The energy balance leads to an average energy for the particle of $\hbar\omega/2$. Planck's constant enters through the overall strength of the vacuum field and its value is determined by experiment. The vacuum field is found as a Lorentz boost invariant solution of the Maxwell equations in free space [31]. In this sense, which is not widely recognized, SED is a theory that is independent of quantum mechanics.

The vacuum modes that are used cover the damped harmonic oscillator resonance width. No other vacuum modes are used in the simulation. The results converge as the width of the spectrum is increased. The vacuum field as a function of time can be approximated by a single mode for less than a coherence time τ, which is the reciprocal of the vacuum field bandwidth. The particle's response to this field is that of a damped harmonic oscillator driven by one frequency and has the usual double peaked classical probability distribution. Averaging over the particle motion over many coherence times leads to a Gaussian probability distribution that agrees with the quantum mechanical distribution (Fig. 9). Also, the product of the widths of the momentum and position distributions satisfies the equality in Heisenberg uncertainty relation in accordance with a minimum uncertainty packet.

The major challenge for simulating other physical systems, and in particular double slit diffraction, is to incorporate the appropriate SED vacuum field. A representative selection of vacuum field modes is thus the key to successful SED simulations. To push the limits of SED, and with the intention to exceed its validity regime we simulated a harmonic oscillator excited with a pulse of which the carrier frequency can be varied. The 1-D equation of motion in the x-direction used for the simulation is Newton equation for a charged particle damped by radiation and driven by the vacuum field and a pulsed field,

$$m\ddot{x} = -m\omega_0^2 x - m\Gamma\omega_0^2\dot{x} + q\left[\left(E_p^{(x)} + E_{vac}^{(x)}\right) + \left(\vec{v} \times \left(\vec{B}_p + \vec{B}_{vac}\right)^{(x)}\right)\right]. \quad (8)$$

The field and parameters are described in ref [32]. The result of the simulation is shown in Fig. 10.

The surprise was that the quantum mechanical result agrees well with the SED results, both for numerical simulation and perturbation theory. The 2ω and 3ω overtones, that are absent for a pure classical oscillator, can be explained by parametric excitation [32]. The more detailed agreement of the peak heights in the discrete excitation spectrum was not expected.

Returning to electron double slit diffraction, it may appear straightforward to run a simulation. However, a free electron can interact with all modes. Even ignoring infrared and ultraviolet divergences, an unbounded vacuum spectrum is hard to simulate. To sidestep this problem, Avendano is attempting to use a guiding equation based on a particle model to obtain

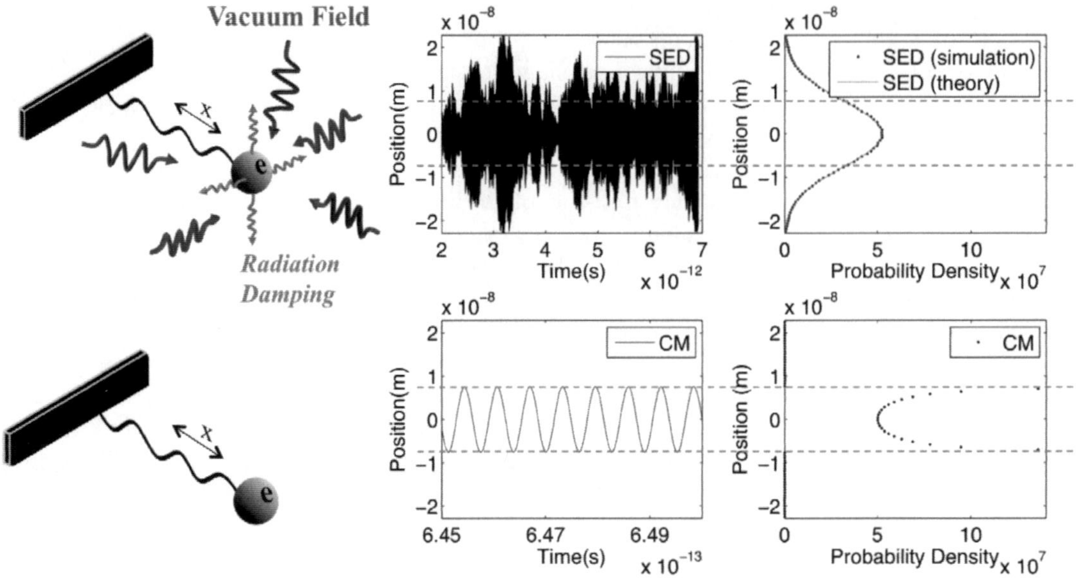

Figure 9. Harmonic oscillators with and without the vacuum field. (Top) The SED harmonic oscillator undergoes an oscillatory motion with modulating oscillation amplitude. The oscillation amplitude modulates at the time scale of the coherence time and is responsible for the resulting Gaussian probability distribution. (Bottom) In the absence of the vacuum field or any external drive, a harmonic oscillator, that is initially displaced from equilibrium, performs simple harmonic oscillation with constant oscillation amplitude. The resulting probability distribution has peaks at the two turning points.

electron trajectory in the vacuum [23]. The results are bundles of trajectories (Fig. 6) that may support diffraction peaks. At this point no probability distributions have been reported, and this work should also be considered unfinished.

It may appear that we can rule out SED as a viable theory as it may be real and local. Perhaps it cannot violate Bell's inequalities and should thus ultimately fail. This discussion, whether or not SED could be considered an emergent quantum theory, is interesting but not important for our present discussion. Even if SED is only an approximate theory with a limited validity range capable of reproducing only some quantum effects, the specific prediction that the vacuum field is responsible for diffraction is worth investigating in its own right. The alternative of using a full QED calculation for electron double slit diffraction, that includes the slits and the vacuum field, is technically more challenging and has never been done. The use of SED for this problem is then one of the options that is available, to address the main question raised in this paper: "By what physical process does the electron exchange momentum with the double slit?"

Acknowledgments

The authors acknowledge helpful discussions with Yves Couder, Ana Maria Cetto, Jaime Avendeno. This work is supported by the National Science Foundation under Grant No. 1306565.

References

[1] Bach R, Pope D, Liou S-H and Batelaan H 2013 *New J. Phys.* **15** 033018
[2] Boyer T H 1975 *Phys. Rev. D* **11** 790–808

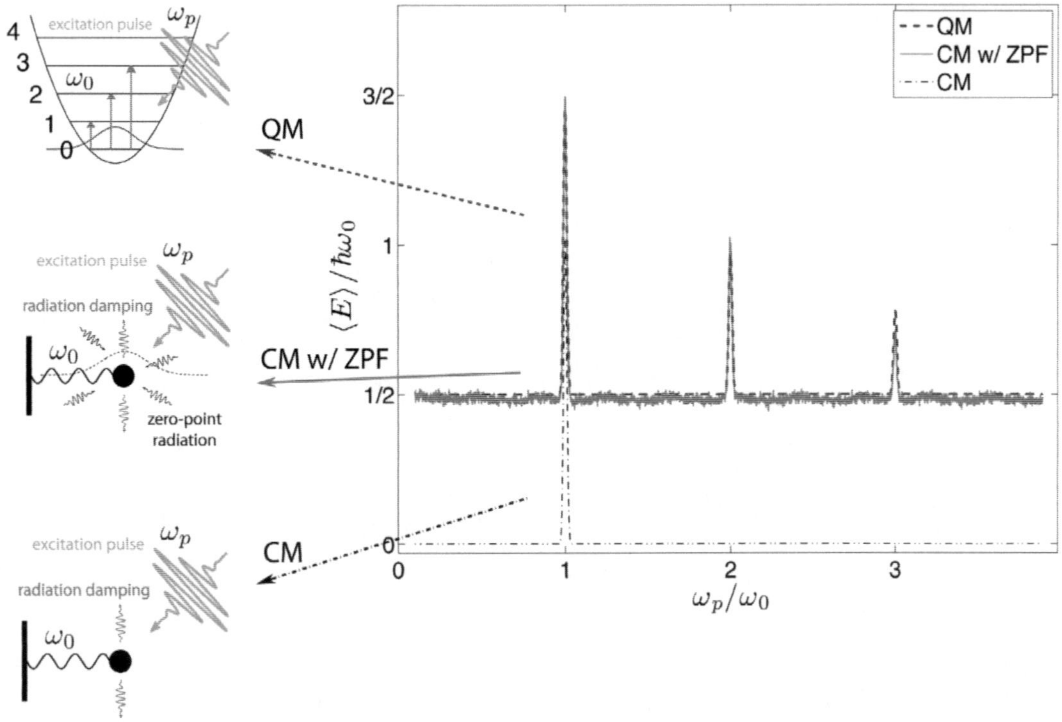

Figure 10. Excitation spectra of harmonic oscillators for different theories. (Left) Schematics of harmonic oscillators are shown for quantum theory (left-top), for SED (left-middle), and for classical theory (left-bottom). The red arrows in the quantum system represents the one-step transition at different pulse frequencies ω_p. (Right) The average value of energy $\langle E \rangle$ after excitation is plotted as a function of pulse frequency ω_p. For the classical theory, the ensemble average is computed. For the quantum theory, the expectation value is computed. The classical oscillator in the vacuum field (red solid line) exhibits an excitation spectrum in agreement with the quantum result (blue dashed line). In the absence of the vacuum field, the classical oscillator has only one single resonance peak at the natural frequency ω_0 (black dash-dot line). The excitation peak heights and the relative ratio are confirmed by classical and quantum perturbation theory [32].

[3] Hestenes D 1985 *Found. Phys.* **15** 63–87
[4] Kracklauer A F 2013 *Found. Phys.* Lett. **12** 441–453
[5] Grössing G, Fussy S, Mesa Pascasio J and Schwabl H 2012 *Ann. Phys.* **327** 421–437
[6] Cavalleri G, Barbero F, Bertazzi G, Cesaroni E, Tonni E, Bosi L, Spavieri G and Gillies G T 2009 *Front. Phys. China* **5** 107–122
[7] Marshall T W 1963 *Proc. R Soc. Lond. Math. Phys. Eng. Sci.* **276** 475–491
[8] Schilpp P A 1998 Ed., Albert Einstein, Philosopher-Scientist: The Library of Living Philosophers Vol. VII, 3rd edition (Open Court).
[9] Wheeler J A and Zurek W H 2014 Quantum Theory and Measurement (Princeton University Press).
[10] Feynman 2015 Feynman Lectures on Physics, Vol. III: Quantum Mechanics (CreateSpace Independent Publishing Platform).
[11] Wootters W K and Zurek W H 1979 *Phys. Rev. D* **19** 473–484
[12] Kapitza P L and Dirac P A M 1933 *Math. Proc. Camb. Philos. Soc.* **29** 297–300
[13] Freimund D L, Aflatooni K and Batelaan H 2001 *Nature* **413** 142–143
[14] Batelaan H 2007 *Rev. Mod. Phys.* **79** 929–941

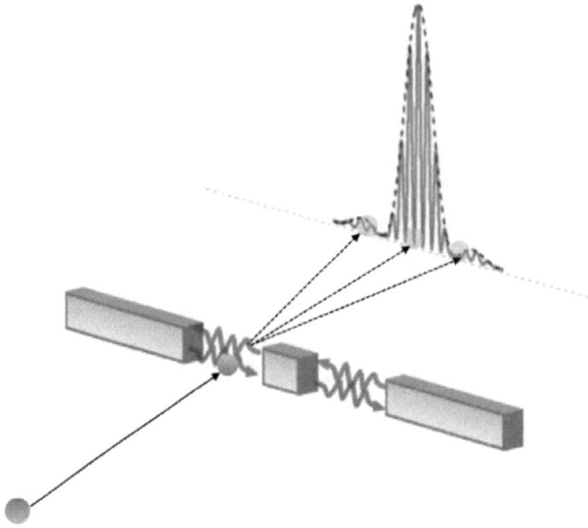

Figure 11. Diffraction mechanism. Vacuum field modes are selected by the double-slit structure, which in turn guide the electron particle motion.

[15] McGregor S, Huang W C-W, Shadwick B A and Batelaan H 2015 *Phys. Rev. A* **92** 023834
[16] Davisson C J and Germer L H 1928 *Proc. Natl. Acad. Sci. USA* **14** 317–322
[17] Gronniger G, Barwick B, Batelaan H, Savas T, Pritchard D and Cronin A 2005 *Appl. Phys. Lett.* **87** 124104
[18] Barwick B, Gronniger G, Yuan L, Liou S-H and Batelaan H 2006 *J. Appl. Phys.* **100** 074322
[19] Spruch L 2008 *Phys. Today* **39** 37–45
[20] Milonni P W 1983 *Int. J. Theor. Phys.* **22** 323–328
[21] Boyer T H 1973 *Phys. Rev. A* **7** 1832–1840
[22] Wayne H B and Huang C-W 2012 *Found. Phys.* **45**
[23] de la Peña L, Cetto A M and Valdés Hernández A 2015 The Emerging Quantum (Springer International Publishing).
[24] Hensen B, Bernien H, Dreau A E, Reiserer A, Kalb N, Blok M S, Ruitenberg J, Vermeulen R F L, Schouten R N, Abellan C, Amaya W, Pruneri V, Mitchell M W, Markham M, Twitchen D J, Elkouss D, Wehner S, Taminiau T H and Hanson R 2015 *Nature* **526** 682–686
[25] Shalm L K, Meyer-Scott E, Christensen B G, Bierhorst P, Wayne M A, Stevens M J, Gerrits T, Glancy S, Hamel D R, Allman M S, Coakley K J, Dyer S D, Hodge C, Lita A E, Verma V B, Lambrocco C, Tortorici E, Migdall A L, Zhang Y et al. 2015 *Phys. Rev. Lett.* **115** 250402
[26] Giustina M, Versteegh M A M, Wengerowsky S, Handsteiner J, Hochrainer A, Phelan K, Steinlechner F, Kofler J, Larsson J-Å, Abellán C, Amaya W, Pruneri V, Mitchell M W, Beyer J, Gerrits T, Lita A E, Shalm L K, Nam S W, Scheidl T et al. 2015 *Phys. Rev. Lett.* **115** 250401
[27] Couder Y and Fort E 2006 *Phys. Rev. Lett.* **97** 154101
[28] Bush J W M 2015 *Phys. Today* **68** 47–53
[29] Andersen A, Madsen J, Reichelt C, Rosenlund Ahl S, Lautrup B, Ellegaard C, Levinsen M T and Bohr T 2015 *Phys. Rev. E* **92** 013006
[30] Huang H C-W and Batelaan H 2013 *J. Comput. Methods Phys.* **2013** 308538
[31] Boyer T H 1985 *Sci. Am.* **253** 70–78
[32] Huang H C-W and Batelaan H 2015 *Found. Phys.* **45** 333–353

Electron system correlated by the zero-point field: physical explanation for the spin-statistics connection

A M Cetto and L de la Peña

Instituto de Física, Universidad Nacional Autónoma de México,
Ap. postal 20-364, 01000 México DF, Mexico

E-mail: `ana, luis@fisica.unam.mx`

Abstract. Which is the physical agent behind the antisymmetry of the electron state vectors? With the purpose to find an answer to this key question, we analyze the stationary states of a system containing two noninteracting electrons, using the tools of stochastic electrodynamics. In previous work, the resonant response of two particles to common modes of the random zero-point field has been shown to lead to the nonfactorizability of the composite state vector. Here we extend the analysis to particles with spin. When two electrons constitute a single system, a correlation is established between their dynamical variables through the common relevant modes of the zero-point field, which acts as a mediator. An analysis of the exchange properties of the bipartite state vectors obtained is shown to lead to the connection between spin and symmetry. The conclusion is that due consideration of the vacuum field in first quantization leads to the corresponding statistics for an assembly of electrons.

Keywords: Zero-point radiation field, foundations of quantum mechanics, stochastic electrodynamics, spin-statistics, Pauli Principle

1. Introduction

One of the most fruitful tenets of quantum mechanics is the link between spin and statistics for an assembly of equal particles, and the related exclusion principle for fermions. The spin-statistics theorem constitutes a core support for some of the most successful applications of quantum theory, including the whole of atomic, molecular and nuclear physics, and of course, quantum statistical physics. Yet despite its central importance, it is conventionally taken in quantum mechanics as a given law of nature, an empirical ingredient distinctive of quantum systems. Pauli himself, the author of the first 'rigorous' proof of the spin-statistics theorem, expressed his concern and dissatisfaction with the state of affairs on this matter in his Nobel lecture and subsequently [1]-[3] —already over six decades ago. Various explanations have been proposed to justify the postulate, most of which resort to mathematical assumptions that replace the usual arguments with some even more formal ones, such as topological properties of 3D space,[4] or conditions on the state vectors imposed by permissible representations of the permutation group.[5] A particular case of the latter, of relevance for the present work, involves the inclusion of the internal (spin) coordinates among the parameters affected by the exchange operation. [6] There are numerous other proposals, both relativistic and nonrelativistic (see e.g. Refs. [7]-[14], and additional references in Ref. [6]), but none has gained general acceptance up to now.

Here we tackle the problem of the spin-symmetry relation from of a physical perspective, as opposed to a formal one. The task is accomplished by making recourse to results derived on the basis of stochastic electrodynamics (SED). [15]-[18] We study in detail the case of a system of two electrons, yet the results are seen to extend naturally to an arbitrary number of particles and to any spin value.

We recall that according to SED as presented extensively in Ref. [18],[1] (nonrelativistic) quantum mechanics emerges as a result of the permanent interaction of otherwise classical particles with the random electromagnetic field at zero temperature (the zero-point field, ZPF). It should be stressed that at the formal level, the final descriptions of SED and usual quantum mechanics are indistinguishable (see Ref. [18]); yet the SED approach serves to disclose key physical elements that are not apparent in the usual formalism, thereby providing interesting clues about the physics behind quantum mechanics. The present work represents a further step in this direction.

A basic finding of SED is that, when the particle-plus-field system has reached a stationary quantum regime, the particle responds linearly and resonantly to a given set of field modes, precisely the modes involved in the transitions between states. As has been shown in Ref. [18], in the two-particle case it is the interaction between the particles mediated by common (transition) ZPF modes what correlates their motion and leads to entangled states. A further key result to be used in this paper, is the emergence of the electron spin as a consequence of the interaction of the particle with polarized modes of the vacuum field, as described in Refs. [18] and [19]. This allows us to address the problem of the two electrons with spin and find the corresponding correlated states. The symmetry properties of the states are thus found to be intimately associated with the particle spin.

The structure of the paper is as follows. We start by briefly recalling the main elements of SED that are needed for subsequent derivations. The description of the random ZPF must now include an angular momentum component, with its corresponding degree of freedom. First we present the solutions for the one-particle stationary states. Next, we show for the two-particle stationary state how the nonfactorizable state vectors result from the correlations generated by the ZPF-mediated interaction between (otherwise noninteracting) particles, due to their common transition frequencies. Finally, an analysis of the exchange properties of these bipartite state vectors is shown to lead to the connection between spin and symmetry. The extension of this result to systems with more than two particles is then discussed. The paper ends with a discussion of the physical implications of the results obtained.

2. Stationary SED solutions. The single-particle case

Let us briefly recall that linear SED studies the dynamics of a charged particle embedded in the random ZPF and acted on by an arbitrary external conservative (binding) force $\boldsymbol{f}(\boldsymbol{x})$. [15], [18] The standard starting point is the Abraham-Lorentz equation,

$$m\ddot{\boldsymbol{x}} = \boldsymbol{f}(\boldsymbol{x}) + m\tau\dddot{\boldsymbol{x}} + e\boldsymbol{E}(\boldsymbol{x},t) + \frac{e}{c}\boldsymbol{v}\times\boldsymbol{B}(\boldsymbol{x},t), \tag{1}$$

where $m\tau\dddot{\boldsymbol{x}}$ (with $\tau = 2e^2/3mc^3$) represents the radiation reaction force, and the Lorentz force is due to the random ZPF. The electric component of this field is written as

$$\boldsymbol{E}(t) = \sum_{\mathbf{k},\lambda} \tilde{E}(\omega)\boldsymbol{\epsilon}_{\kappa}^{\lambda} a^0(\omega) e^{i(\omega t - \boldsymbol{k}\cdot\boldsymbol{x})} + c.c., \tag{2}$$

where the value of the (real) coefficient $\tilde{E}(\omega)$ corresponds to a field having an energy $\mathcal{E}(\omega) = \hbar\omega/2$ per mode, and the $a^0(\omega)$ are complex coefficients of the form

$$a^0(\omega) = e^{i\zeta}, \tag{3}$$

[1] A more detailed knowledge of Ref. ([18]) is convenient for a full understanding of the present paper.

with statistically independent random phases ζ corresponding to different realizations of the field. As usual, λ stands for the polarization index, $\boldsymbol{\epsilon}_{\kappa}^{\lambda} \cdot \boldsymbol{k} = 0$ and $|\boldsymbol{k}| \equiv k = \omega/c$.

Since electrons are known to interact with electromagnetic field modes of circular polarization, it is convenient to decompose the ZPF into such modes, which means that the index λ refers to the two possible circular polarizations, left and right (or clockwise and counter clockwise), associated with the spin angular momentum of the electromagnetic field. We may therefore write the factor multiplying the coefficient $\tilde{E}(\omega)$ as

$$a(\omega, \phi, t) = a^0(\omega) e^{i(\omega t - \boldsymbol{k}\cdot\boldsymbol{x} + \gamma\phi)}. \tag{4}$$

In the long-wavelength (dipole) approximation $\exp i(\boldsymbol{k} \cdot \boldsymbol{x}) \simeq 1$ —which has been shown to be sufficient to establish contact with quantum mechanics, see [18]— this reduces to

$$a(\omega, \phi, t) = a^0(\omega) e^{i\omega t} e^{i\gamma\phi}. \tag{5}$$

As for the rotation angle ϕ, notice that it changes by 2π during a time interval equal to $2\pi/\omega$ so that

$$\gamma = \pm 1 \tag{6}$$

for the full rotation to take place in a full period (or along a full wavelength $\lambda = 2\pi/k$), the plus and minus signs referring to the left and right polarizations, respectively.

In previous work [15], [18] it has been shown that as a result of the interplay between the action of this field and the particle's own radiation reaction, as described by Eq. (1), the system reaches eventually a condition of energy balance. In this regime the particle attains any one of a set of possible stationary states (say, state α), described by the stationary solutions of Eq. (1), namely

$$x_\alpha(t, \phi) = \sum_\beta x_{\alpha\beta} a_{\alpha\beta} = \sum_\beta x_{\alpha\beta} a^o_{\alpha\beta} e^{i\omega_{\alpha\beta} t} e^{i\gamma_{\alpha\beta}\phi}, \tag{7}$$

which indicates that the particle responds linearly to a well-defined set of (polarized) field modes $\{\alpha\beta\}$, with amplitudes given by

$$a_{\alpha\beta} = a^o_{\alpha\beta} e^{i\omega_{\alpha\beta} t} e^{i\gamma_{\alpha\beta}\phi}, \tag{8}$$

where β denotes any other state attainable by means of a transition from α. This linear response to the field is the reason for the name linear SED (or LSED, for short) given to the theory. The corresponding relevant frequencies $\omega_{\alpha\beta}$ turn out to be given precisely by the frequencies of transition between states α and β according to Bohr's formula, [15]

$$\omega_{\alpha\beta} = \Omega_\alpha - \Omega_\beta = (\mathcal{E}_\alpha - \mathcal{E}_\beta)/\hbar, \tag{9}$$

$\mathcal{E}_\alpha$ ($\mathcal{E}_\beta$) being the energy of the partilce in state α (β). The response of the particle becomes extremely selective, as it leaves out all the modes contained in Eq. (2) except for those appearing in Eq. (7). Equation (7) is but a particular case of the general expression for a dynamical variable $F_\alpha(t)$ in state α, namely (using 1D notation) [18]

$$F_\alpha(t, \phi) = \sum_\beta F_{\alpha\beta} a_{\alpha\beta} = \sum_\beta F_{\alpha\beta} a^o_{\alpha\beta} e^{i\omega_{\alpha\beta} t} e^{i\gamma_{\alpha\beta}\phi}. \tag{10}$$

Notice that the stochasticity is contained solely in the random phases of the relevant field variables, $a^o_{\alpha\beta} = e^{i\zeta_{\alpha\beta}}$. These phases $\zeta_{\alpha\beta}$ vary from realization to realization, whence the treatment is necessarily statistical. Under the condition of ergodicity, assumed to hold in the stationary regime, the stochastic amplitudes must obey the *chain rule* $a^o_{\alpha\beta} a^o_{\beta\gamma} = a^o_{\alpha\gamma}$, which demands that [18]

$$a^o_{\alpha\beta} = e^{i\zeta_{\alpha\beta}} = e^{i(\varphi_\alpha - \varphi_\beta)}. \tag{11}$$

Journal of Physics: Conference Series **701** (2016) 012008 doi:10.1088/1742-6596/701/1/012008

When the polarization of the field variables does not play a relevant role, the dependence of F_α on the internal rotation angle ϕ may be omitted and the expression for the dynamical variable becomes simply

$$F_\alpha(t) = \sum_\beta F_{\alpha\beta} a^o_{\alpha\beta} e^{i\omega_{\alpha\beta} t}. \tag{12}$$

This description has been sufficient to develop the LSED theory so far; [18] as we shall see below, however, in dealing with spin the dependence on ϕ will need to be taken into account.

The (nonrandom) coefficients $F_{\alpha\beta}$ appearing in Eq. (12) turn out to be the matrix elements in the energy representation of the (quantum) operator associated with the variable F, and they satisfy the corresponding matrix algebra. In particular, the statistical (ensemble) average of the dynamical variable (12) coincides with the average calculated with the usual quantum formula, i.e.,

$$\overline{F_\alpha} = \sum_\beta F_{\alpha\beta} a^o_{\alpha\beta} \delta_{\alpha\beta} = F_{\alpha\alpha}. \tag{13}$$

In summary, the description of the particle state in the time-asymptotic, ergodic regime, which is identified as the *quantum regime,* is given by the formulas and rules familiar from quantum mechanics in its matrix form. Equations such as (12) for the (stochastic) dynamical variable $F(t)$ in state α endow this description with a concrete physical meaning, in which the ZPF plays a central role in ensuring that such stationary state is maintained.

For practical reasons, and to establish contact with the usual quantum formalism, it is convenient to introduce an appropriate Hilbert space for the description; this will prove particularly useful for the analysis of the entangled states to be made below. With this purpose we observe that the matrix $\hat{F}(t)$, represented by the set of elementary oscillators $F_{\alpha\beta} e^{i\omega_{a\beta} t}$, can be expanded as

$$\hat{F}(t) = \sum_\beta F_{\alpha\beta} e^{i\omega_{a\beta} t} \left| e_\alpha \right\rangle \left\langle e_\beta \right| = \sum_\beta F_{\alpha\beta} e^{i(\Omega_a - \Omega_\beta) t} \left| e_\alpha \right\rangle \left\langle e_\beta \right| \tag{14}$$

where $\{|e_\alpha\rangle\}$ represents an orthonormal basis spanning the Hilbert space of states of the system. The corresponding matrix elements are then given by

$$F_{\alpha\beta}(t) = \left\langle e_\alpha \right| \hat{F}(t) \left| e_\beta \right\rangle = \left\langle \alpha(t) \right| \hat{F} \left| \beta(t) \right\rangle, \tag{15}$$

with $\hat{F}$ having matrix elements $F_{\alpha\beta}$ and

$$\left| \alpha(t) \right\rangle = e^{-i\Omega_\alpha t} \left| e_\alpha \right\rangle. \tag{16}$$

In particular, the mean value given by (13) can be written according to (15) as $F_{\alpha a} = \langle \alpha | \hat{F} | \alpha \rangle$, which is the usual quantum formula.

3. Inclusion of spin into the description

In recent work (see Refs. [18], [19]) it has been shown that the electron spin emerges as an internal angular momentum of value $\hbar/2$ acquired by the particle, due to its interaction with modes of the ZPF of a given circular polarization. In other words, it is precisely the internal (or spin) angular momentum of the polarized modes of the ZPF what gives rise to and sustains the spin angular momentum of the electrons.

To include the spin in our analysis we need to use the complete expression for the dynamical variable $F_\alpha(t)$ with the angular factor included as given by (10), instead of (12). In order to establish contact with the usual quantum description, we propose to write

$$\gamma_{\alpha\beta} = \gamma_\alpha - \gamma_\beta, \tag{17}$$

and hence,

$$F_\alpha(t,\phi) = \sum_\beta F_{\alpha\beta} a^0_{\alpha\beta} e^{i(\gamma_\alpha - \gamma_\beta)\phi} e^{i(\Omega_\alpha - \Omega_\beta)t}, \tag{18}$$

with γ_α, γ_β taking half-integral, positive or negative values. This is a natural way to include the spin as a contribution to the total angular momentum of the particle into the description. The separation of $\gamma_{\alpha\beta}$ into individual components α, β implies that the full amplitude $a_{\alpha\beta}$ given by Eq. (8) complies with the chain rule, a property that will prove to be fundamental in what follows. In writing $F_\alpha(\phi,t)$ in the above form, the dependence on ϕ can be transferred to the state vectors, in the same way as was done in the previous section with the time dependence, by means of a unitary transformation applied to (16),

$$|\alpha(\phi,t)\rangle = e^{-i\gamma_\alpha\phi} |\alpha(t)\rangle = e^{-i\gamma_\alpha\phi} e^{-i\Omega_\alpha t} |e_\alpha\rangle, \tag{19}$$

so that the matrix elements become, in analogy with (15),

$$F_{\alpha\beta}(\phi,t) = \langle e_\alpha | \hat{F}(\phi,t) | e_\beta\rangle = \langle \alpha(\phi,t)| \hat{F} |\beta(\phi,t)\rangle. \tag{20}$$

The factor $e^{i\gamma_{\alpha\beta}\phi}$, associated originally with the amplitude of the field mode $(\alpha\beta)$, has thus been incorporated into the (particle) state vectors $|\alpha\rangle$, $|\beta\rangle$. Now these vectors include the γ space, so they span the product Hilbert space $\mathcal{H}_{\text{old}} \otimes \mathcal{H}_\gamma$. As in previous LSED treatments, the ZPF has disappeared altogether from the description; yet the effect of its action on the particle is evident once more, now through the appearance of the angular momentum transferred by the field.

Notice that ϕ is the azimuthal angle of rotation with respect to an internal quantization axis, the orientation of which is in general independent of the coordinates in 3D space. It is only when a privileged direction is used for the angular momentum projection —e.g., in the presence of an external magnetic field— that this axis may be made to coincide with the z-direction (say) in the configuration space.

4. The bipartite system

Let us now consider a system of two noninteracting particles under the action of a common binding potential. Here one has to take into account that if the two particles belong to one and the same system and they are sufficiently near one to another, they are subject to the common background field; i.e, to the same realization of the random ZPF. It is under this condition that we say that the two particles constitute a single system.

From (10), the product of any two (stochastic) dynamical variables pertaining each to one particle is given in general by

$$F_\alpha(t,\phi) G_{\alpha'}(t,\phi) = \sum_{\beta,\beta'} F_{\alpha\beta} G_{\alpha'\beta'} a^0_{\alpha\beta} a^0_{\alpha'\beta'} e^{i\gamma_{\alpha\beta}\phi} e^{i\gamma_{\alpha'\beta'}\phi} e^{i\omega_{\alpha\beta}t} e^{i\omega_{\alpha'\beta'}t}. \tag{21}$$

On both sides of this expression, the first factor refers always to particle 1 and the second one to particle 2; e.g., $F_\alpha(\phi,t)G_{\alpha'\beta'}(t,\phi) = F_{1\alpha}(\phi,t)G_{2\alpha'\beta'}(t,\phi)$; $a^0_{\alpha\beta}a^0_{\alpha'\beta'} = a^0_{1\alpha\beta}a^0_{2\alpha'\beta'}$, a.s.o. For simplicity in the reading we shall consistently omit the corresponding subindices $1, 2$.

Now it is the composite system that eventually reaches a stationary state, characterized by a total energy $\mathcal{E}_A$. Let us write this state as $A = (\alpha, \alpha')$, where α, α' represent stationary (single-particle) states of particles 1 and 2, respectively. Then in concise form, the product FG in state A writes as

$$\begin{aligned} (FG)_A(\phi,t) &= \sum_B (FG)_{AB} b_{AB} e^{i\gamma_{\alpha\beta}\phi} e^{i\gamma_{\alpha'\beta'}\phi} e^{i\omega_{AB}t} \\ &= \sum_B (FG)_{AB} b_{AB} e^{i(\gamma_\alpha - \gamma_\beta)\phi} e^{i(\gamma_{\alpha'} - \gamma_{\beta'})\phi} e^{i(\Omega_A - \Omega_B)t} \end{aligned} \tag{22}$$

with $(FG)_{AB}$ the elements of a matrix in the product Hilbert space $\mathcal{H}_1 \otimes \mathcal{H}_2$ of the two single-particle Hilbert spaces involved,

$$(FG)_{AB} = F_{\alpha\beta} G_{\alpha'\beta'}. \tag{23}$$

The index B refers to any other accessible state of the compound system, with total energy $\mathcal{E}_B$, and hence it is, like A, a compound index of the form $B = (\beta, \beta')$. The coefficient b_{AB} in (22) is the product of the random variables of the common ZPF appropriate for describing the linear response in the bipartite case,

$$b_{AB} = a^0_{\alpha\beta} a^0_{\alpha'\beta'}. \tag{24}$$

Because of (11), these coefficients satisfy the following relations

$$b^*_{AB} = b_{BA}, \quad |b_{AB}|^2 = b_{AA} = b_{BB} = 1, \quad b_{AC} = b_{AB} b_{BC}, \tag{25}$$

where C denotes an arbitrary state accessible from A (or B). Further, because of (9), the (composite) transition frequencies are given by

$$\omega_{AB} = \hbar^{-1}(\mathcal{E}_A - \mathcal{E}_B). \tag{26}$$

Now the condition of ergodicity implies that the time average of $(FG)_A$, which according to (22) is

$$\overline{FG}^t_A = \sum_B (FG)_{AB} b_{AB} e^{i\gamma_{\alpha\beta}\phi} e^{i\gamma_{\alpha'\beta'}\phi}\Big|_{\omega_{AB}=0}, \tag{27}$$

must be independent of the realization, which means that for every $\omega_{AB} = 0$, the corresponding b_{AB} is nonrandom. Let us call λ_{AB} these nonrandom coefficients, i.e.,

$$\lambda_{AB} = b_{AB}|_{\omega_{AB}=0} = \lambda^*_{BA}, \tag{28}$$

where for the last equality we used (11) and (24). Then the (statistical) average of (22) is given by

$$\overline{FG}_A = \sum_B (FG)_{AB} \lambda_{AB} e^{i\gamma_{\alpha\beta}\phi} e^{i\gamma_{\alpha'\beta'}\phi} \quad (\omega_{AB} = 0). \tag{29}$$

If there is no energy degeneracy, i.e., if to every state A there corresponds a different energy $\mathcal{E}_A$, the condition $\omega_{AB} = 0$ is satisfied only for $A = B$, i.e. $\alpha = \beta$ and $\alpha' = \beta'$, and Eq. (29) reduces to the single term representing the product of the diagonal elements of F and G

$$\overline{FG}_A = (FG)_{AA} = F_{\alpha\alpha} G_{\alpha'\alpha'}, \tag{30}$$

since by virtue of (25) $\lambda_{AA} = b_{AA} = 1$. When the two particles are different, their stationary energy levels will be different in general so that Eq. (30) applies.

However, let us consider two *identical* particles, with different single energy eigenstates α, α' such that $\mathcal{E}_\alpha \neq \mathcal{E}_{\alpha'}$ (i.e., nondegenerate single states). The composite states will always be degenerate in this case, because to every state $A(\alpha, \alpha')$ there corresponds a different state $B(\alpha', \alpha)$ having the same total energy,

$$\mathcal{E}_A = \mathcal{E}_B = \mathcal{E}_\alpha + \mathcal{E}_{\alpha'} \quad (\mathcal{E}_\alpha \neq \mathcal{E}_{\alpha'}). \tag{31}$$

There are therefore two terms in the sum (29) that contribute to $\overline{FG}_A$, namely the first one with $\beta = \alpha$, $\beta' = \alpha'$ (i.e. $A = B$), and the second one with $\beta = \alpha'$, $\beta' = \alpha$ $(\alpha \neq \alpha')$, [16]-[18]

$$\overline{FG}_A = (FG)_{AA} + (FG)_{AB} \lambda_{AB} e^{i\gamma_{\alpha\alpha'}\phi} e^{i\gamma_{\alpha'\alpha}\phi}. \tag{32}$$

Journal of Physics: Conference Series **701** (2016) 012008 doi:10.1088/1742-6596/701/1/012008

where according to (23),

$$(FG)_{AB} = F_{\alpha\alpha'}G_{\alpha'\alpha} \tag{33}$$

and from (24) and (28),

$$\lambda_{AB} = a^0_{\alpha\alpha'}\, a^0_{\alpha'\alpha}. \tag{34}$$

Analogously we find for $\overline{FG}_B$, where B is the compound state having the same energy as A but with the states α, α' interchanged,

$$\overline{FG}_B = (FG)_{BB} + (FG)_{BA}\lambda_{BA}e^{i\gamma_{\alpha'\alpha}\phi}e^{i\gamma_{\alpha\alpha'}\phi}. \tag{35}$$

Let us now use the standard notation for the matrix elements (as before the first factor corresponding consistently to particle 1 and the second one to particle 2)

$$(FG)_{AA} = \langle A|\,\widehat{FG}\,|A\rangle = \langle\alpha|\,\langle\alpha'|\,\widehat{FG}\,|\alpha\rangle\,|\alpha'\rangle$$

and

$$(FG)_{AB} = \langle A|\,\widehat{FG}\,|B\rangle = \langle\alpha|\,\langle\alpha'|\,\widehat{FG}\,|\alpha'\rangle\,|\alpha\rangle\,.$$

Eqs. (32) and (35) take then the form

$$\overline{FG}_A = \sqrt{2}e^{i\gamma_\alpha\phi}e^{i\gamma_{\alpha'}\phi}\,\langle A|\,\widehat{FG}\,|\psi_{AB}\rangle\,, \tag{36a}$$

$$\overline{FG}_B = \sqrt{2}\lambda_{BA}e^{i\gamma_{\alpha'}\phi}e^{i\gamma_\alpha\phi}\,\langle B|\,\widehat{FG}\,|\psi_{AB}\rangle\,, \tag{36b}$$

with

$$\begin{aligned}|\psi_{AB}\rangle &= \tfrac{1}{\sqrt{2}}\Big(e^{-i\gamma_\alpha\phi}e^{-i\gamma_{\alpha'}\phi}\,|A\rangle + \lambda_{AB}e^{-i\gamma_{\alpha'}\phi}e^{-i\gamma_\alpha\phi}\,|B\rangle\Big)\\ &= \tfrac{1}{\sqrt{2}}\Big(e^{-i\gamma_\alpha\phi}e^{-i\gamma_{\alpha'}\phi}\,|\alpha\rangle\,|\alpha'\rangle + \lambda_{AB}e^{-i\gamma_{\alpha'}\phi}e^{-i\gamma_\alpha\phi}\,|\alpha'\rangle\,|\alpha\rangle\Big).\end{aligned} \tag{37}$$

By combining the two expressions (36a), (36b) we obtain for the average $\overline{FG}$ in the degenerate state with $\mathcal{E}_A = \mathcal{E}_B$ $(A \neq B)$

$$\overline{(FG)}_{AB} = \langle\psi_{AB}|\,\widehat{FG}\,|\psi_{AB}\rangle = \tfrac{1}{2}\left[\overline{FG}_A + \overline{FG}_B\right], \tag{38}$$

with $|\psi_{AB}\rangle$ given by (37).

The vector $|\psi_{AB}\rangle$, describing the state of the bipartite system with energy $\mathcal{E}_A = \mathcal{E}_B$ $(A \neq B)$, represents an entangled state. Crucial for this entanglement is the presence of the parameter λ_{AB}, which according to (34) is different from zero. Entanglement appears therefore as an unavoidable consequence of the intervention of ZPF modes that are common to the two particles, with frequency $\omega_{\alpha\alpha'} = -\omega_{\alpha'\alpha}$ equal to the frequency of transition between the two single-particle states. This is to say that the resonant response of both particles to modes of the ZPF of the same frequency is what correlates the motions of the particles.[2] In the usual quantum formalism, the possibility of adding two (or more) state vectors to construct a third one (which is key for the existence of entanglement) is taken as a (formal) consequence of the linearity of the Schrödinger equation. Here we have identified a physical origin and a *need* for this (linear) combination, as a consequence of the mediation of the common modes of the background field. This is an additional example of the contributions offered by the SED formulation of quantum mechanics to a deeper physical understanding of the theory.

[2] As of recently, a distinction is made between entanglement and nonfactorizability in the case of identical particles. This distinction, which departs from the traditional use of the term entanglement since its introduction by Schrödinger in 1935, [20] is still under debate; for a discussion and an extensive list of references see, e.g., Ref. [21]. Since here such distinction is irrelevant, we shall use freely the term entanglement as equivalent to nonfactorabizability.

5. Spin-symmetry relation

In order to explore the connection between spin and symmetry, let us analyze the properties of the entangled state vector $|\psi_{AB}\rangle$ given by Eq. (37) under different permutations. We shall first perform an exchange of particles, i.e., $(1,2) \rightarrow (2,1)$, which means putting particle 1 in the position of particle 2 and vice versa. When performing this exchange operation one must apply it to the entire state vector, including the angular factors; this means that in switching the positions of particles 1 and 2 in the equation, one must take care that the rotations that take particle 1 to the azimuthal position of particle 2 and vice versa, are both made in the same sense —say clockwise— as discussed e.g. in Ref. ([6]).

Assume first that $\phi_2 > \phi_1$, then ϕ_1 transforms into ϕ_2 (if the rotation is made in the clockwise sense) and ϕ_2 transforms into $2\pi + \phi_1$, i.e., with $\Delta\phi = \phi_2 - \phi_1$:

$$\phi_1 \rightarrow \phi_2 = \phi_1 + \Delta\phi. \tag{39a}$$

$$\phi_2 \rightarrow 2\pi + \phi_1 = \phi_2 + 2\pi - \Delta\phi. \tag{39b}$$

Therefore, as a result of the exchange applied to Eq. (37) we get

$$\begin{aligned} |\psi_{AB}\rangle_{1\leftrightarrow 2} &= \tfrac{1}{\sqrt{2}}\Big(e^{-i2\pi\gamma_{\alpha'}}e^{-i\gamma_{\alpha'}\phi}e^{-i\gamma_{\alpha}\phi}\,|\alpha'\rangle\,|\alpha\rangle \\ &\quad +\lambda_{BA}e^{-i2\pi\gamma_{\alpha}}e^{-i\gamma_{\alpha}\phi}e^{-i\gamma_{\alpha'}\phi}\,|\alpha\rangle\,|\alpha'\rangle\Big). \end{aligned} \tag{40}$$

Now γ_α and $\gamma_{\alpha'}$ have always a half-integral value, as discussed in Sect. 3. Therefore $e^{-i2\pi\gamma_\alpha} = e^{-i2\pi\gamma_{\alpha'}} = (-1)^{2\gamma_\alpha}$ regardless of the value of $\gamma_\alpha, \gamma_{\alpha'}$, and one may take out this common factor from the parentheses in the above expression:

$$|\psi_{AB}\rangle_{1\leftrightarrow 2} = \tfrac{1}{\sqrt{2}}(-1)^{2\gamma_\alpha}\Big(e^{-i\gamma_{\alpha'}\phi}e^{-i\gamma_{\alpha}\phi}\,|\alpha'\rangle\,|\alpha\rangle + \lambda_{BA}e^{-i\gamma_{\alpha}\phi}e^{-i\gamma_{\alpha'}\phi}\,|\alpha\rangle\,|\alpha'\rangle\Big). \tag{41}$$

If instead, $\phi_2 < \phi_1$, then ϕ_2 transforms into ϕ_1, ϕ_1 transforms into $2\pi + \phi_2$, and the exchange applied to Eq. (37) leads to

$$\begin{aligned} |\psi_{AB}\rangle_{1\leftrightarrow 2} &= \tfrac{1}{\sqrt{2}}\Big(e^{-i2\pi\gamma_{\alpha}}e^{-i\gamma_{\alpha'}\phi}e^{-i\gamma_{\alpha}\phi}\,|\alpha'\rangle\,|\alpha\rangle \\ &\quad +\lambda_{BA}e^{-i2\pi\gamma_{\alpha'}}e^{-i\gamma_{\alpha}\phi}e^{-i\gamma_{\alpha'}\phi}\,|\alpha\rangle\,|\alpha'\rangle\Big), \end{aligned} \tag{42}$$

which gives the same result, Eq. (41). Therefore we have in both cases

$$|\psi_{AB}\rangle_{1\leftrightarrow 2} = \lambda^*_{AB}(-1)^{2\gamma_\alpha}\,|\psi_{AB}\rangle\,. \tag{43}$$

Since the particles are identical, their interchange should have no effect whatsoever; stated explicitly, the interchange of particle positions denoted by $(1,2) \rightarrow (2,1)$ should have no consequence on the state vector. This means that $\lambda^*_{AB}(-1)^{2\gamma_\alpha} = 1$. Further, given that the sign of $(-1)^{2\gamma_\alpha}$ is determined solely by the (half-integer) contribution due to the spin σ, we may write $(-1)^{2\gamma_\alpha} = (-1)^{2\sigma}$, whence

$$\lambda_{AB} = \lambda^*_{AB} = (-1)^{2\sigma} = -1. \tag{44}$$

Let us now take the state vector given by Eq. (37) and apply to it a different operation, consisting of the interchange of states $A \Leftrightarrow B$ —which is equivalent to replacing (α, α') by (α', α). Since the states A and B are different by construction, the state vector is in principle

Journal of Physics: Conference Series **701** (2016) 012008 doi:10.1088/1742-6596/701/1/012008

not invariant under this interchange. Indeed, the state vector (40) transformed by the operation $A \Leftrightarrow B$ (or $\alpha \Leftrightarrow \alpha'$) is readily seen to be

$$|\psi_{BA}\rangle = \lambda^*_{AB} |\psi_{AB}\rangle, \tag{45}$$

which indicates that its transformation property depends on the sign of the entanglement parameter λ_{AB}, given by Eq. (44). Introducing this value into Eq. (45) we obtain the well-known result

$$|\psi_{BA}\rangle = -|\psi_{AB}\rangle, \tag{46}$$

and Eq. (37) becomes the antisymmetric state vector

$$|\psi_{AB}\rangle = \frac{1}{\sqrt{2}}\left(e^{-i\gamma_\alpha\phi}e^{-i\gamma_{\alpha'}\phi}|A\rangle - e^{-i\gamma_{\alpha'}\phi}e^{-i\gamma_\alpha\phi}|B\rangle\right). \tag{47}$$

We recall that by construction, the states A, B must be different for the two particles to constitute a single system. If we were to take $A = B$, Eq. (47) would reduce to $|\psi_{AA}\rangle = 0$, which is just an expression of the well-known Pauli exclusion principle.

5.1. The many-electron system

The above discussion can be extended to a system composed of more than two electrons with spin, subject again to a common ZPF. To determine the resulting state of the system, in which the total energy is again a constant, we must consider the various possible configurations A, B, C, ... of stationary states of the individual particles corresponding to the same total energy $\mathcal{E}_A$. Specifically, we are interested in the question whether in general the symmetrization principle finds an explanation in terms of the correlations established by the ZPF between the electronic motions.

Direct application of the above procedure to a larger number of electrons becomes rather cumbersome, as the degeneracy of the composite state increases as $N!$ with the number of particles N. A convenient approach would be to apply the above procedure firstly to any pair of electrons of the system, say those in states α, α'. By taking successively every possible pair, all relevant frequencies of resonance will be accounted for, and all the respective symmetries will thus be included.[6] Since as a result no pair of electrons can be in the same (single-particle) state, the state of the entire system will be described by a totally antisymmetric, multiply entangled state vector built of different single-particle states that carry the factor $(-1)^{2k\sigma}$ in front of each term, where k stands for the number of transpositions needed to reach the corresponding exchanged state, starting from the initial state, and $\sigma = 1/2$. If two states were made to coincide (say, $\alpha' = \alpha$) the entire state vector would reduce to zero; we have again a direct expression of the exclusion principle, here applied to more than two electrons.

6. Final comments

The results presented here allow us to give an answer to two important questions that have remained open in quantum mechanics along the years. The first one refers to the physics behind the structure of the state vectors for a bipartite or multipartite system of electrons: When its components resonate to a common frequency of the common background field, it is no longer possible to describe the state of the compound system as if every constituent particle were isolated and independent from the others. Particles and field constitute a physical unit of interconnected parts, which explains why this state is represented by a nonfactorizable vector, that peculiar quantum entity that entails correlations between the particles' dynamical variables. It is the permanent presence of the vacuum field, resonantly and coherently connecting all parts among themselves, that unveils the mystery of the spin-statistics association in matter-plus-field behavior. The corollary is that the consideration of the complete physical system in first quantization is sufficient to explain the statistics of assemblies of electrons.

The second question is how it is that in almost ninety years of quantum mechanics no definitive *physical* argument has been advanced (to our knowledge) to account for the restriction to antisymmetric states for fermions. An answer to this question is provided by just the above discussion: the *existence* of the ZPF is responsible for the correlations, but this field is outside the scope of usual quantum mechanics. In ignoring the ZPF, quantum mechanics misses the possibility in principle to delve into the source of this (anti)symmetry.

It is important to stress that the procedure followed here to determine the symmetry properties of the two-electron state vector differs in a subtle but essential way from usual quantum mechanics, where the interchange of identical (or 'indistinguishable') particles $1 \Longleftrightarrow 2$ is considered equivalent to the interchange of states $\alpha, \alpha' \Leftrightarrow \alpha', \alpha$. According to the present discussion, by contrast, whereas the interchange of identical particles means an interchange of labels and should therefore leave the state vector unchanged, the interchange of individual states takes the compound state from $A(\alpha, \alpha')$ to a new state $B(\alpha', \alpha)$; although the respective state vectors are physically equivalent and yield the same values for the observables, they happen to differ by an overall minus sign. This minus sign comes from the value of the parameter λ (determined in its turn by the half-integral nature of the spin variable) that correlates the motions of the two electrons via the common ZPF modes.

We should like to add a comment on the extension of the present treatment to particles with different intrinsic angular momentum values. The fact that the sign of the correlation parameter $\lambda_{AB} = (-1)^{2\sigma}$ is determined by the integral or half-integral value of σ, suggests that the results of Sect. 5 may be applied not just to fermions, but equally well to bosons or higher-spin particles.

Finally, a brief comment on the difference between identity and indistinguishability is in place. The notion of distinguishability implies our capacity to distinguish, and thus involves a subjective element. By contrast, in dealing with identical particles, one must consider the indifference between them, in the statistical context of the quantum description. The point may seem irrelevant for some, but it acquires importance for at least two reasons. One is that, contrary to indistinguishability, the notion of statistical indifference does not contain unphysical elements. The second is more technical and refers to classical physics. In the whole of thermodynamics, classical particles are treated as distinguishable, yet a radical change is made in the derivation of the Maxwell–Boltzman law, which requires to consider them as indistinguishable in order to avoid contradictions such as Gibbs' paradox. [23] Introducing the notion of statistical indifference when the *statistical* behavior of classical particles is being considered leads to the correct results without putting at stake the objectivity of the treatment.

Acknowledgments
The authors acknowledge helpful criticism and suggestions from Andrea Valdés-Hernández and Jaime Avendaño. This work has been supported by DGAPA-UNAM through projects PAPIIT IN112714 and IN104816.

References

[1] Pauli W 1945 *Exclusion Principle and Quantum Mechanics* (Nobel lecture)
[2] Pauli W 1950 On the connection between spin and statistics, *Prog. Theor. Phys.* **5** 526–543
[3] Kaplan L G 2013 The Pauli exclusion principle. Can it be probed? *Found. Phys.* **43** 1233–1251 (2013)
[4] Bernstein H J and Phillips A V 1981 Fiber bundles and quantum theory *Sci. Am.* **245**(7) 123–137
[5] Imbo T D, Imbo Ch Sh and Sudarshan E C D 1990 Identical particles, exotic statistics, and braid groups *Phys. Lett. B* **234** 103–107
[6] Jabs A 2010 Connecting spin and statistics in quantum mechanics *Found. Phys.* **40** 776–792 (*Preprint* `arXiv:0810.2399v4 [quant-ph]` 3 Feb 2014)
[7] Hilborn R C 1995 Answer to Question 7, 'The spin-statistics theorem' *Am. J. Phys.* **63** 298–299
[8] Duck I and Sudarshan E C G 1998 *Pauli and the Spin-Statistics Theorem* (Singapore: World Scientific); Toward an understanding of the spin-statistics theorem *Am. J. Phys.* **66** 285–303
[9] McMurry S M 1993 *Quantum Mechanics* (Reading: Addison-Wesley)

[10] Bennett A F 2015 Spin-Statistics connection for relativistic quantum mechanics *Found. Phys.* **45** 370–381
[11] Leinaas J M and Myrheim J 1977 On the theory of identical particles *Il Nuovo Cim. B* **37** 1–23
[12] Broylos A A 1999 Derivation of the pauli exchange principle (*Preprint* `arXiv:quant-ph/9906046v1`
[13] Reginatto M 1998 Derivation of the Pauli equation using the principle of minimum Fisher information *Phys. Lett. A* **249** 355
[14] Santamato E and De Martini F 2015 Proof of the spin-statistics theorem *Found. Phys.* **45** 858
[15] de la Peña L and Cetto A M 2006 The foundations of linear stochastic electrodynamics *Found. Phys.* **36** 350
[16] de la Peña L, Valdés-Hernández A and Cetto A M 2010 Entanglement of particles as a result of their coupling through the common background zero-point radiation field *Physica E* **42** 308
[17] Valdés-Hernández A, de la Peña L and Cetto A M 2011 Bipartite entanglement induced by a common background (zero-point) radiation field *Found. Phys.* **41** 843
[18] de la Peña L, Cetto A M and Valdés-Hernández A 2015 *The emerging quantum. The Physics behind Quantum Mechanics* (Berlin: Springer)
[19] Cetto A M, de la Peña L and Valdés-Hernández A 2014 Emergence of quantization: the spin of the electron *JPCS* **504** 012007; See also Ref. [22]
[20] Schrödinger E 1935 Die gegenwärtige Situation in der Quantenmechanik I, II, III *Naturwiss.* **23** 807, 823, 844; English translation in Wheeler J A and Zurek W H (eds.) 1983 *Quantum theory and measurement* (Princeton, Princeton University Press)
[21] Benatti F, Floreanini R and Titimbo K 2014 Entanglement of identical particles *Open Syst. Inf. Dyn.* **21** 1440003
[22] de la Peña L and Cetto A M 1996 *The Quantum Dice. An Introduction to Stochastic Electrodynamics* (Dordrecht: Kluwer)
[23] McQuarrie D A 2000 *Statistical Mechanics* (Sausalito, CA: University Science Books)

Journal of Physics: Conference Series **701** (2016) 012009 doi:10.1088/1742-6596/701/1/012009

Trading drift and fluctuations in entropic dynamics: quantum dynamics as an emergent universality class

Daniel Bartolomeo and Ariel Caticha

Department of Physics, University at Albany-SUNY, Albany, NY 12222, USA.

E-mail: ariel@albany.edu

Abstract. Entropic Dynamics (ED) is a framework that allows the formulation of dynamical theories as an application of entropic methods of inference. In the generic application of ED to derive the Schrödinger equation for N particles the dynamics is a non-dissipative diffusion in which the system follows a "Brownian" trajectory with fluctuations superposed on a smooth drift. We show that there is a family of ED models that differ at the "microscopic" or sub-quantum level in that one can enhance or suppress the fluctuations relative to the drift. Nevertheless, members of this family belong to the same universality class in that they all lead to the same emergent Schrödinger behavior at the "macroscopic" or quantum level. The model in which fluctuations are totally suppressed is of particular interest: the system evolves along the smooth lines of probability flow. Thus ED includes the Bohmian or causal form of quantum mechanics as a special limiting case. We briefly explore a different universality class – a non-dissipative dynamics with microscopic fluctuations but no quantum potential. The Bohmian limit of these hybrid models is equivalent to classical mechanics. Finally we show that the Heisenberg uncertainty relation is unaffected either by enhancing or suppressing microscopic fluctuations or by switching off the quantum potential.

1. Introduction

Entropic Dynamics (ED) is a framework in which quantum theory is derived as an application of entropic methods of inference.[1] As in any theory of inference, establishing the subject matter is the first and most crucial step; this amounts to a choice of microstates, that is, a choice of *beables*. Once that choice is made the dynamics is driven by entropy subject to constraints which reflect the information needed for making physical predictions [2–5].

ED naturally leads to an epistemic view of the quantum state ψ but with an added twist. Within an inferential framework it is not sufficient to merely interpret the probability $|\psi|^2$ as a state of knowledge.[2] It is just as important to require that the dynamics, that is the updates of ψ, be consistent with the established rules for updating probability distributions. This is where the method of maximum entropy and the Bayesian methods enters. Furthermore, the entropic dynamics must include *both* the unitary time evolution described by the Schrödinger equation *and* the collapse of the wave function during measurement. As a result the ED framework turns out to be very restrictive. In a fully *entropic* dynamics we do not postulate an underlying

[1] The principle of maximum entropy as a method for inference can be traced to the pioneering work of E. T. Jaynes. For a pedagogical overview of Bayesian and entropic inference and further references see [1].
[2] For a review with references on the epistemic vs ontic interpretations of the quantum state see [6].

mechanics with an action principle that operates at some deeper level. Instead, at the sub-quantum level there is only inference and at the quantum level the emergent dynamics is described by an action principle that is derived rather than posited.

There is a vast literature on the attempts to reconstruct quantum mechanics and it is inevitable that the ED approach will resemble them in one aspect or another. Indeed, to the extent that any of these approaches are successful they must sooner or later converge to the same Schrödinger equation. However, there are important differences. For example, the central concern with the notion of time makes ED significantly different from other approaches that are also based on information theory (such as *e.g.*, [7–14]). And ED also differs from those approaches (see *e.g.*, [15–21]) that aim to explain the emergence of quantum behavior as the effective statistical mechanics of some underlying sub-quantum mechanics which might possibly include some additional stochastic element. Indeed, ED makes no reference to any sub-quantum action principles whether classical, deterministic, or stochastic.

As stated above in ED inferences are carried out on the basis of information introduced in the form of constraints. In the particular case of the ED of N particles the microstates are the positions of the particles. The basic physical input — that the particles follow continuous trajectories — is implemented through N constraints, one for each particle. The multiple roles played by the corresponding Lagrange multipliers α_n $(n = 1 \ldots N)$ are by now well understood. The multipliers regulate the flow of time and they serve to unify the concept of mass with that of quantum fluctuations.

If these were the only constraints each particle would experience its own independent dynamics and the resulting motion would be an isotropic diffusion. It turns our that one can obtain more interesting forms of dynamics by imposing just one single additional constraint. This constraint acts on configuration space and involves a "drift potential," an epistemic tool that plays a role somewhat analogous to that of a pilot wave. The drift potential contributes to the phase of the wave function, it correlates the motion of particles, and it causes such quintessential quantum effects as interference and entanglement. The corresponding multiplier α' is not nearly as well understood and the purpose of this paper is to fill this gap [22].

We begin with a brief overview of ED following the presentation in [1]. We show that the role of the multiplier α' is to control the relative magnitudes of drift and fluctuations. Our main result is simple: ED models with different values of α' lead to the same Schrödinger equation. In other words, different "microscopic" or sub-quantum models can lead to the same emergent quantum behavior — they belong to the same "universality" class. The limit of large α' deserves particular attention. In this limit fluctuations are suppressed, the drift motion prevails, and the particles tend to move along the smooth lines of probability flow. This means that ED includes the Bohmian form of quantum mechanics as a special limiting case [23–25].

We briefly explore one alternative family of ED models — a different universality class. At the microscopic level these models also describe particles that follow Brownian trajectories but without the non-local correlations induced by the quantum potential [26]. The multiplier α' also acts to suppress microscopic fluctuations and the Bohmian, or large α' limit, approaches classical mechanics. All these microscopic models lead to the same dynamics at the macroscopic level — the emergent dynamics is an essentially classical mechanics.

The value of the ED approach to quantum theory lies in part in the conceptual clarity it brings to issues of interpretation. In [26] the methods originally developed in the context of stochastic mechanics [27–30] were adapted to derive the Heisenberg uncertainty relation for position and momentum within the context of ED.

Finally, the fact that in ED we can enhance or suppress microscopic fluctuations relative to the drift and that we can switch on or off the quantum potential leads us to raise the question of how these changes affect the uncertainty relations. Are there potential violations of the uncertainty principle? We find that the uncertainty relations are unaffected by changes in the microscopic fluctuations, even when suppressing them to the Bohmian limit, or by switching off the quantum potential.

2. Entropic Dynamics — a brief overview

We consider the ED of N particles living in a flat Euclidean space $\mathbf{X}$ with metric δ_{ab}. In the ED framework the particles have definite positions x_n^a and it is their unknown values that we wish to infer. (The index $n = 1 \ldots N$ labels the particle and $a = 1, 2, 3$ its spatial coordinates.) The position of the system in configuration space $\mathbf{X}_N = \mathbf{X} \times \ldots \times \mathbf{X}$ will also be denoted x^A where $A = (n, a)$.

The main dynamical assumption is that motion is continuous which means that it can be analyzed as a sequence of many infinitesimally short steps. Thus we first find the probability $P(x'|x)$ that the system takes a short step from x^A to $x'^A = x^A + \Delta x^A$ and then we determine how such short steps accumulate. To find $P(x'|x)$ we maximize the (relative) entropy,

$$\mathcal{S}[P, Q] = -\int dx'\, P(x'|x) \log \frac{P(x'|x)}{Q(x'|x)} \,, \tag{1}$$

where we adopt the notation $dx' = d^{3N}x'$ and the prior distribution $Q(x'|x)$ describes the state of knowledge (of an ideally rational agent) *before* any information about the motion is taken into account. A priori we shall assume extreme ignorance which is expressed by a uniform distribution: we assume that $Q(x'|x)dx'$ is proportional to the volume element.[3] Since the space $\mathbf{X}_N$ is flat we can set $Q(x'|x) = 1$.

The physical information about the motion is introduced through constraints. The fact that particles take infinitesimally short steps from x_n^a to $x_n'^a = x_n^a + \Delta x_n^a$ is imposed through N separate constraints,

$$\langle \Delta x_n^a \Delta x_n^b \rangle \delta_{ab} = \kappa_n \,, \qquad (n = 1 \ldots N) \tag{2}$$

where κ_n are constants. The κ_n's are chosen to be constant to reflect the translational symmetry of the space $\mathbf{X}$; they are n-dependent in order to accommodate non-identical particles; and eventually we take $\kappa_n \to 0$ to implement infinitesimally short steps.

There is one additional constraint that leads to correlations among the particles,

$$\langle \Delta x^A \rangle \partial_A \phi = \sum_{n=1}^{N} \langle \Delta x_n^a \rangle \frac{\partial \phi}{\partial x_n^a} = \kappa' \,, \tag{3}$$

which introduces the "drift" potential ϕ and $\partial_A = \partial/\partial x^A = \partial/\partial x_n^a$. κ' is another small but for now unspecified position-independent constant. Eq.(3) is a single constraint that acts on the $3N$-dimensional configuration space.[4]

Maximizing the entropy (1) subject to (2), (3), and normalization leads to

$$P(x'|x) = \frac{1}{\zeta} \exp[-\sum_n (\frac{1}{2}\alpha_n \, \Delta x_n^a \Delta x_n^b \delta_{ab} - \alpha' \Delta x_n^a \frac{\partial \phi}{\partial x_n^a})] \,, \tag{4}$$

where ζ is a normalization constant and α_n and α' are the Lagrange multipliers associated to (2) and (3). The limit of infinitesimally short steps $\kappa_n \to 0$ is achieved as $\alpha_n \to \infty$.

[3] Strictly uniform non-normalizable priors are mathematically problematic. This difficulty can be avoided by adopting a physically reasonable normalizable prior. By "uniform" we actually mean any distribution that is essentially flat over macroscopic scales.

[4] Elsewhere, in the context of particles with spin, we will see that the potential $\phi(x)$ can be given a natural geometric interpretation as an angular variable. Its integral over any closed loop is $\oint d\phi = 2\pi n$ where n is an integer.

Already at this early stage we can see that ED exhibits a rather trivial symmetry: Imposing the constraint (3) with the pair (ϕ, κ') leads to the same transition probability $P(x'|x)$ as a constraint with the pair $(\tilde{\phi}, \tilde{\kappa}') = (C\phi, C\kappa')$ where C is some arbitrary constant. In previous work [4] we took advantage of the symmetry and rescaled with $C = 1/\alpha'$. This amounts to setting $\alpha'\phi \to \phi$ which eliminates α'. Here we wish to examine the effect of α' for given ϕ so we will keep it explicit.

To find how these short steps accumulate to produce a finite change we introduce a book-keeping device — this is how the notion of time enters dynamics. As discussed in [2–5] entropic time is measured by the fluctuations themselves (see eq.(8) below) which leads to the choice

$$\alpha_n = \frac{m_n}{\eta \Delta t} , \tag{5}$$

where Δt is the interval over which the short step $x \to x'$ occurs, the m_n are particle-specific constants that will be called "masses", and η is a constant that fixes the units of time relative to those of length and mass. With this choice of α_n a generic displacement can be expressed as an expected drift plus a fluctuation,

$$\Delta x^A = b^A \Delta t + \Delta w^A , \tag{6}$$

where $b^A(x)$ is the drift velocity,

$$\langle \Delta x^A \rangle = b^A \Delta t \quad \text{with} \quad b^A = \frac{\eta\alpha'}{m_n}\delta^{AB}\partial_B\phi = \eta\alpha' m^{AB}\partial_B\phi . \tag{7}$$

$m_{AB} = m_n\delta_{AB}$ is the "mass" tensor and $m^{AB} = \delta^{AB}/m_n$ is its inverse. The fluctuations Δw^A satisfy,

$$\langle \Delta w^A \rangle = 0 \quad \text{and} \quad \langle \Delta w^A \Delta w^B \rangle = \frac{\eta}{m_n}\delta^{AB}\Delta t = \eta m^{AB}\Delta t . \tag{8}$$

Comparing equations (7) and (8) for short steps, as $\Delta t \to 0$, we see that the fluctuations are much larger than the drift ($\Delta w^A \sim \Delta t^{1/2}$ while $\langle \Delta x^A \rangle \sim \Delta t$) which leads to non-differentiable trajectories characteristic of a Brownian motion. They also show that for fixed ϕ the effect of the multiplier α' is to enhance or suppress the drift $b^A \Delta t$ relative to the fluctuations Δw^A.

Having introduced a convenient notion of time through (5), the accumulation of many short steps leads to a probability distribution $\rho(x,t)$ in configuration space that obeys a Fokker-Planck equation (FP), [1–3]

$$\partial_t \rho = -\partial_A \left(\rho v^A\right) . \tag{9}$$

In this equation v^A is the velocity of the probability flow in configuration space or *current velocity*. It is given by

$$v^A = b^A + u^A \quad \text{where} \quad u^A = -\eta m^{AB}\partial_B \log \rho^{1/2} \tag{10}$$

is called the *osmotic velocity*. The interpretation of u^A follows immediately from looking at its contribution to the probability flux,

$$\rho u^A = -\frac{1}{2}\eta m^{AB}\partial_B\rho . \tag{11}$$

This is recognized as the analogue of Fick's law of diffusion with a diffusion tensor $\eta m^{AB}/2$. Since both b^A and u^A are gradients the current velocity v^A is a gradient too,

$$v^A = m^{AB}\partial_B\Phi \quad \text{where} \quad \Phi = \eta\alpha'\phi - \eta \log \rho^{1/2} \tag{12}$$

will be called the *phase*. Thus, the phase also has a drift component and a diffusive or osmotic component.

The dynamics described by the FP equation (9) is a standard diffusion. To describe a "mechanics" we require that the diffusion be "non-dissipative" which is achieved by an appropriate readjustment of the constraint (3) after each step Δt. The net effect is that the drift potential ϕ, or equivalently the phase Φ, is promoted to a fully dynamical degree of freedom. The diffusion is said to be "non-dissipative" when the actual updating of Φ is implemented by imposing that a certain functional $\tilde{H}[\rho,\Phi]$ be conserved. In order to offset the entropic change $\rho \to \rho + \delta\rho$, one requires that Φ changes $\Phi \to \Phi + \delta\Phi$ in such a way that

$$\tilde{H}[\rho + \delta\rho, \Phi + \delta\Phi] = \tilde{H}[\rho,\Phi] \ . \tag{13}$$

As shown in [4] the requirement that $\tilde{H}$ be conserved for arbitrary choices of ρ and Φ implies that the coupled evolution of ρ and Φ is given by a conjugate pair of Hamilton's equations,

$$\partial_t \rho = \frac{\delta \tilde{H}}{\delta \Phi} \quad \text{and} \quad \partial_t \Phi = -\frac{\delta \tilde{H}}{\delta \rho} \ . \tag{14}$$

The "ensemble" Hamiltonian $\tilde{H}$ is chosen so that the first equation reproduces the FP equation (9). Then the second equation becomes a Hamilton-Jacobi equation (HJ). Further arguments from information geometry can then be invoked to fully specify the form of the functional $\tilde{H}[\rho,\Phi]$ [4]. They suggest that the natural choice of $\tilde{H}$ is

$$\tilde{H}[\rho,\Phi] = \int dx\, \rho \left[\frac{1}{2} m^{AB} \partial_A \Phi \partial_B \Phi + V + \xi m^{AB} \frac{1}{\rho^2} \partial_A \rho \partial_B \rho \right] \ . \tag{15}$$

The first term in the integrand is the "kinetic" term that reproduces the FP equation (9). The second term represents the simplest non-trivial interaction, a potential energy that is linear in ρ and introduces the standard potential $V(x)$. The third term, motivated by information geometry, is the trace of the Fisher information and is called the "quantum" potential. The parameter ξ turns out to be crucial: it controls the relative contributions of the two potentials. When $\xi > 0$ we write $\xi = \hbar^2/8$. Thus ξ defines the value of what we call Planck's constant $\hbar$, and sets the scale that separates quantum from classical regimes. The case $\xi = 0$ will be addressed below; the case $\xi < 0$ leads to instabilities and will not be discussed further.

To conclude this brief review of ED we note that at this point the dynamics is fully specified by equations (14) and (15). Nothing prevents us however from combining ρ and Φ into a single complex function,

$$\Psi = \rho^{1/2} \exp(i\Phi/\hbar) \ . \tag{16}$$

Then the pair of Hamilton's equations (14) can be tremendously simplified and written as a single complex linear equation,

$$i\hbar \partial_t \Psi = -\frac{\hbar^2}{2} m^{AB} \partial_A \partial_B \Psi + V\Psi \ , \tag{17}$$

which we recognize as the Schrödinger equation.

3. Trading drift and fluctuations

According to ED at the microscopic sub-quantum level the dynamics is very irregular. The particles perform a Brownian motion, eq.(6), with an expected drift and fluctuations given by (7) and (8). From these equations we see that the effect of α' is to enhance the drift relative to the fluctuations. This means that different values of the multiplier α' correspond different types of microscopic dynamics.

We can study the sub-quantum effect of α' directly through eqs.(7) and (8). However, it may be more instructive to rescale η and write $\eta = \tilde{\eta}/\alpha'$. Under such rescaling the α' dependence migrates from the drift to the fluctuations,

$$\langle \Delta x^A \rangle = \tilde{\eta} m^{AB} \partial_B \phi \, \Delta t \quad \text{and} \quad \langle \Delta w^A \Delta w^B \rangle = \frac{\tilde{\eta}}{\alpha'} m^{AB} \Delta t \; , \tag{18}$$

and we see that increasing α' at fixed $\tilde{\eta}$ has the effect of suppressing the fluctuations.

In contrast, at the "macroscopic" or quantum level the dynamics is very smooth. It is a non-dissipative diffusion described by Hamilton's equations (14) or by the Schrödinger equation (17). The quantum dynamics is clearly independent of α' which means that there is a whole family of microscopic models — one could call it a universality class — that leads to the same emergent quantum dynamics.

From eq.(12) we have

$$v^A = m^{AB} \partial_B \Phi \quad \text{with} \quad \Phi = \tilde{\eta}\phi - \frac{\tilde{\eta}}{\alpha'} \log \rho^{1/2} \; . \tag{19}$$

Here too we see that when we change α' the phase Φ remains unchanged but the relative contributions of the drift and osmotic components change.

The Bohmian limit For large α' the fluctuations are suppressed; osmotic or diffusion effects become negligible. As α' increases the particles follow smoother trajectories which resemble a Brownian motion only at increasingly shorter spatial scales. In the limit $\alpha' \to \infty$ we have

$$\Phi \to \tilde{\eta}\phi \quad \text{so that} \quad v^A \to b^A \; . \tag{20}$$

The current and the drift velocities coincide and particles follow smooth trajectories that coincide with the lines of probability flow. This is exactly the kind of motion postulated by Bohmian mechanics [23–25].

But a word of caution is necessary. ED is driven by entropy; it is essentially non-causal and indeterministic. This is what led us to introduce probabilities and allowed us to write down a Fokker-Planck equation. Therefore the $\Delta t \to 0$ limit or, going back to (5), the $\alpha_n \to \infty$ limit, which is the limit that enforces the continuity of trajectories, must be taken at fixed α', before we consider the effect of $\alpha' \to \infty$. Only then, once we recognize that we are dealing with a tricky singular limit, can we claim that entropic dynamics includes Bohmian mechanics as a special limiting case. Thus, no matter how large the (fixed) value of α', entropic dynamics remains "entropic". Even for large α' the dynamics is still driven by fluctuations and at sufficiently microscopic scales the expected motion is Brownian.

We must also emphasize that it is only with respect to the mathematical formalism that ED includes Bohmian mechanics as a special case. The philosophical differences constitute an unbridgeable gap. Bohmian mechanics attempts to provide an actual description of reality, a description of the ontology of the universe as it "really" is and as it "really" happens. In the Bohmian view the universe consists of real particles that have definite positions and their trajectories are guided by a real field, the wave function Ψ [23].

On the other hand ED is a purely epistemic theory. Its pragmatic goal is less ambitious: to make the best possible predictions on the basis of very incomplete information. In ED the particles also have definite positions and its formalism includes a function Φ that behaves as a wave. But Φ is a tool for reasoning; it is not meant to represent anything real. There is no implication that the particles move the way they do because they are pushed around by a pilot wave or by some stochastic force. In fact ED is silent on the issue of what causative power is responsible for the peculiar motion of the particles. What the probability ρ and the phase Φ are designed to do is not to guide the particles but to guide our inferences. They guide our expectations of where and when to find the particles but they do not exert any causal influence on the particles themselves.

4. Another universality class and its Bohmian limit

ED as a non-dissipative diffusion is defined by Hamilton's equations (14). The ensemble Hamiltonian, eq.(15), includes a parameter ξ that regulates the strength of the quantum potential. Any non-zero value $\xi > 0$ yields a fully *quantum* mechanics, albeit with differing values of $\hbar$. We can also treat ξ and $\hbar$ as independent parameters and set $\xi = 0$.[5] This leads us to a qualitatively different theory — a different universality class.

At the microscopic level the particles follow irregular Brownian trajectories described by eqs.(6-8). For $\xi = 0$, α' has the same effect of enhancing the drift relative to the fluctuations, and therefore different values of α' correspond to different types of microscopic dynamics.

At the "macroscopic" level the emergent behavior is smooth. According to equations (14) and (15) for $\xi = 0$ the probability ρ follows the gradient of Φ,

$$\partial_t \rho = \frac{\delta \tilde{H}}{\delta \Phi} = -\partial_A \left(\rho v^A\right) \quad \text{with} \quad v^A = m^{AB} \partial_B \Phi \ , \tag{21}$$

and Φ evolves according to

$$\partial_t \Phi = -\frac{\delta \tilde{H}}{\delta \rho} = -\frac{1}{2} m^{AB} \partial_A \Phi \partial_B \Phi - V \ , \tag{22}$$

which we recognize as the classical Hamilton-Jacobi equation. Therefore the probability ρ flows along the classical path. In fact, these are exactly the classical equations of motion in a Liouville representation. We conclude that the emergent macroscopic dynamics is essentially classical mechanics. There is, however, no implication that the particles themselves follow the classical paths. Indeed, at any instant of time the particles undergo the same fluctuations, eq.(18), that we would expect for any non-zero value of ξ.

For $\xi = 0$ ED is a hybrid theory; it resembles classical mechanics in some respects and quantum mechanics in others. Just as in quantum mechanics the particles follow Brownian paths and the dynamics is a non-dissipative diffusion; they even satisfy an uncertainty principle [26]. On the other hand, just as in classical mechanics, the probability flows according to paths described by the classical Hamilton-Jacobi equation. One can even combine ρ and Φ into a single complex function,[6] $\Psi = \rho^{1/2} \exp(i\Phi/\hbar)$, and write the coupled evolution of ρ and Φ in terms of a single complex equation that resembles a Schrödinger equation,

$$i\hbar \partial_t \Psi_k = -\frac{\hbar^2}{2} m^{AB} \partial_A \partial_B \Psi_k + V \Psi_k + \frac{\hbar^2}{2} m^{AB} \frac{\partial_A \partial_B |\Psi_k|}{|\Psi_k|} \Psi_k \ . \tag{23}$$

But this equation is not linear which means that a central feature of quantum behavior, the superposition principle, has been lost.

Within the family of microscopic models with $\xi = 0$ we can also take the "Bohmian" limit, $\alpha' \to \infty$. Increasing α' at fixed $\tilde{\eta}$ suppresses the fluctuations so the particles follow smoother

[5] In the hybrid theory ξ and $\hbar$ are independent parameters. ξ is set to 0 and $\hbar$ is defined as the constant with the appropriate units of action that is needed to define a wave function $\Psi = \rho^{1/2} e^{i\Phi/\hbar}$.

trajectories that increasingly approximate the lines of probability flow determined by the classical Hamilton-Jacobi equation (22). Therefore for $\alpha' \to \infty$ the particles follow classical trajectories. We conclude that the $\alpha' \to \infty$ limit of the hybrid theory is classical mechanics.

Here too a word of caution is needed. We say the emergent entropic dynamics in the $\xi = 0$ case is "essentially" classical mechanics. The point is that here too, entropic dynamics remains "entropic". Even for very large α', at sufficiently microscopic scales the expected motion remains Brownian.

To summarize: Within the universality class of quantum dynamics, suppressing microscopic fluctuations yields the Bohmian mechanics. Similarly, within the universality class of hybrid dynamics, suppressing microscopic fluctuations yields classical mechanics.

5. Momentum and its uncertainty relations

In the ED of particles there is one set of beables — their positions. This immediately raises questions about the nature of other observables: are they beables or are they created in the act of measurement? The case of momentum and the Heisenberg uncertainty relation is discussed in [26]; other related matters in [1, 31].

In this paper we have seen that ED allows the construction of a variety of models that can differ both at the microscopic and at the macroscopic level — we can enhance or suppress microscopic fluctuations by tuning α' and we can turn the quantum potential on and off by setting $\xi > 0$ or $\xi = 0$. Our goal here is to revisit the issue of momentum and its uncertainty relations to find out how they are affected by the choices of α' and ξ.

First, let us recall the notion of momentum. For simplicity we consider a single particle. Since it follows a non-differentiable trajectory it is clear that the classical momentum $md\vec{x}/dt$ tangent to the trajectory cannot be defined. One can however introduce two notions of momentum. One is the usual quantum operator,

$$\hat{p}^a = -i\hbar\delta^{ab}\partial_b \ , \tag{24}$$

which acts on the space of wave functions Ψ and generates infinitesimal translations. The other momentum is a local quantity associated to the current velocity, $v^A = m^{AB}\partial_B\Phi$ given in eq.(19), which leads to

$$p_A(x) = m_{AB}v^B(x) = \partial_A\Phi(x) \ . \tag{25}$$

For a single particle this "local" or "current" momentum is

$$p^a(x) = mv^a(x) = \delta^{ab}\partial_b\Phi(x) \ . \tag{26}$$

As we see from (19), the local momentum can be decomposed into drift and osmotic components, $p^a = p^a_d + p^a_o$ where

$$p^a_d = mb^a = \delta^{ab}\tilde{\eta}\partial_b\phi \ , \tag{27}$$

$$p^a_o = mu^a = -\delta^{ab}\frac{\tilde{\eta}}{\alpha'}\partial_b \log \rho^{1/2} \ . \tag{28}$$

Two points deserve to be emphasized. First, the two notions of momentum, (24) and (26), are not unrelated. As shown in [5] the infinitesimal displacement of a functional $f[\rho, \Phi]$ is given by its Poisson bracket with the "ensemble" momentum

$$\tilde{P}^a[\rho, \Phi] = \int d^3x\, \rho\delta^{ab}\partial_b\Phi = \langle p^a \rangle \ . \tag{29}$$

The second point is that the local momentum $p^a(x)$ is expressed in terms of the probability ρ and the drift potential ϕ. It is not a beable; it is an epistemic concept. *It is not an attribute of the particle but of the wave function.*

Next we revisit the uncertainty relation. We recall that the relation between expected values follows from

$$\langle \partial_a \log \rho \rangle = 0 \Rightarrow \langle p_o^a \rangle = 0 \quad \text{and} \quad \langle p^a \rangle = \langle p_d^a \rangle \ . \tag{30}$$

Also, using $\Psi = \rho^{1/2} \exp(i\Phi/\hbar)$, we have

$$\langle \hat{p}^a \rangle = \int dx \, \Psi^* \frac{\hbar}{i} \delta^{ab} \partial_b \Psi = \langle p^a \rangle \ . \tag{31}$$

The corresponding covariances are related by

$$\mathrm{Cov}(\hat{p}^a, \hat{p}^b) = \langle \hat{p}^a \hat{p}^b \rangle - \langle \hat{p}^a \rangle \langle \hat{p}^b \rangle = \mathrm{Cov}(p^a, p^b) + \frac{\hbar^2}{4} I^{ab} \ , \tag{32}$$

where $I^{ab} = \delta^{ac} \delta^{bd} I_{cd}$ and

$$I_{cd} = \int dx \, \rho \partial_c \log \rho \partial_d \log \rho = \mathrm{Cov}(\partial_c \log \rho, \partial_d \log \rho) \ , \tag{33}$$

is the Fisher information matrix. The corresponding variances of $\hat{p}^a$ and p^a are related by

$$\mathrm{Cov}(\hat{p}^a, \hat{p}^a) = \mathrm{Var}(\hat{p}^a) = \mathrm{Var}(p^a) + \frac{\hbar^2}{4} \mathrm{Var}(\partial^a \log \rho) \ . \tag{34}$$

(No sum on repeated a's.) The uncertainty relation is obtained by multiplying (34) by $\mathrm{Var}(x^b)$,

$$\mathrm{Var}(\hat{p}_q^a)\mathrm{Var}(x^b) = \mathrm{Var}(p^a)\mathrm{Var}(x^b) + \frac{\hbar^2}{4} \mathrm{Var}(\partial^a \log \rho) \, \mathrm{Var}(x^b) \ . \tag{35}$$

and invoking the Cauchy-Schwartz inequality,

$$\mathrm{Var}(A)\mathrm{Var}(B) \geq \mathrm{Cov}^2(A, B) \ , \tag{36}$$

to get

$$\mathrm{Var}(\hat{p}^a)\mathrm{Var}(x^b) \geq \mathrm{Cov}^2(p^a, x^b) + \frac{\hbar^2}{4} \mathrm{Cov}^2(\partial^a \log \rho, x^b) \ . \tag{37}$$

Finally, an explicit calculation of the covariances on the right gives

$$\mathrm{Cov}(\hat{p}^a, x^b) = \frac{1}{2} \langle \hat{p}^a \hat{x}^b + \hat{x}^b \hat{p}^a \rangle - \langle \hat{p}^a \rangle \langle \hat{x}^b \rangle = \mathrm{Cov}(p^a, x^b) \ , \tag{38}$$

and

$$\mathrm{Cov}(\partial^a \log \rho, x^b) = -\delta^{ab}. \tag{39}$$

Substituting into (37) leads to

$$\mathrm{Var}(\hat{p}^a)\mathrm{Var}(x^b) \geq \mathrm{Cov}^2(\hat{p}_q^a, x^b) + \left(\frac{\hbar}{2}\right)^2 , \tag{40}$$

which is the version of the uncertainty relation proposed originally by Schrödinger. The weaker version due to Heisenberg follows immediately,

$$\mathrm{Var}(\hat{p}^a)\mathrm{Var}(x^b) \geq \left(\frac{\hbar}{2}\right)^2 \quad \text{or} \quad \Delta \hat{p}^a \Delta x^b \geq \frac{\hbar}{2} \ . \tag{41}$$

The derivation above shows that we can set $\xi = 0$ and switch the quantum potential off without affecting either the momentum or the uncertainty relation. It also shows that the value of α' never entered the argument. We can enhance or suppress microscopic fluctuations without affecting the momentum or the uncertainty relation. α' can affect the relative drift and osmotic components of the local momentum, but not the local momentum itself.

Acknowledgments

We would like to thank M. Abedi, C. Cafaro, N. Caticha, S. DiFranzo, A. Giffin, S. Ipek, D.T. Johnson, K. Knuth, S. Nawaz, C. Rodríguez, K. Vanslette, and very specially M. Reginatto for many discussions on entropy, inference and quantum mechanics.

References

[1] For a pedagogical review see Caticha A 2012 *Entropic Inference and the Foundations of Physics* (Monograph commissioned by the 11th Brazilian Meeting on Bayesian Statistics, EBEB-2012) URL `http://www.albany.edu/physics/ACaticha-EIFP-book.pdf`
[2] Caticha A 2011 *J. Phys. A: Math. Theor.* **44** 225303 (*Preprint* arXiv:1005.2357 [quant-ph])
[3] Caticha A 2014 *J. Phys.: Conf. Ser.* **504** 012009 (*Preprint* arXiv:1403.3822 [quant-ph])
[4] Caticha A, Bartolomeo D and Reginatto M 2015 Bayesian inference and maximum entropy methods in science and engineering *AIP Conf. Proc.* **1641** 155 (*Preprint* arXiv:1412.5629 [quant-ph])
[5] Caticha A 2015 *Entropy* **17** 6110 (*Preprint* arXiv:1509.03222 [quant-ph])
[6] Leifer M S 2014 Is the quantum state real? an extended review of ψ-ontology theorems *Quanta* **3** 67–155 (*Preprint* arXiv:1409.1570 [quant-ph])
[7] Wootters W K 1981 *Phys. Rev.* D **23** 357
[8] Caticha A 1998 *Phys. Lett.* A **244** 13
Caticha A 1998 *Phys. Rev.* A **57** 1572
Caticha A 2000 *Found. Phys.* **30** 227
[9] Brukner C and Zeilinger A 2003 Information and fundamental elements of the structure of quantum theory, ed Castell L and Ischebeck O *Time, Quantum, Information* (Springer) (*Preprint* arXiv:quant-ph/0212084)
[10] Spekkens R 2007 *Phys. Rev.* A **75** 032110
[11] Goyal P, Knuth K and Skilling J 2010 *Phys. Rev.* A **81** 022109
[12] Hardy L 2011 Reformulating and reconstructing quantum theory (*Preprint* arXiv:1104.2066 [quant-ph])
[13] Hall M J W and Reginatto M 2002 *J. Phys.* A **35** 3289
Hall M J W and Reginatto M 2002 Quantum mechanics from a heisenberg-type equality *Fortschr. Phys.* **50** 646–651 (*Preprint* arXiv:quant-ph/0201084)
[14] Reginatto M 1312 From information to quanta: a derivation of the geometric formulation of quantum theory from information geometry (*Preprint* arXiv:1312.0429 [quant-ph])
[15] Nelson E 1985 *Quantum Fluctuations* (Princeton: Princeton University Press)
[16] de la Peña L, Cetto A M and Valdez-Hernández A 2015 *The Emerging Quantum: the Physics Behind Quantum Mechanics* (Springer)
[17] Smolin L 2006 Could quantum mechanics be an approximation to another theory? (*Preprint* arXiv:quant-ph/0609109)
[18] 't Hooft G 2002 Determinism beneath quantum mechanics (*Preprint* arXiv:quant-ph/0212095)
't Hooft G 2007 Emergent quantum mechanics and emergent symmetries (*Preprint* arXiv:0707.4568 [hep-th])
[19] Elze H T and Schipper O 2002 *Phys. Rev.* D **66** 044020
Elze H T 2003 *Phys. Lett.* A **310** 110
[20] Adler S 2004 *Quantum Theory as an Emergent Phenomenon* (Cambridge: Cambridge University Press)
[21] Grössing G 2008 *Phys. Lett.* A **372** 4556 (*Preprint* arXiv:0711.4954 [quant-ph])
Grössing G *et al.* 2012 *J. Phys.: Conf. Ser.* **361** 012008 (*Preprint* arXiv:1205.3393 [quant-ph])
[22] Some of these ideas have been briefly reported in Bartolomeo D and Caticha A 2015 Entropic Dynamics: The Schrödinger equation and its Bohmian limit (*Preprint* **arXiv:1512.09084**)
[23] Bohm D 1952 *Phys. Rev.* **85** 166 and 180
[24] Bohm D and Hiley B J 1993 *The undivided universe – An ontological interpretation of quantum theory* (New York: Routledge)
[25] Holland P R 1993 *The Quantum Theory of Motion* (Cambridge: Cambridge University Press)
[26] Nawaz S and Caticha A 2012 Momentum and uncertainty relations in the entropic approach to quantum theory, ed Knuth K *et al. AIP Conf. Proc.* vol 1443 (Am. Inst. Phys.) p 112 (*Preprint* arXiv:1108.2629 [quant-ph])
[27] de Falco D, Martino S D and De Siena S 1982 *Phys. Rev. Lett.* **49** 181
[28] De Martino S and De Siena S 1984 *Nuovo Cimento* **79b** 175
[29] Golin S 1985 *J. Math. Phys.* **26** 2781
[30] Golin S 1986 *J. Math. Phys.* **27** 1549
[31] Johnson D T and Caticha A 2012 Entropic dynamics and the quantum measurement problem, ed Knuth K *et al. AIP Conf. Proc.* vol 1443 (Am. Inst. Phys.) p 104 (*Preprint* arXiv:1108.2550 [quant-ph])

Journal of Physics: Conference Series **701** (2016) 012010 doi:10.1088/1742-6596/701/1/012010

Structure Process, Weak Values and Local Momentum

B. J. Hiley

Physics Department, University College, London, Gower Street, London WC1E 6BT

E-mail: b.hiley@bbk.ac.uk.

Abstract. We explain how weak values and the local momentum can be better understood in terms of Bohm's notion of structure process. The basic ideas of this approach can be expressed in a fully algebraic way, generalising Heisenberg's original matrix mechanics. This approach leads to questions that are now being experimentally investigated by our group at University College London.

1. Introduction

1.1. Act not Fact

It is almost a century since quantum mechanics was discovered but there remains the unresolved problem of the collapse of the wave function. Could it be that we have been focusing on the wrong mathematical structure to provide an adequate understanding of quantum phenomena? David Finkelstein [1] thinks so and argues that "to speak about the *wave function of the system* is a syntactic error". I agree, so let me explain what he means.

When we set up a Stern-Gerlach magnet with its axis in a certain direction and find the spin is, say up, the wave function contains information about what the experimenter is *doing* to get such a result. Reorienting the magnet will produce a different result on an identically produced system. This involves a double act, Act 1, as Finkelstein calls it, is the preparation and Act 2, is mistakenly called the 'measurement'. However there is no difference between the two Acts, they are just carried out in a different order. Act 2 could be used for the preparation of another experiment and therefore could be called 'preparation 2', but today we use the terminology 'pre- and post-selection'.

Before embarking on a development of a formal algebraic approach to quantum phenomena it is important to realise that the context in which we are working involves a radical reappraisal of underlying physical ideas forming the basis of our approach. These ideas go back to the 1960s when David Bohm, Roger Penrose and I were exploring the possibility of changing the geometrical picture of the space-time structure to accommodate quantum processes. Bohm's proposal [2] was that we should not assume that these quantum processes can be analysed as particles moving in an *a priori* given space-time, but rather *all* is a modern form of Heraclitian flux so that what is *is* the process of becoming itself. All objects and particles are then ultimately quasi-stable, quasi-local forms that can be abstracted from this underlying process which we called the *holomovement* and hence they are themselves inseparable from the underlying activity. In a little known paper, Bohm [3] showed that this activity could be described by a set of

Journal of Physics: Conference Series **701** (2016) 012010 doi:10.1088/1742-6596/701/1/012010

elemental structure-processes (see Figure 1) whose meaning we will bring out as we go along. This structure of elemental processes undergoes a discrete or continuous movement as it unfolds in a process of development. Penrose explored these ideas in his theory of *spin networks* [4], which ultimately led to the development of twistors [5]. Bohm and I developed a more general algebraic structure-process, originally based on orthogonal Clifford algebras, but now extended to symplectic Clifford algebras [6, 7].

Within this context, let me develop the mathematics in a heuristic manner and show that the formal mathematical structure mentioned above is closely related to the conventional formalism but sufficiently different to offer new insights. To do this, let us make use of a formalism already introduced by Dirac [8], but generally ignored.

We will denote Act 2 by $\psi\rangle$ and Act 1 by $\phi\rangle$; we are then looking at the process $\phi\rangle \rightarrow \psi\rangle$[1]. Following Feynman, we define the transition probability amplitude, [TPA], for the process,

$$TPA = \langle\psi|\phi\rangle. \tag{1}$$

But this brings the process to an end by forming a complex number. Indeed forming $\langle x|\psi\rangle = \psi(x)$, produces the wave function. But this is simply another transition probability amplitude, $\psi\rangle \rightarrow x\rangle$, which tells you more about the *act* of producing an effect, rather than telling us about the *state of the system.*

In the standard approach eigenvalues are assumed to be associated with *properties* of the system, giving these properties a *value* which is *assumed* to describe the *state of the system.* We have argued above that the wave function is a TPA and so how are we to understand eigenvalues in terms of TPAs? They are simply TPAs in which Act 2 is the same as Act 1, $\psi\rangle \rightarrow \psi\rangle$, that is the pre- and post-selections are identical. Therefore they are values that persist unchanging when you subject the system to the *same process* – nothing more, nothing less. How then are we to develop a mathematics of *process*?

1.2. Clifford Algebras and Process

As I was exploring the origins of Clifford algebras and their relations to quaternions (essentially Pauli spinors), my attention was drawn to a lecture by Hamilton [9] in which he was discussing the metaphysical aspects of algebra. He writes:

> In algebra, the relations .... are between successive states of some *changing thing or thought.* Numbers are the names or nouns of the algebra; they are the marks or signs, by which one of these successive states may be remembered and distinguished from another..... relations between successive thoughts thus viewed as successive states of one or more general changing thought, are the primary relations of algebra.

In attempting to generalise the quaternions to higher dimensional space, Clifford [10] points out that there are two sides to the notion of a product. The relation 2×3 equals 6, may be regarded as the product of the two numbers 2 and 3; or it may be regarded as the act of doubling the number 3. In the latter, 3 is a number but 2 is an operation, and the two factors play very distinct parts. Thus Hamilton's quaternions are likewise operations which transform one vector into another but now, when we a dealing with quantum phenomena, we have an important new feature, namely, non-commutativity, a feature which plays a crucial role in distinguishing between classical and quantum mechanics. For more details see Hiley [7, 11].

Let us return to the quaternion and consider its action of rotating and stretching a vector ρ to produce a vector σ,

$$q\rho = \sigma.$$

[1] This is not a typo. The absence of '|' will be explained below.

But this multiplication can be non-commutative, so that we may find that $\rho q = \eta \neq \sigma$. The precise relationships between the various elements appearing here are determined, not individually, but by the overall structure of the algebra.

We want to effect a further change and make the vectors themselves part of the algebra; we therefore introduce ρ^*, a dual to ρ so that

$$q = \sigma\rho^* \quad \text{with} \quad \rho\rho^* = 1. \tag{2}$$

Notice that q is an *action* and therefore the product $\sigma\rho^*$ must also be an *action*.

Let us consider re-writing equation (2) in the form

$$q = \sigma\rangle\langle\rho. \tag{3}$$

Now we are in a position to explain the *standard ket*, a new notation first introduced by Dirac [8] and subsquently discussed in more detail in the 3rd edition of his classic text on quantum mechanics [12]. Originally the symbol, $|$, in front of the ket $|\ \rangle$ was introduced to separate operators from the labels used to specify the wave function. But the content of the wave function is determined by the operators themselves, simply by replacing the operators by their eigenvalues. Dirac thereby shows that nothing is lost by removing the distinction between operator and wave function.

What Dirac had effectively done was to remove the distinction between an element of the operator algebra and the wave function without losing any information about the content of what is carried by the wave function. Thus everything now is contained in the algebra itself. This is exactly what we require if we want to obtain an algebraic description of process. In fact Dirac had formed an element of a left ideal in the algebra by the symbol $\rangle$. Only multiplication from the left is allowed. He called the object $\rangle$ the *standard ket* and its dual $\langle$ the *standard bra*. We then have a way of distinguishing between left and right translations, a distinction that is essential in non-commutative geometry.

We have now reached the position in which all elements are within the same algebra except one, namely $\rangle\langle$. To identify this symbol, notice that in a non-nilpotent algebra, there always exists an idempotent, i.e., an element ϵ such that $\epsilon^2 = \epsilon$. Then we can always write an element of a left ideal in the form $\Psi_L = \psi\epsilon$ where ϵ is an idempotent satisfying $\epsilon^2 = \epsilon$. Thus $\rangle\langle$ is a symbol for an idempotent, which is, unfortunately, not in the nilpotent Heisenberg algebra and therefore must be added.

The fact that $\rangle\langle$ is an idempotent has already been noticed by Kauffman [13] and its role has been discussed in more detail in Hiley [7]. There we point out that in order to introduce spin and relativity, we must introduce the orthogonal Clifford algebra. That algebra is not nilpotent and many idempotents exist. They are used to describe the break in the space-time symmetry when a homogeneous magnetic field is introduced by the experimenter.

In a series of insightful papers on quantum geometry, Schönberg [14] has already introduced the required idempotent to complete the Heisenberg algebra through the defining relations

$$E^2 = E, \quad PE = EX = 0$$

where $E \equiv \rangle\langle$ and X and P are elements of the Heisenberg algebra. To make the role of E clearer, Frescura and Hiley [15] showed that in the usual transformation to the bosonic $a, a^\dagger$ algebra, E is replaced by the projection onto the vacuum state $V = |0\rangle\langle 0|$ so that $aV = Va^\dagger = 0$.

What we have done can easily be understood in terms of a matrix representation. A general way of writing a rank one matrix, q, is to product a column matrix, σ with a row matrix ρ^*, which is done in equation (2). Thus we have returned to a "matrix mechanics", where our matrices now have an actual physical meaning in terms of a structure-process. In fact we do

Journal of Physics: Conference Series **701** (2016) 012010 doi:10.1088/1742-6596/701/1/012010

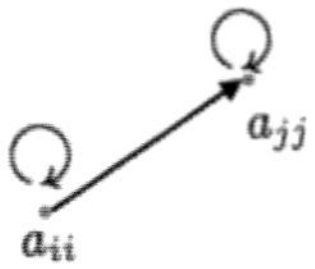

Figure 1. Elementary Process

not need to resort to a matrix representation once we realise that a simple element of a general algebra can be written as a product of an element of a left ideal Ψ_L and an element of a right ideal Ψ_R.

It should be noted that the mathematical structure we have highlighted here is of physical significance in the context of structure process. Furthermore it can be shown that this algebraic structure can be *mapped* isomorphically onto the standard Hilbert space formalism. However if that is done and the standard interpretation is adopted, we are right back into the measurement problem and interpretational difficulties.

2. Structure-Process

2.1. The Elementary Process and Groupoids

Let us take a step back and try to examine in more detail the mathematical description we have introduced heuristically in the previous section. Recall that *becoming* was taken as primary, whereas *being*, a natural primary term in the classical paradigm, is a special case of becoming, a repeated becoming that produces no change. In order to put these words into a firmer mathematical context, let us represent becoming by an element a_{ij} and take multiplication to be the *order of succession*. Each process has a source i and a target j. The sources and targets can be regarded as the idempotents a_{ii} and a_{jj}. A useful image is shown in Figure 1.

To give the abstract notion of an elementary process more physical meaning, recall that in relativity we are forced to take the notion of a *point event* as basic. Then the elementary process would be the relation between a pair of events, the primary connection being a light ray. This is the basic idea that Penrose [5] adopts in his twistor theory. However the meaning of the symbol is not directly relevant. It is only the relationships and the operations in which these symbols take part that gives the whole structure its meaning. In other words the single algebraic symbol is similar to a word, in the sense that its implicit meaning only comes out in the way in which the language as a whole is used [2].

If a process g_{ij} is followed by another process $g'_{j'k}$, then 'succession' is defined only when $j = j'$. In other words we have assumed that our structure process forms a groupoid. In symbols

$$g, g' \in G \text{ and } j = j', \quad \text{then} \quad g_{ij}g'_{j'k} = g''_{ik}$$

where G is some group.

What then is a 'being', a particle, in a process philosophy? Eddington [16] has suggested that we do not want to introduce a particle as a "lump of matter", matter being a metaphysical term that has no place in a process philosophy. Rather, as we have hinted above, a natural structural concept of existence can be represented by an idempotent, $a_{ii}a_{ii} = a_{ii}$. This is a process that continuously turns into itself. This idempotent could, itself, be a complicated substructure of processes, a notion that is exactly what is needed for the concept of a particle in relativity, where there is a fundamental difficulty with the idea of extended matter. In relativity, an extended object must be represented by a world tube of a complex structure of events in space-time.

The assumption that structure process can be described by a groupoid fits in naturally with Heisenberg's original idea of replacing each of position and momentum by two-indexed symbols in order to explain the Ritz-Rydberg combination principle $\nu(nm) = \nu(nk) + \nu(km)$. Then if we replace the position q by $q(nm)$ and write it in the form

$$q \to q(nm) \to a(nm) \exp[2\pi i\nu(nm)t]$$

we see immediately that the product produces the correct combination of frequencies since $h\nu(nm) = (E_n - E_m)$.

Heisenberg actually went further by writing $E(nm) = E_n\delta_{nm}$ so that

$$Eq = \sum_k E(mk)q(kn) = \sum_k E_m\delta_{mk}q(kn) = E_m\delta_{mn}$$
$$qE = \sum_k q(mk)E(kn) = \sum_k q(mk)E_k\delta_{kn} = E_m\delta_{mn}$$

giving Heisenberg's equation of motion,

$$\dot{q} = \frac{i}{\hbar}(Eq - qE).$$

Thus by introducing addition to capture the *order of coexistence*, we are led to a non-commutative algebra, which is at the heart of quantum mechanics. Our approach is not merely to accept formalism as an abstract structure but to provide a physical structure lying behind the formalism, leading to the need for a clearer understanding of non-commutative geometry.

2.2. New Concept of Motion

Notice that in a non-commutative structure, a key factor is the recognition that we must distinguish between left and right translations. In other words, *order is vital.* The distinction gives rise to the possibility of a new type of motion, namely, that an old *structure* may evolve into a new *structure* via the inner automorphism:

$$e' = M_1^{-1}(\tau)eM_2(\tau) \tag{4}$$

where τ parameterises the order of succession. This is what Bohm [2] calls an enfolding-unfolding movement that lies behind the path integral method introduced by Feynman [17]. We will assume that $M_1 = M_2 = \exp[iH\tau]$ so that for small τ, equation (4) can be written in the form

$$e' = (1 - iH\tau)e(1 + iH\tau) \quad \Rightarrow \quad i\frac{(e' - e)}{\tau} = He - eH. \tag{5}$$

In the limit this becomes the Heisenberg equation of motion, provided we identify τ with time. Thus we directly make contact with the historical starting point of quantum mechanics in terms of structure-process.

We can say more. e and e' are elements of the algebra so that they can be written in the form $e = |\psi\rangle\langle\phi|$ in the way we have suggested above[2]. We find equation (5) becomes

$$i\frac{d|\psi\rangle}{d\tau}\langle\phi| + i|\psi\rangle\frac{d\langle\phi|}{d\tau} = (H|\psi\rangle)\langle\phi| - |\psi\rangle(\langle\phi|H)$$

[2] We retain the $|$ in order to make immediate contact with the usual notation.

which splits into two equations

$$i\frac{d|\psi\rangle}{d\tau} = H|\psi\rangle \quad \text{and} \quad -i\frac{d\langle\phi|}{d\tau} = \langle\phi|H.$$

So the Schrödinger equation *and* its dual emerge from this process [18]. Since the dual is considered as the complex conjugate of the wave function, it is usually assumed that this adds no new content and therefore the dual equation is of no consequence. However approaching through the algebraic point of view where we are distinguishing between the left and right translation, it does have a significance. For more details see Hiley [11].

2.3. Equations of Motion from the Lagrangian

Notice however that we also obtain two equations from the Heisenberg Lagrangian

$$\mathcal{L} = -\frac{1}{2m}\nabla\psi^* \cdot \nabla\psi + \frac{i}{2}[(\partial_t\psi)\psi^* - (\partial_t\psi^*)\psi] - V\psi^*\psi.$$

We then use the Euler-Lagrange equations treating ψ and ψ^* as independent so that

$$\frac{\partial\mathcal{L}}{\partial\psi} - \partial_\mu\left(\frac{\partial\mathcal{L}}{\partial\psi_\mu}\right) = 0 \quad \text{and} \quad \frac{\partial\mathcal{L}}{\partial\psi^*} - \partial_\mu\left(\frac{\partial\mathcal{L}}{\partial\psi^*_\mu}\right) = 0$$

where $\psi_\mu = \partial_\mu\psi$ and $\psi^*_\mu = \partial_\mu\psi^*$. Evaluating the two equations of motion, we find

$$i\hbar\partial_t\psi = H\psi \quad \text{and} \quad -i\hbar\partial_t\psi^* = H\psi^*$$

or more suggestively

$$i\hbar\overrightarrow{\partial}_t|\psi\rangle = \overrightarrow{H}|\psi\rangle \quad \text{and} \quad -i\hbar\langle\psi|\overleftarrow{\partial}_t = \langle\psi|\overleftarrow{H}.$$

Now we can rewrite the Lagrangian in terms of the two real fields R and S where as usual $\psi = R\exp^{iS/\hbar}$ so that

$$\mathcal{L} = -R^2\left(\frac{\partial S}{\partial t} + \frac{(\nabla S)^2}{2m} + \frac{\hbar^2}{2m}\frac{(\nabla R)^2}{R^2} + V\right). \tag{6}$$

Treating R and S as independent variables, we find the two Euler-Lagrange equations now become

$$\partial_t\rho + \nabla.(\rho\nabla S/m) = 0$$

and

$$\frac{\partial S}{\partial t} + \frac{1}{2m}(\nabla S)^2 - \frac{\hbar^2}{2m}(\nabla^2 R)/R + V(x) = 0. \tag{7}$$

These equations are, of course, the equations central to the Bohm approach but notice they are also central to the quantum formalism.

Returning to the Lagrangian (6), we see that there are two kinetic energy terms, $(\nabla S)^2/2m$ and $\hbar^2(\nabla R)^2/2mR^2$. This is confirmed by noting that the diagonal terms in the usual expression for the energy-momentum tensor, evaluated using the Lagrangian given in equation (6), can be written as

$$T^{kk} - \mathcal{L}\delta^{kk} = \frac{R^2}{2m}\left[(\partial^{x_k}S)^2 + \hbar^2\frac{(\partial^{x_k}R)^2}{R^2}\right] + VR^2.$$

Journal of Physics: Conference Series **701** (2016) 012010 doi:10.1088/1742-6596/701/1/012010

Furthermore we find two canonical momenta, namely,

$$p_1(x,t) = \nabla S(x,t) \quad \text{and} \quad p_2(x,t) = \hbar \nabla R(x,t)/R(x,t). \tag{8}$$

The first term is simply the real part of the local momentum, sometimes called the Bohm momentum, while the second term is called the 'osmotic' momentum after it appeared in Nelson's [19, 20] stochastic model used in his derivation of the Schrödinger equation from Brownian motion. However we are talking about a single particle, and therefore such an approach requires the existence of some form of sub-quantum medium, an old idea which has dropped out of favour.

In our approach the two momenta arise because the order of succession is vital and gives rise to a non-commutative structure. In such a structure it is necessary to distinguish between left and right translations and it is this difference that gives rise to the 'strange' features of quantum processes.

In an early important paper by Hirschfelder *et al* [21] it is shown that

$$\frac{\langle x|P|\psi\rangle}{\langle x|\psi\rangle} = p_1 + ip_2 \tag{9}$$

where p_1 and p_2 are exactly the same terms as appear in equation (8). But the LHS of equation (9) is what is now called the weak value of the momentum for a post-selected state that is the final position.

Our approach to the behaviour of what we can call a single particle is dictated by equation (7) which ensures the conservation of energy provided we regard the term

$$Q = -\frac{\hbar^2}{2m}(\nabla^2 R)/R$$

as a new quality of energy arising from the additional 'osmotic' process. Traditionally this has been called the quantum potential energy because it has, at times, been regarded as giving rise to an 'internal force'. This is not the way we regard it in the structure process approach. Rather it is an internal energy arising from an extra degree of freedom in which the process acts.

In the above equations, the properties are assumed to be local, hence the term 'local momentum'. If we regard equation (7) as a local particle equation, then where are the *global* properties which are required if the experimental conditions are to play a key role as Bohr has claimed? They are encoded in the quantum potential energy which enables us to work with a *local* model.

This is in contrast to standard quantum field theory which works with *global* properties of both energy and momentum with the momentum of the particle defined through

$$P^j = \int T^{0j}(x,t)d^3x \quad \text{and} \quad E = \int T^{00}(x,t)d^3x.$$

Thus we are left with the question of whether we should regard a quantum particle as a global or local object, a conflict that Colosi and Rovelli [22] have already discussed at some length and so we will not repeat their arguments here but strongly recommend their paper to the interested reader.

3. The Problem of Stationary States

In order to illustrate one of the advantages of the new analysis through structure-process, let us return to the objection that Einstein [23] raised against the original Bohm proposals. In a stationary state, the spatial part of the wave function can be real so that $p_B = \nabla S = 0$. This implies that the 'particle' is stationary, a result that violates our physical intuition.

A way to see what is going on is to analyse a simple example, namely the eigenstate of a particle in a one-dimensional box with infinite potential delta function walls. Here, as every undergraduate learns, the eigenfunctions, $\psi_n(x,t)$, are all real

$$\psi_n(x,t) = \sqrt{\frac{2}{a}}\sin\left(\frac{n\pi x}{a}\right)e^{-\frac{iE_n t}{\hbar}}$$

where a is the length of the box. In terms of a polar decomposition of the wave function

$$R_n = \sqrt{\frac{2}{a}}\sin\left(\frac{n\pi x}{a}\right) \quad \text{and} \quad S_n = -E_n t.$$

Then

$$\partial_x S_n = 0; \quad \partial_t S_n = -E_n \quad \text{and} \quad Q_n = -\frac{\hbar^2}{2m}\nabla^2 R_n/R_n = \frac{n^2\hbar^2\pi^2}{2ma^2}.$$

Thus $p_1 = 0$. However notice that equation (8) gives $p_2 \neq 0$ so there is activity present even in the stationary state, a point that was not apparent in the original Bohm model. Thus the appearance of the quantum potential energy seemed mysterious, simply arising in the real part of the Schrödinger equation, a feature that led Heisenberg [24] to declare that the potential was *ad hoc*. However Hirschfelder *et al* [21] constructed the local value (ie weak value) of the kinetic energy showing

$$(2m\psi)^{-1}\hat{\boldsymbol{p}}.\hat{\boldsymbol{p}}\psi = (2\psi)^{-1}\hat{\boldsymbol{p}}.(\boldsymbol{v}_1 + i\boldsymbol{v}_2)\psi = mv_1^2/2 + Q + i(m\boldsymbol{v_1}.\boldsymbol{v_2} - \hbar\boldsymbol{\nabla}.\boldsymbol{v_2}/2).$$

Then we find

$$2Q = -mv_2^2 + \hbar\boldsymbol{\nabla}.\boldsymbol{v}_2 = -\frac{\hbar^2}{m}\frac{\nabla^2 R}{R}. \tag{10}$$

This gives a very different insight as to the origins of the quantum potential energy. It also shows that the kinetic energy of the particle cannot be negative, thus contradicting an earlier result of Aharonov and Rohrlich [25] who concluded from an argument using weak values that the kinetic energy must be negative.

In terms of our model of structure process, the particle is an invariant feature of an underlying process, which comprises two components arising from the non-commutative aspect of the process. What this shows is the condition $p_1 = 0$ does not imply that there is no activity in the box. On the contrary there is still activity present, but it is in the form of what has been called 'osmotic' activity. This term originates from early attempts to understand quantum processes in terms of some underlying Brownian-type process induced by a sub-quantum medium subjecting the particle to random stochastic forces. Nelson showed that such stochastic background processes could lead to the Schrödinger equation provided the diffusion parameter was replaced by a term including Planck's constant.

The main step in his argument that is of relevance to our work involves the two derivatives he introduces, the 'backward derivative' and the 'forward derivative'. These are defined through the following simplified equations

$$b_*(x,t) = \frac{x(t) - x(t-\Delta t)}{\Delta t} \quad \text{and} \quad b(x,t) = \frac{x(t+\Delta t) - x(t)}{\Delta t}.$$

Journal of Physics: Conference Series **701** (2016) 012010 doi:10.1088/1742-6596/701/1/012010

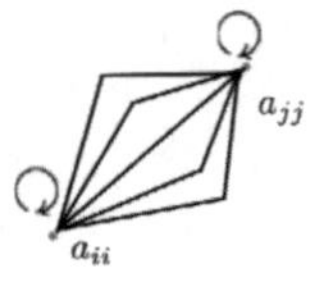

Figure 2. Many Elementary Processes between two Idempotents

These are the analogues of the left and right translations referred to above. In the appropriate limit, these form derivates from which we can construct two velocities

$$v_1(x,t) = [b(x,t) + b_*(x,t)]/2 \quad \text{and} \quad v_2(x,t) = [b(x,t) - b_*(x,t)]/2.$$

Nelson [19] also shows that the resulting Fokker-Planck equation gives rise to a continuity equation

$$\partial_t \rho + \boldsymbol{\nabla}.(\boldsymbol{v}\rho) = 0$$

where ρ is the density of the fluid. This equation is the analogue of the conservation of probability equation, the imaginary part of the Schrödinger equation under polar decomposition of the wave function. Thus we can identify $m\boldsymbol{v}$ with the local momentum, the Bohm momentum p_1 of equation (8).

4. Weak Values

4.1. Statistics

The key equation in the argument of Hirschfelder *at al* [21] is (9). There we see that the LHS is a particular example of what Aharonov *et al* [27] called a 'weak value'. In our approach, however, the form of the expression tells us that it is a ratio of two transition probability amplitudes. So the question is how are the various TPAs related to each other?

Before we can answer that question we need to take a closer look at how statistics enters the algebraic structure. There may be a number of different ways to connect a pair of idempotents as shown in Figure 2. In a simple algebra there is a Wedderburn theorem which can be written in the form

$$\epsilon A \epsilon = \langle A \rangle \epsilon$$

where $\langle A \rangle$ is the expectation value. We can see this result very quickly if we write

$$\epsilon = |\psi\rangle\langle\psi| \quad \text{so that} \quad |\psi\rangle\langle\psi|A|\psi\rangle\langle\psi| = \langle A \rangle |\psi\rangle\langle\psi|.$$

This can also be written as

$$\langle A \rangle = tr(\rho A) \quad \text{where} \quad \rho = |\psi\rangle\langle\psi| = \Psi_L \Psi_R = \psi_L \epsilon \psi_R.$$

Thus probability arises naturally in the context of structure-process.

4.2. Relationship of Weak Values to Expectation Values

To see how the weak value fits into our structure-process approach, we write

$$tr(A\rho) = \sum_{\phi_j} \rho(\phi_j) \frac{\langle \phi_j | A | \psi \rangle}{\langle \phi_j | \psi \rangle} = \sum_{\phi_j} \rho(\phi_j) \langle A_{(\phi_j;\psi)} \rangle_w$$

with $\rho(\phi_j) = |\langle\phi_j|\psi\rangle|^2$. If we choose the following basis

$$|\phi_j\rangle = \sum_{j,n} d_{jn}|a_n\rangle; \quad |\psi\rangle = \sum_k c_k|a_k\rangle \quad \text{and} \quad A|a_n\rangle = a_n|a_n\rangle,$$

we find that

$$\langle A_{(\phi_j;\psi)}\rangle_w = \frac{\langle\phi_j|A|\psi\rangle}{\langle\phi_j|\psi\rangle} = \frac{\sum_{n,k} d^*_{jn}c_k\langle a_n|A|a_k\rangle}{\sum_{n.k} d^*_{jn}c_k\langle a_n|a_k\rangle} = \frac{\sum_n d^*_{jn}c_n a_n}{\sum_n d^*_{jn}c_n}.$$

Thus in a two-state system, if we choose the following coefficients

$$a_1 = -1,\ a_2 = 1,\ d_{j1} = -99,\ d_{j2} = 101, \quad \text{and} \quad c_1 = c_2 = 1,$$

we find

$$\langle A_{(\phi_j;\psi)}\rangle_w = \frac{\langle\phi_j|A|\psi\rangle}{\langle\phi_j|\psi\rangle} = 100.$$

So a two-state system with eigenvalues ± 1 can easily give rise to a weak value of 100. Thus there is nothing surprising about the result we have obtained, once we realise that we are talking about TPAs and *not* about properties the *system possesses.*

The usual result focussed on in the standard interpretation is

$$\langle\psi|A|\psi\rangle = \sum_n \rho'_n a_n,$$

with $\rho'_n = |\langle a_n|\psi\rangle|^2$. This can be obtained from the weak value by first choosing $|\phi_j\rangle = |a_j\rangle$ and $|\psi\rangle = \sum_k c_k|a_k\rangle$ and finding

$$\langle A_{(\phi_j;\psi)}\rangle_w = \frac{\sum_k c_k\langle a_j|A|a_k\rangle}{\sum_k c_k\langle a_j|a_k\rangle} = \frac{c_j a_j}{c_j} = a_j.$$

Thus we see that the eigenvalue is a special case of the weak value.

The example we have chosen might seem somewhat artificial but it is easily reproduced by a more realistic system for the weak value of the z-component of a spin-half system defined by

$$\langle\sigma_{z(\phi;\psi)}\rangle_w = \frac{\langle\phi|\sigma_z|\psi\rangle}{\langle\phi|\psi\rangle}.$$

This gives a whole spectrum of weak values from $0 \to \infty$ simply by noting that with $\langle\phi| = \langle\uparrow_z| + \langle\downarrow_z|$ and $\sigma_z|\psi\rangle = d_+|\uparrow_z\rangle + d_-|\downarrow_z\rangle)$ and choosing

$$\sqrt{2}d_+ = \cos\alpha/2 + \sin\alpha/2 \quad \text{and} \quad \sqrt{2}d_- = \cos\alpha/2 - \sin\alpha/2,$$

we find $\langle\sigma_{z(\phi;\psi)}\rangle_w = \tan\alpha/2$, a result that has already been obtained by Duck *et al* [26]. Note that $\langle\sigma_{z(\phi;\psi)}\rangle_w = \tan\alpha/2$ gives the eigenvalues $+1$ or -1 when $\alpha = \pi/2$ or $\alpha = 3\pi/2$.

By highlighting what appears to be an anomalous result, Aharonov, Albert and Vaidman [27] have drawn attention to a weakness in the standard interpretation. It is clear that the important change in interpretation can now be tested experimentally in a number of ways [28, 29].

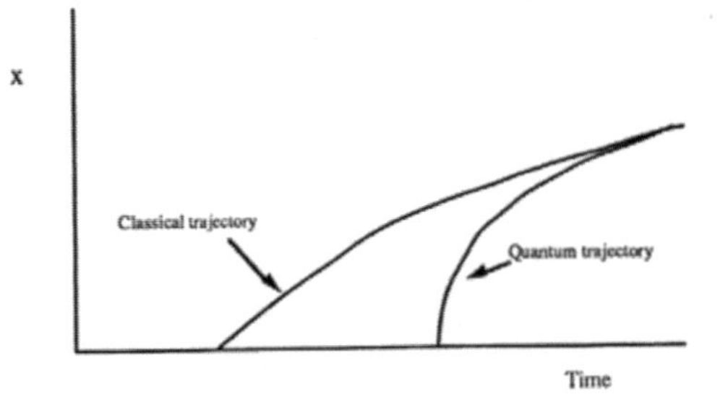

Figure 3. Pair of Merging Trajectories

4.3. Why doesn't the 'Osmotic' Velocity show up in the Standard Approach?

Where does the osmotic velocity appear in the standard approach? Consider the momentum expectation values defined by

$$\langle\psi|P|\psi\rangle = \int \left[\rho\nabla S + \frac{\hbar}{2}\nabla\rho\right] d^3x = \int \left[\rho\nabla S + \frac{\hbar\rho}{2}\frac{\nabla\rho}{\rho}\right] d^3x$$
$$= m\int [\rho\boldsymbol{v}_1 + \rho\boldsymbol{v}_2]\, d^3x.$$

Since $\rho \to 0$ as $x \to \pm\infty$, the second term, which contains the osmotic component, vanishes as $\int \nabla\rho\, d^3x = 0$. Thus only the local velocity contributes to the mean momentum.

Does this mean that there are no consequences of the existence of the osmotic component in the standard approach? As we have already seen from equation (10), it is the osmotic component that gives rise to the quantum potential energy which already appears in the real part of the Schrödinger equation. In terms of the present discussion we should note that the mean value of the kinetic energy is

$$\langle\psi|P^2|\psi\rangle = \int \rho[(mv)^2 + 2mQ]d^3x,$$

while the weak value of the kinetic energy is

$$\langle P^2_{x,\psi}(x,t)\rangle_w/2m = mv^2/2 + Q.$$

Thus once again we cannot escape the appearance of the quantum potential energy. It is an essential component of quantum processes; it is essential for the conservation of energy as seen in equation (7).

4.4. Emergence of Classical Mechanics

The local momentum has been used to calculate sets of streamlines [21, 30, 31] which were then interpreted as 'particle trajectories'. This has always been a contentious intrepretation. The claim originates from the similarity of the form of equation (7) and the classical Hamilton-Jacobi equation. If $Q \to 0$ and $S \to S_a$, the classical action, one has the possibility of a smooth transition of a quantum trajectory to a classical trajectory. When will such a transition take place?

It is possible to illustrate how this could happen using a toy-model. Noticing that Guth and Pi [32] had constructed a simple model to explain an inflationary universe using an up-side-down harmonic potential, Hiley and Mufti [33] showed that by starting with a Gaussian wave packet on the top of the potential, a set of trajectories which started out as 'quantum' smoothly transformed into a set of 'classical' trajectories as the 'particles' accelerated down the potential. A pair of converging trajectories is shown in Figure 3.

The explanation is simple at one level. As the particle accelerates down the potential the quantum potential becomes negligible compared with the kinetic energy so that the classical motion results. However in the context of our approach, where the 'rock-like' image of the particle is being seriously challenged, all we can conclude is that this suggests that it is not inconsistent to regard the flow lines as particle trajectories.

Within this simplified context, we see a novel form of energy flow

$$\frac{1}{2m}(\nabla S)^2 \leftrightarrow \frac{1}{2m}\left[(\nabla S)^2 - \hbar^2\frac{\nabla^2 R}{R}\right].$$

As a metaphor, one could liken this process to the KERS device used in F-1 car racing where internal energy can be stored before being released into the classical world!

5. Conclusions

The concept of a local momentum has been around for many years, but it seems to have been ignored and dismissed as not being meaningful in a quantum context. Indeed Fritz London [34] as long ago as 1945 wrote:

> The local mean velocity has no true quantum mechanical significance since it cannot be expressed as the expectation value of any linear operator.

In other words it has been excluded on the grounds that it does not fit into standard quantum mechanics. In this paper we have shown that the standard quantum formalism is a fragment of a larger non-commutative algebraic structure. Further, it is within this wider context that the local momentum can be given a physical meaning.

More importantly, weak values can be measured experimentally, as has been shown by Ritchie, Story and Hulet [35], Parkes, Cullen and Stoudt [36] and Kocsis *et al* [37]. This last experiment has been the most relevant for us since it measured the local momentum of photons in a two-slit experiment from which photon 'trajectories' can be constructed. The exact meaning of these momentum flow lines is still openly debated. The Bohm approach treats photons very differently from finite rest mass particles such as atoms. For example in the Bohm theory applied to photons [38] there are no photon trajectories. This makes the experimental determination of local momenta using atoms of vital significance. Such an experiment is being developed by our group and is reported in the paper by Morley, Edmunds and Baker, [39].

Acknowledgments

The author would like to thank the Fetzer Franklin Fund for their support. Thanks also to Robert Callaghan, Glen Dennis, Robert Flack and David Robson for their many stimulating discussions.

References

[1] Finkelstein D 1987 All is Flux, in *Quantum Implications: Essays in Honour of David Bohm* ed by Hiley B J and Peat D (London: Routledge and Kegan Paul) pp 289–294
[2] Bohm D 1980 *Wholeness and the Implicate Order* (London: Routledge)
[3] Bohm D 1965 Space *Proc. Int. Conf. on Elementary Particles 1965 Kyoto* 252
[4] Penrose R 1971 *Quantum Theory and Beyond* ed. Bastin T (Cambridge: Cambridge Uni Press) 151
[5] Penrose R 1967 *J. Math. Phys.* **8** 345
[6] Bohm D 1986 in *Physics and the Ultimate Significance of Time* ed. Griffin D R (New York: SUNY Press) 172–6 & 177–208

[7] Hiley B J 2011 Process, Distinction, Groupoids and Clifford Algebras: an Alternative View of the Quantum Formalism (*Lecture Notes in Physics vol 813*), ed Coecke B *New Structures for Physics* (Springer) pp 705–750; (*Preprint* `arXiv:1211.2098`
[8] Dirac P A M 1939 *Mathematical Proceedings of the Cambridge Philosophical Society* **35** 416
[9] Hamilton W R 1837 *Trans. Royal Irish Acad.* **17** 293–422
[10] Clifford W 1878 *Am. J. Maths. Pure and Applied* **1** 350
[11] Hiley B J 2015 *Journal of Computational Electronics* **14** 869
[12] Dirac P A M 1947 *The Principles of Quantum Mechanics* (Oxford: Oxford University Press, 3rd ed.) p 79
[13] Kauffman L H 2002 *Contem. Maths.* **305** 101
[14] Schönberg M 1957 *Il Nuovo Cimento, Supp.* **VI** 356–380; id. 1958 II *An. Acad. Brasil. Cien.* **30** 1–20; id. 1958 III *An. Acad. Brasil. Cien.* **30** 117; id. 1958 IV *An. Acad. Brasil. Cien.* **30** 259–280; id. 1958 V *An. Acad. Brasil. Cien.* **30** 429–446
[15] Frescura F A M and Hiley B J 1984 *Revista Brasileria de Fisica*, Vol. Especial Os 70 anos de Mario Schonberg, 49
[16] Eddington A 1958 *The Philosophy of Physical Science* (University of Michigan Press) p 162
[17] Feynman R P 1948 *Rev. Mod. Phys.* **20** 367
[18] Dirac P A M 1928 *Proc. Roy. Soc.* **117** 610–24
[19] Nelson E 1967 *Dynamical Theories of Brownian Motion* (Princeton, NJ: Princeton University Press)
[20] Bohm D and Hiley B J 1989 *Phys. Reps.* **172** 92
[21] Hirschfelder J O, Christoph A C and Palke W E 1974 *J. Chem. Phys.* **61** 5435
[22] Colosi D and Rovelli C 2009 *Class. Quant. Gravity* **26** 025002
[23] Einstein A 1953 in *Scientific Papers Presented to M. Born on his Retirement from the University of Edinburgh* (Edinburgh: Oliver and Boyd) p 33
[24] Heisenberg W 1958 *Physics and Philosophy: the revolution in modern science* (London: George Allen and Unwin) p 116
[25] Aharonov Y and Rohrlich D 2008 *Quantum paradoxes: quantum theory for the perplexed* (Weinheim: Wiley) p 234
[26] Duck I M, Stevenson P M and Sudarshan E C G 1989 *Phys. Rev.* **40** 2112
[27] Aharonov Y, Albert D Z and Vaidman L 1988 *Phys. Rev. Lett.* **60** 1351
[28] Sponar S, Denkmayr T, Geppert H, Lemmel H, Matzkin A and Hasegawa Y 2014 (*Preprint* `arXiv:1404.2125`)
[29] Monachello V and Flack R 2016 The weak Value of Spin for Atomic Systems *This publication*
[30] Dewdney C, Philippidis C and Hiley B J 1979 *Nuovo Cimento* **52B** 15
[31] Dewdney C and Hiley B J 1982 *Found. Phys.* **12** 27
[32] Guth A H and So-Young Pi 1985 *Phys. Rev.* **D 32** 1899
[33] Hiley B J and Mufti A 1995 in *Fundamental Problems in Quantum Physics*, ed. Ferrero and van der Merwe (Dordrecht: Kluwer) pp 141–156
[34] London F 1945 *Rev. Mod. Phys.* **17** 310
[35] Ritchie N W, Story J G and Hulet R G 1991 *Phys. Rev. Lett.* **66** 1107
[36] Parkes A D, Cullen D W and Stoudt D C 1998 *Proc. R. Soc. Lond.* **A 454** 2997
[37] Kocsis S, Braverman B, Ravets S, Stevens M J, Mirin R P, Shalm L K and Steinberg M A 2011 *Science* **332** 1170
[38] Bohm D J, Hiley B J and Kaloyerou P N 1987 *Phys. Rep.* **144** 349
[39] Morley J, Edmunds P and Baker B 2016 *This publication*

Journal of Physics: Conference Series **701** (2016) 012011 doi:10.1088/1742-6596/701/1/012011

Weak Values, the Reconstruction Problem, and the Uncertainty Principle

Charlyne de Gosson and Maurice de Gosson
Universität Wien, NuHAG, Fakultät für Mathematik, A-1090 Wien

Abstract. Closely associated with the notion of weak value is the problem of reconstructing the post-selected state: this is the so-called reconstruction problem. We show that the reconstruction problem can be solved by inversion of the cross-Wigner transform, using an ancillary state. We thereafter show, using the multidimensional Hardy uncertainty principle, that maximally concentrated cross-Wigner transforms corresponds to the case where a weak measurement reduces to an ordinary von Neumann measurement.

1. Introduction

In time-symmetric quantum mechanics (TSQM) the state of a system is represented by a two-state vector $\langle\phi|\ |\psi\rangle$ where the state $\langle\phi|$ evolves backwards from the future and the state $|\psi\rangle$ evolves forwards from the past [1–6]. The interpretation of the two-state vector goes as follows: assume that at an initial time $t_{\rm i}$ an observable $\widehat{A}$ is measured and a non-degenerate eigenvalue was found: $|\psi(t_{\rm i})\rangle = |\widehat{A} = \alpha\rangle$; similarly at a later (final) time $t_{\rm f}$ a measurement of another observable $\widehat{B}$ yields $|\phi(t_{\rm f})\rangle = |\widehat{B} = \beta\rangle$. The two-time state $\langle\phi|\ |\psi\rangle$ can be constructed as follows: Alice prepares the state $|\psi(t_{\rm i})\rangle$ at initial time $t_{\rm i}$. She then sends the system to Bob (the observer), who is free to perform any measurement he wants. The system is then returned to Alice, who then performs a strong measurement with the state $|\phi(t_{\rm f})\rangle$ as one of the outcomes. Only if this outcome is obtained, does Bob keep the results of his measurement. Let now t be some intermediate time: $t_{\rm i} < t < t_{\rm f}$. Following the time-symmetric approach to quantum mechanics at this intermediate time the system is described by the *two* wavefunctions

$$\psi = U_{\rm i}(t, t_{\rm i})\psi(t_{\rm i}) \quad , \quad \phi = U_{\rm f}(t, t_{\rm f})\phi(t_{\rm f}) \tag{1}$$

where $U_{\rm i}(t, t') = e^{-i\widehat{H}_{\rm i}(t-t')/\hbar}$ and $U_{\rm f}(t, t') = e^{-i\widehat{H}_{\rm f}(t-t')/\hbar}$ are the unitary operators governing the evolution of the state before and after time t; $|\psi\rangle$ is the pre-selected state and $|\phi\rangle$ the post-selected state. By definition, the complex number

$$\langle\widehat{A}\rangle_{\phi,\psi} = \frac{\langle\phi|\widehat{A}|\psi\rangle}{\langle\phi|\psi\rangle} \tag{2}$$

is the *weak value* of $\widehat{A}$ in the two-time state $\langle\phi|\ |\psi\rangle$; it is well-defined as long as the vectors ϕ and ψ are non-orthogonal (which we assume from now on). Notice that when $\phi = \psi$ then $\langle\widehat{A}\rangle_{\phi,\psi}$ is just the usual mean $\langle\widehat{A}\rangle_\psi$ value of $\widehat{A}$ in the pure state $|\psi\rangle$. It is easy to show, using Wigner's phase space formalism [7–10], that formula (2) can be written [11–13]

$$\langle\widehat{A}\rangle_{\phi,\psi} = \frac{1}{\langle\phi|\psi\rangle}\iint a(x,p)W(\psi,\phi)(x,p)d^n p d^n x \tag{3}$$

where $a = a(x,p)$ is the dequantization of the operator $\widehat{A}$ (i.e. the classical observable corresponding to $\widehat{A}$) and $W(\psi,\phi)$ is the cross-Wigner transform of ψ and ϕ:

$$W(\psi,\phi)(x,p) = \left(\tfrac{1}{2\pi\hbar}\right)^n \int e^{-\frac{i}{\hbar}py}\psi(x+\tfrac{1}{2}y)\phi^*(x-\tfrac{1}{2}y)d^n y \tag{4}$$

(it reduces to the usual Wigner distribution [8, 10] $W\psi$ when $\psi = \phi$.

Now, as was already shown by Lundeen et al. [14] in 2011 (also see Lundeen and Bambe [15]), weak values can be used to reconstruct the state $|\psi\rangle$; they considered the following experiment on a particle: a weak measurement of the real position variable x is performed; this amounts to applying the projection operator $\widehat{\Pi}_x = |x\rangle\langle x|$ to the pre-selected state $|\psi\rangle$; thereafter they perform a strong measurement of momentum, which yields the value p_0, that is $\phi(x) = \left(\frac{1}{2\pi\hbar}\right)^{1/2} e^{\frac{i}{\hbar}p_0 x}$. The result of the weak measurement is thus

$$\langle\widehat{\Pi}_x\rangle_{\psi,\phi} = \frac{\langle p_0|x\rangle\langle x|\psi\rangle}{\langle p_0|\psi\rangle} = \left(\frac{1}{2\pi\hbar}\right)^{1/2}\frac{e^{-\frac{i}{\hbar}p_0 x}\psi(x)}{\widehat{\psi}(p_0)} \tag{5}$$

($\widehat{\psi}$ the Fourier transform of ψ). Since the value of p_0 is known we get

$$\psi(x) = \frac{1}{k}e^{\frac{i}{\hbar}p_0 x}\langle\widehat{\Pi}_x\rangle_{\psi,\phi} \tag{6}$$

where $k = (2\pi\hbar)^{n/2}\widehat{\psi}(p_0)$; formula (6) thus allows to determine $\psi(x)$ by scanning through the values of x. In our previous work [11–13] we have shown that this procedure boils down to the following question:

Knowing $W(\psi,\phi)$ and the postselected state ϕ can we determine the preselected state ψ?

(The effective determination of $W(\psi,\phi)$ is essentially a problem of quantum tomography, see [16–25]).

For instance, in the case of Lundeen's experiment we have $a(x,p) = \delta(x - x_0)$ and

$$W(\psi,\phi)(x,p) = \frac{1}{2\pi\hbar}\int_{-\infty}^{\infty} e^{-\frac{i}{\hbar}py}\psi(x+\tfrac{1}{2}y)e^{-\frac{i}{\hbar}p_0(x-\frac{1}{2}y)}dy$$

hence, by formula (3),

$$\langle\widehat{\Pi}_{x_0}\rangle_{\psi,\phi} = \left(\frac{1}{2\pi\hbar}\right)^{1/2}\frac{1}{\widehat{\psi}(p_0)}e^{-\frac{i}{\hbar}p_0 x}\psi(x_0)$$

which is (6) if one replaces x with x_0. We are going to show that it is indeed always possible to reconstruct the preselected state $|\psi\rangle$ (in Lundeen et al.'s example, the wavefunction) if one knows the classical observable a of which the operator $\widehat{A}$ is the (Weyl) quantization together with the cross-Wigner distribution of the two-time state $\langle\phi|\ |\psi\rangle$; a symmetry argument will show that one can actually reverse the situation and determine the post-selected state as well; more intriguingly we will see that, in fact, that the whole two-time state can be reconstructed (up to a factor) if one knows $W(\psi,\phi)$. This actually raises another question, intimately related to the uncertainty principle.

Notation. We will work with systems having n degrees of freedom. Position (*resp.* momentum) variables are denoted $x=(x_1,...,x_n)$ (*resp.* $p=(p_1,...,p_n)$). The corresponding phase space variable is $z=(x,p)$. The scalar product $p_1x_1+\cdots+p_nx_n$ is denoted by px. When integrating we will use, where appropriate, the volume elements $d^nx=dx_1\cdots dx_n$, $d^np=dp_1\cdots dp_n$. The unitary $\hbar$-Fourier transform of a square-integrable function ψ of x is by definition

$$F\psi(p)=\widehat{\psi}(p)=\left(\tfrac{1}{2\pi\hbar}\right)^{n/2}\int e^{-\frac{i}{\hbar}px}\psi(x)d^nx. \qquad (7)$$

We denote by $\widehat{x}=(\widehat{x}_1,...,\widehat{x}_n)$ and $\widehat{p}=(\widehat{p}_1,...,\widehat{p}_n)$ the (vector) operators defined by $\widehat{x}_j\psi=x_j\psi$, $\widehat{p}_j\psi=-i\hbar\partial_{x_j}\psi$. We will assume from now on that $|\psi\rangle$ and $|\phi\rangle$ are two normalized non-orthogonal states: $\langle\psi|\psi\rangle=\langle\phi|\phi\rangle=1$, $\langle\phi|\psi\rangle\neq 0$.

2. A General Reconstruction Formula

We begin by recalling some well-known results from harmonic analysis. Let $\widehat{T}(x_0,p_0)=e^{-\frac{i}{\hbar}(p_0\widehat{x}-x_0\widehat{p})}$ be the Heisenberg operator [7, 9, 26]; it is a unitary operator whose action on a wavefunction ψ is given by

$$\widehat{T}(x_0,p_0)\psi(x)=e^{\frac{i}{\hbar}(p_0x-\frac{1}{2}p_0x_0)}\psi(x-x_0). \qquad (8)$$

An associated operator is the Grossmann–Royer reflection operator $\widehat{\Pi}(x_0,p_0)$ whose definition is due to Grossmann [27] and Royer [28] (see [7, 29] for a detailed study of these operators):

$$\widehat{\Pi}(x_0,p_0)=\widehat{T}(x_0,p_0)\widehat{\Pi}\widehat{T}(x_0,p_0)^\dagger \qquad (9)$$

where $\widehat{\Pi}$ is the parity operator: $\widehat{\Pi}\psi(x)=\psi(-x)$; the explicit action of $\widehat{\Pi}(x_0,p_0)$ on wavefunctions is easily obtained using formula (8) and one finds

$$\widehat{\Pi}(x_0,p_0)\psi(x)=e^{\frac{2i}{\hbar}p_0(x-x_0)}\psi(2x_0-x). \qquad (10)$$

Now, a straightforward calculation shows that the cross-Wigner transform is given by[1]

$$W(\psi,\phi)(x,p)=\left(\tfrac{1}{\pi\hbar}\right)^n\langle\widehat{\Pi}(x,p)\phi|\psi\rangle. \qquad (11)$$

This formalism allows us to prove a basic reconstruction result. Before that we note that, using the Fourier inversion formula and the definition (4) of $W(\psi,\phi)$, we get the formulas

$$\psi(x')\phi^*(x'')=\int e^{\frac{i}{\hbar}p(x'-x'')}W(\psi,\phi)(\tfrac{1}{2}(x'+x''),p)d^np. \qquad (12)$$

$$\widehat{\psi}(p')\widehat{\phi}^*(p'')=\int e^{-\frac{i}{\hbar}(p'-p'')x}W(\psi,\phi)(x,\tfrac{1}{2}(p'+p''),p)d^nx \qquad (13)$$

(the second formula is obtained by applying the first to the Fourier transforms of ψ,ϕ and noting that $W(\widehat{\psi},\widehat{\phi})(x,p)=W(\psi,\phi)(-p,x)$). Observe that when $\psi=\phi$ these formulas reduce, setting respectively $x'=x''$ and $p'=p''=p$ to the familiar [7, 8, 10] marginal properties

$$\int W\psi(x,p)d^np=|\psi(x)|^2 \ , \ \int W\psi(x,p)d^nx=|\phi(x)|^2$$

of the Wigner distribution $W\psi=W(\psi,\psi)$.

[1] Hence the cross-Wigner transform can be viewed as a transition amplitude.

Proposition 1 *Let λ be an arbitrary vector in $L^2(\mathbb{R}^n)$ such that $\langle\phi|\lambda\rangle \neq 0$. We have*

$$\psi(x) = \frac{2^n}{\langle\phi|\lambda\rangle} \iint W(\psi,\phi)(y,p)\widehat{\Pi}(y,p)\lambda(x)d^n p d^n y. \tag{14}$$

Similarly, if $\langle\phi|\lambda'\rangle \neq 0$ then

$$\phi(x) = \frac{2^n}{\langle\phi|\lambda'\rangle} \iint W(\psi,\phi)^*(y,p)\widehat{\Pi}(y,p)\lambda'(x)d^n p d^n y. \tag{15}$$

Proof. (Cf. de Gosson and de Gosson [11]). Multiplying both sides of (12) by $\lambda(x'')$ and integrating, we have

$$\psi(x)\langle\phi|\lambda\rangle = \iint e^{\frac{i}{\hbar}p(x-x')} W(\psi,\phi)(\tfrac{1}{2}(x+x'),p)\lambda(x')d^n p d^n x'.$$

Setting $y = \frac{1}{2}(x+x')$ and using the explicit formula (10) for the Grossmann–Royer parity operator we get (14). Formula (15) follows, swapping ψ and ϕ and noting that $W(\psi,\phi)^* = W(\phi,\psi)$. ■

This result is remarkable, because it shows that the knowledge of the cross-Wigner transform suffices to determine the two-state vector $\langle\phi|\ |\psi\rangle$ (and hence the weak value $\langle\widehat{A}\rangle_{\phi,\psi}$).

We are going to apply this result to study the case where $W(\phi,\psi)$ is maximally concentrated around a phase space point.

3. Hardy's Uncertainty Principle

3.1. Statement of the HUP

Hardy's uncertainty principle (HUP) is a precise statement of the folk wisdom following which a function and its Fourier transform cannot be simultaneously sharply located. Hardy [30] proved that if ψ and its Fourier transform $\widehat{\psi}$ satisfy estimates of the type

$$|\psi(x)| \leq C_a e^{-\frac{a}{2\hbar}x^2} \quad , \quad |\widehat{\psi}(p)| \leq C_b e^{-\frac{b}{2\hbar}p^2} \tag{16}$$

where a, b, C_a, C_b are positive constants then the following holds true:

- *If $ab > 1$ then we must have $\psi = 0$;*
- *If $ab = 1$ we have $\psi(x) = Ce^{-\frac{a}{2\hbar}x^2}$ for some complex constant C;*
- *If $ab < 1$ there exists a whole space of $\mathcal{S}$ solutions, containing the functions $\psi(x) = Q(x)e^{-\frac{a}{2\hbar}x^2}$ where Q is a polynomial (equivalently, $\mathcal{S}$* contains the finite linear combinations of Hermite polynomials).

Note that in the limiting case $ab = 1$ the function ψ is a squeezed Gaussian, hence a minimum-uncertainty state (It saturates the Robertson–Schrödinger inequalities).

We are going to use here the following multi-dimensional variant of Hardy's result (de Gosson and Luef [31, 32]) which we state for functions defined on phase space:

Proposition 2 *Let A and B be two real positive definite $2n \times 2n$ matrices and Ψ a square integrable function on phase space $\mathbb{R}^{2n}$, $\Psi \neq 0$. Assume that*

$$|\Psi(z)| \leq C_A e^{-\frac{1}{2\hbar}Az^2} \quad \textit{and} \quad |F\Psi(\zeta)| \leq C_B e^{-\frac{1}{2\hbar}B\zeta^2} \tag{17}$$

for some constants $C_A, C_B > 0$. Then:

(i) The eigenvalues λ_j, $j = 1, ..., 2n$, of the $2n \times 2n$ matrix AB are all ≤ 1;

(ii) If $\lambda_j = 1$ for all j, then $\psi(x) = Ce^{-\frac{1}{2\hbar}Ax^2}$ for some some complex constant C;

(iii) If $\lambda_j < 1$ for some j then the space of functions satisfying (17) contains every $\psi(x) = Q(x)e^{-\frac{1}{2\hbar}Ax^2}$ where Q is a complex polynomial.

In (17) $F\Psi$ is the $\hbar$-Fourier transform (7) where n is replaced with $2n$. Notice that the eigenvalues of AB are real since AB is equivalent to the symmetric matrix $A^{1/2}BA^{1/2}$.

This result allows us to prove the following Lemma, which we will need in next Subsection to prove our main result:

Lemma 3 *Assume that there exists a real positive-definite $2n \times 2n$ matrix M such that*

$$|W(\psi,\phi)(z)| \leq Ce^{-\frac{1}{\hbar}Mz^2} \tag{18}$$

for some constant $C > 0$. Assume that ϕ is an even state: $\phi(-x) = \phi(x)$. Then the eigenvalues $\mu_1, ..., \mu_{2n}$ of M are all ≤ 1.

Proof. Set $\Psi = W(\psi,\phi)$ and consider the Fourier transform $F\Psi$; the function

$$A(\psi,\phi)(x,p) = F\Psi(p,-x)$$

is the cross-ambiguity function familiar from radar theory [26]; it is related to the Wigner transform by [7, 26, 33] the formula

$$A(\psi,\phi)(z) = W(\psi,\phi^\vee)(\tfrac{1}{2}z)$$

where $\phi^\vee(x) = \phi(-x)$. Since we are assuming that ϕ is even we thus have here

$$A(\psi,\phi)(z) = W(\psi,\phi)(\tfrac{1}{2}z)$$

and hence

$$F\Psi(x,p) = W(\psi,\phi)(-\tfrac{1}{2}p, \tfrac{1}{2}x)$$

so that the inequality (18) implies that

$$|F\Psi(x,p)| \leq Ce^{-\frac{1}{4\hbar}Mz^2}.$$

Setting $A = 2M$ and $B = \frac{1}{2}M$ Proposition 2 implies that the eigenvalues $\mu_1^2, ..., \mu_{2n}^2$ of $AB = M^2$ must be ≤ 1, hence the result since M is positive definite. ■

This result implies – in particular – that the (cross-)Wigner transform $W(\psi,\phi)$ can never be a compactly supported phase space function, because if it were the case we could dominate $W(\psi,\phi)$ by an arbitrarily sharply peaked phase space Gaussian.

Notice that if all the eigenvalues of M are equal to one, then M is the identity matrix, hence (18) becomes

$$|W(\psi,\phi)(x,p)| \leq Ce^{-\frac{1}{\hbar}(|x|^2+|p|^2)}. \tag{19}$$

This inequality corresponds to *minimum uncertainty in phase space*. We are going to draw an interesting consequence of this below.

3.2. Application to weak values

We are going to see that a true weak measurement can never be associated with a maximally peaked cross-Wigner distribution (19); in fact that such a situation corresponds to an ordinary measurement, in which case we have $\psi = \phi$ and $\langle\widehat{A}\rangle_{\phi,\psi} = \langle\widehat{A}\rangle_\psi$ (the ordinary mean value). More precisely:

Proposition 4 *Assume that the cross-Wigner transform of post- and pre-selected states satisfies (19). Then*

$$\psi(x) = \phi(x) = N\phi_0(x)e^{-\frac{1}{2\hbar}|x|^2}$$

for some constant N, the function

$$\phi_0(x) = (\pi\hbar)^{-n/4}e^{-\frac{1}{2\hbar}|x|^2}$$

being the fiducial coherent state.

Proof. In view of formula (12) we have

$$\psi(x')\phi^*(x'') = \int e^{\frac{i}{\hbar}p(x'-x'')}W(\psi,\phi)(\tfrac{1}{2}(x'+x''),p)d^np$$

for all x', x''. Since ϕ is not equal to zero everywhere there exists a value x'' such that $\phi'(x'') \neq 0$; shifting if necessary ϕ it is no restriction to assume that $x'' = 0$ so that we can rewrite, changing x' into x, the equality above as

$$\psi(x)\phi^*(0) = \int e^{\frac{i}{\hbar}px}W(\psi,\phi)(\tfrac{1}{2}x,p)d^np$$

and hence, taking the inequality (19) into account,

$$\begin{aligned}|\psi(x)| &\leq \frac{1}{|\phi'(0)|}\int |W(\psi,\phi)(\tfrac{1}{2}x,p)|d^np \\ &\leq \frac{C}{|\phi'(0)|}e^{-\frac{1}{\hbar}|x|^2}\int e^{-\frac{1}{\hbar}|p|^2}dp.\end{aligned}$$

hence, since the integral in p is trivially convergent,

$$|\psi(x)| \leq C'e^{-\frac{1}{\hbar}|x|^2} \tag{20}$$

for some constant $C' > 0$. Similarly, using (13), there exists $C'' > 0$ such that

$$|\widehat{\psi}(p)| \leq C''e^{-\frac{1}{\hbar}|p|^2}. \tag{21}$$

In view of the multidimensional Hardy uncertainty principle these inequalities imply that $\psi(x)$ must be proportional to $\phi_0(x)$. ■

4. Discussion

In our treatment of the reconstruction problem we have explicitly assumed that the "right" phase space picture of quantum mechanics is obtained by using the traditional Wigner formalism which is closely associated with Weyl's quantization of classical observables [7, 9]. But life is perhaps not that simple. It turns out that there are strong indications that one should replace Weyl quantization by that introduced by Born and Jordan [34, 35] in 1925, prior to Weyl's work; we refer to de Gosson [29, 36] for a discussion of these rules. If this proved to be the case, the reconstruction formulae in Proposition 1 would no longer hold. It turns out, as we already discussed in [11], that the reconstruction problem in general is much more difficult than in the Wigner case.

Acknowledgments
The second author has been supported by the Austrian Research Foundation (FWF) grant P-27773. Both authors thank Gerhard Grössing for having invited us to prestigious Conference EmQM 2015.

References
[1] Aharonov Y, Bergmann P G and Lebowitz J L 1964 Time symmetry in the quantum process of measurement *Phys. Rev.* **134**(6B) B1410–B1416
[2] Aharonov Y, Cohen E and Elitzur A C 2015 Can a future choice affect a past measurement's outcome? *Ann. Phys.* **355** 258–268 (*Preprint* arXiv:1206.6224 [quant-ph])
[3] Aharonov Y and Vaidman L 1990 Properties of a quantum system during the time interval between two measurements *Phys. Rev.* A **41**(1) 11–20
[4] Aharonov Y and Vaidman L 1991 Complete description of a quantum system at a given time *J. Phys. A: Math. Gen.* **24**(10) 2315–2328
[5] Aharonov Y and Vaidman L 2008 The two-state vector formalism: An updated review (*Lecture Notes in Physics* vol 734), ed Muga J G, Mayato R S and Egusquiza I L *Time in Quantum Mechanics* (Berlin, Heidelberg: Springer Berlin Heidelberg) pp 399–447 ISBN 978-3-540-73472-7
[6] Cohen O and Hiley B J 1995 Retrodiction in quantum mechanics, preferred Lorentz frames, and nonlocal measurements *Found. Phys.* **25**(12) 1669–1698
[7] de Gosson M A 2011 *Symplectic Methods in Harmonic Analysis and in Mathematical Physics* (Basel: Birkhäuser) ISBN 978-3-7643-9991-7
[8] Hillery M, O'Connell R F, Scully M O and Wigner E P 1984 Distribution functions in physics: Fundamentals *Phys. Rep.* **106**(3) 121–167
[9] Littlejohn R G 1986 The semiclassical evolution of wave packets *Phys. Rep.* **138**(4-5) 193–291
[10] Wigner E 1932 On the quantum correction for thermodynamic equilibrium *Phys. Rev.* **40**(5) 749–759
[11] de Gosson C and de Gosson M A 2015 The phase space formulation of time-symmetric quantum mechanics *Quanta* **4**(1) 27–34
[12] de Gosson M A and de Gosson S M 2012 The reconstruction problem and weak quantum values *J. Phys. A: Math. Theor.* **45**(11) 115305
[13] de Gosson M A and de Gosson S M 2011 Weak values of a quantum observable and the cross-Wigner distribution *Phys. Lett.* A **376**(4) 293–296 (*Preprint* arXiv:1109.3665v1 [quant-ph])
[14] Lundeen J S, Sutherland B, Patel A, Stewart C and Bamber C 2011 Direct measurement of the quantum wavefunction *Nature* **474**(7350) 188–191 (*Preprint* arXiv:1112.3575v1 [quant-ph])
[15] Lundeen J S and Bamber C 2012 Procedure for direct measurement of general quantum states using weak measurement *Phys. Rev. Lett.* **108**(7) 070402
[16] Artiles L M, Gill R D and Guta M I 2005 An invitation to quantum tomography *J. R. Statist. Soc. B* **67**(1) 109–134 (*Preprint* arXiv:math/0405595)
[17] Asorey M, Ibort A, Marmo G and Ventriglia F 2015 Quantum tomography twenty years later *Phys. Scr.* **90**(7) 074031
[18] Dunn T J, Walmsley I A and Mukamel S 1995 Experimental determination of the quantum-mechanical state of a molecular vibrational mode using fluorescence tomography *Phys. Rev. Lett.* **74**(6) 884–887
[19] Fischbach J and Freyberger M 2012 Quantum optical reconstruction scheme using weak values *Phys. Rev.* A **86**(5) 052110 (*Preprint* arXiv:1211.2185 [quant-ph])
[20] Leonhardt U 1997 *Measuring the quantum state of light* (*Cambridge studies in modern optics* no 22) (Cambridge: Cambridge Univ. Press) ISBN 978-0-521-49730-5
[21] Leonhardt U and Paul H 1995 Measuring the quantum state of light *Prog. Quantum Electron.* **19**(2) 89–130
[22] Lutterbach L G and Davidovich L 1997 Method for direct measurement of the Wigner function in cavity QED and ion traps *Phys. Rev. Lett.* **78**(13) 2547–2550
[23] Lvovsky A I and Raymer M G 2009 Continuous-variable optical quantum-state tomography *Rev. Mod. Phys.* **81**(1) 299–332
[24] Roy A S and Roy S M 2014 Optimum phase space probabilities from quantum tomography *J. Math. Phys.* **55**(1) 012102 (*Preprint* arXiv:1304.2857 [quant-ph])
[25] Smithey D T, Beck M, Raymer M G and Faridani A 1993 Measurement of the Wigner distribution and the density matrix of a light mode using optical homodyne tomography: Application to squeezed states and the vacuum *Phys. Rev. Lett.* **70**(9) 1244–1247
[26] Folland G B 1989 *Harmonic analysis in phase space* (*Annals of mathematics studies* no 122) (Princeton, NJ: Princeton Univ. Press) ISBN 978-0-691-08527-2
[27] Grossmann A 1976 Parity operator and quantization of δ-functions *Commun. Math. Phys.* **48**(3) 191–194

[28] Royer A 1977 Wigner function as the expectation value of a parity operator *Phys. Rev.* A **15**(2) 449–450
[29] de Gosson M A 2016 *Born–Jordan Quantization: Theory and Applications* (*Fundamental Theories of Physics* no 182) (Springer) ISBN 978-3-319-27902-2
[30] Hardy G H 1933 A theorem concerning Fourier transforms *J. London Math. Soc.* **s1-8**(3) 227–231
[31] de Gosson M and Luef F 2007 Quantum states and Hardy's formulation of the uncertainty principle : a symplectic approach *Lett. Math. Phys.* **80**(1) 69–82
[32] de Gosson M and Luef F 2009 Symplectic capacities and the geometry of uncertainty: The irruption of symplectic topology in classical and quantum mechanics *Phys. Rep.* **484**(5) 131–179
[33] Gray J 2010 An interpretation of Woodward's ambiguity function and its generalization, *2010 IEEE Radar Conference* pp 859–864
[34] Born M and Jordan P 1925 Zur Quantenmechanik *Z. Phys.* **34** 858–888
[35] Born M, Heisenberg W and Jordan P 1925 Zur Quantenmechanik II *Z. Phys.* **35** 557–615
[36] de Gosson M A 2014 Born–Jordan quantization and the equivalence of the Schrödinger and Heisenberg pictures *Found. Phys.* **44**(10) 1096–1106

Journal of Physics: Conference Series **701** (2016) 012012 doi:10.1088/1742-6596/701/1/012012

Models of spontaneous wave function collapse: what they are, and how they can be tested

Angelo Bassi

Department of Physics, University of Trieste, 34151 Miramare-Trieste, Italy

Istituto Nazionale di Fisica Nucleare, Sezione di Trieste, Via Valerio 2, 34127 Trieste, Italy

E-mail: bassi@ts.infn.it

Abstract. There are few proposals, which explicitly allow for (experimentally testable) deviations from standard quantum theory. Collapse models are among the most-widely studied proposals of this kind. The Schrödinger equation is modified by including nonlinear and stochastic terms, which describe the collapse of the wave function in space. These spontaneous collapses are rare for microscopic systems, hence their quantum properties are left almost unaltered. On the other hand, collapses become more and more frequent, the larger the object, to the point that macroscopic superpositions are rapidly suppressed. The main features of collapse models will be reviewed. An update of the most promising experimental tests will be presented.

1. Introduction

Collapse models [1, 2] are nonlinear and stochastic (phenomenological) modifications of the Schrödinger equation, which add the collapse of the wave function to the standard quantum evolution. The fundamental dynamics, stripped from all details, which are important to have a physically consistent model, but which are not necessary to understand the fundamental mechanism, is given by the following stochastic differential equation for the wave function:

$$d|\psi_t\rangle = \left[-\frac{i}{\hbar}H dt + \sqrt{\lambda}(A - \langle A\rangle_t)dW_t - \frac{\lambda}{2}(A - \langle A\rangle_t)^2 dt\right]|\psi_t\rangle, \tag{1}$$

where H is the usual quantum Hamiltonian of the system, $\langle A\rangle_t = \langle\psi_t|A|\psi_t\rangle$ is the quantum expectation of the operator A, here assumed to be self-adjoint, λ is a positive constant, which sets the strength of the collapse mechanism, and W_t is a standard Wiener process.

Eq. (1) is rather easy to understand [3]. It describes a diffusion process on the unit sphere of the Hilbert space, since the norm is conserved. H as usual induces a unitary "rotation" on the sphere, while the remaining terms tend to collapse the wave function towards one of the eigenstates of the operator A, in a stochastic fashion. The composite effect depends on the strength of one term with respect to the other. If the dynamics induced by H is dominant, then the evolution is deterministic and only slightly blurred by the collapse terms, for very long times. If on the other hand the collapse terms are dominant, then very rapidly the wave function randomly collapses, with a probability distribution, which is almost equal to the Born probability rule.

It is interesting to note that all known collapse models, in spite of the differences among them, sometimes also significant, have the same structure, which is captured by Eq. (1). This is not an accident. It can be shown [4] that the requirement of no-faster-than-light signalling, together with norm conservation, almost uniquely fixes the form of the possible modifications of the Schrödinger equation. Therefore, in this sense, one can claim that collapse models are not just a possible way of modifying quantum theory, but are the only way.

2. Collapse models in space

The choice of the operator A in Eq. (1) determines the basis upon which the wave function is localised. If $A = H$, then the collapse will occur in the energy basis, if $A = p$, it will occur on the momentum basis, and so on. Since the main reason for introducing collapse models is to justify the classical properties of macroscopic objects, the most obvious one being that they always occupy a well defined position in space, then the simplest and natural choice is to take A equal to the position operator q, or some function of it. All collapse models considered in the literature are of this kind.

The first model of this type was the GRW (Ghirardi–Rimini–Weber) model [5]. It is defined in terms of a discrete jump process instead of a continuous one, but otherwise the dynamics is equivalent to that of Eq. (1). Subsequently it was generalised to the CSL (Continuous Spontaneous Localisation) model [6], to include identical particles in the description. Parallel to that the QMUPL (Quantum Mechanics with Universal Position Localisation) model [7] was proposed, which can be seen as a short distance approximation of both the GRW and CSL models [8]. To mention also the DP (Diosi–Penrose) model [9], as a first attempt to connect the collapse of the wave function to gravity.

All these models are equivalent from the qualitative point of view: they induce the collapse of the wave function in *space*, and the larger the object, the faster the collapse, as we will see in Section 4. They differ, sometimes significantly, from the quantitative point of view. They share another property: the noise inducing the collapse of the wave function also induces a Brownian type of motion on the system. In momentum space, the wave function picks higher and higher components, meaning that its kinetic energy steadily increases. Although the increase is small and hardly detectable with present-day technology, nevertheless it is rather unpleasant feature of these models.

Recently a partial resolution of the problem was found. It was shown that it is possible to include dissipative effects in collapse models, without altering the collapse properties. This has been successfully accomplished for the QMPUL model [10], the GRW model [11], the CSL model [12], and in this case one speaks of dissipative models (dQMUPL, dGRW, dCSL). Interestingly enough, the same procedure turned out to be more problematic for the DP model [13]. In all these cases, the energy does not increase anymore, but thermalises to a finite temperature, which can be associated to the noise inducing the collapse. This means not only that the energy increase is not an unavoidable feature of collapse models (a naive argument was often told, that by collapsing the wave function in space, one increases the uncertainty in momentum—because of Heisenberg's uncertainty principle—thus increasing the kinetic energy) but also that, actually, the energy can decrease (if initially higher than that of the noise), while the system's wave function collapses in space.

Dissipative collapse models suggest the following picture: the noise causing the collapse is a *real* field filling space, at thermal equilibrium at some temperature T. It couples nonlinearly to quantum systems, causing the localisation of the wave function in space. At the same time, it thermalises them to its own temperature. This second effect is similar to what happens to a classical particle in a thermal bath. Energy is not conserved of course, but now the path towards restoring energy conservation is clear: besides the action of the noise on a quantum system, one should consider the reaction of the system on the noise. The combination of both effects might lead to energy conservation, as classically the fundamental equations of motion (Newton's laws), out of which the phenomenological description of a particle in a gas emerges, are energy conserving. In the case of collapse models, this requires the definition of an underlying theory, out of which they emerge as phenomenological models. This is not available yet (see [4] for a first attempt); our argument here shows that its existence is not unreasonable.

All models so far considered are defined in terms of a noise field, which is white in time. This assumption is very convenient from the mathematical point of view, because the associated dynamics is given in terms of a standard stochastic differential equation, which is relatively easy to analyse. But white noises are only a mathematical idealisation; real noises necessarily have a frequency cut off. Collapse models defined in terms of coloured noises have been introduced [14, 15, 16, 17, 18]. It has been shown that they share the same properties as the original models. Details could be different, especially for cut offs at low frequencies.

3. The mass proportional CSL model

The most widely analysed model in the literature is the mass-proportional version of the CSL model [19]. Its dynamics is given by the following stochastic differential equation:

$$d|\psi_t\rangle = \left[-\frac{i}{\hbar}Hdt + \frac{\sqrt{\gamma}}{m_0}\int d^3x(M(\mathbf{x}) - \langle M(\mathbf{x})\rangle_t)dW_t(\mathbf{x}) \right.$$
$$\left. - \frac{\gamma}{2m_0^2}\int\int d^3x d^3y G(\mathbf{x}-\mathbf{y})(M(\mathbf{x}) - \langle M(\mathbf{x})\rangle_t)(M(\mathbf{y}) - \langle M(\mathbf{y})\rangle_t)\right]|\psi_t\rangle, \quad (2)$$

where γ (having the dimensions of $[\mathrm{L}^3\mathrm{T}^{-1}]$) is the collapse strength of the model, m_0 is a reference mass, which is taken equal to that of a nucleon, and $M(\mathbf{x}) = ma^\dagger(\mathbf{x})a(\mathbf{x})$ is the *mass density* operator ($a^\dagger(\mathbf{x})$ and $a(\mathbf{x})$ are, respectively, the creation and annihilation operator of a particle at point $\mathbf{x}$ of space—here we are ignoring spin), which is a suitable replacement for the position operator in the second-quantized language. $W_t(\mathbf{x})$ is a family of Wiener processes, one for for each point of space, with average equal to zero. They are "white" in time, while they have a Gaussian correlation function in space:

$$\mathbb{E}[W_t(\mathbf{x})W_s(\mathbf{y})] = \delta(t-s)G(\mathbf{x}-\mathbf{y}), \qquad G(\mathbf{x}-\mathbf{y}) = \frac{1}{(4\pi r_C)^{3/2}}\exp[-(\mathbf{x})^2/4r_C^2]. \quad (3)$$

As we can see, the CSL model is defined in terms of two phenomenological parameters: the collapse strength γ and the correlation length of the noise r_C. Usually, one replaces γ with $\lambda = \gamma/(4\pi r_C^2)^{3/2}$, which has the dimensions of a rate $[\mathrm{T}^{-1}]$. Experimental analysis is placing stronger and stronger constraints on these parameters.

4. The amplification mechanism and the numerical choice of the parameters

The CSL equation (2) does what all collapse equations do: it tries to localizes the wave function in space, depending on the relative strength with respect to the Hamiltonian H. More specifically, the collapse terms work as follow [20].

Consider a system (in particular, a rigid object) whose center of mass is in a state, which is delocalized over a distance $\Delta x \ll r_C$. One can show that in this case the collapse is ineffective, almost absent. In a way, the delocalization is too small and the noise, which can resolve only distances greater than r_C, sees it already as well-localized.

If on the other hand the delocalization is lager than r_C, then the noise forces the wave function to collapse with a rate, which is roughly given by:

$$\Lambda = \lambda n^2 N, \quad (4)$$

where λ, introduced before, is the collapse rate for a single nucleon, n is the number of nucleons within a volume r_C^3 (a measure of the density of matter) and N counts how many such volumes can be accommodated in the space occupied by the system (a measure of the total size of the system).

Eq. (4) is the mathematical expression of the *amplification mechanism*: the larger the system, the faster the collapse. This means that it is possible to choose a unique value for λ and r_C such that, at the microscopic level, CSL differs little from ordinary quantum mechanics, while at the macroscopic level it forces wave functions to be well localized in space. In this way, the model is capable of providing a unique description of the quantum properties of macroscopic systems, and the classical properties of macroscopic objects—in particular, a resolution of the quantum measurement problem—within a unique dynamical framework.

In the literature, two values for the CSL parameters have been proposed. The first proposal was [6]: $\lambda \simeq 10^{-17}$ s^{-1} and $r_C = 10^{-7}$ m. More recently, a strongly enhanced value for λ has been suggested by Adler [20], motivated by the requirement of making the wave-function collapse effective at the level of latent image formation in photographic process. The values suggested by Adler are $\sim 10^{9\pm2}$ times larger than standard values at $r_C = 10^{-7}$ m, and $\sim 10^{11\pm2}$ times larger at $r_C = 10^{-6}$ m.

5. Experimental tests

Collapse models explicitly modify the Schrödinger equation, therefore they make predictions, which differ—in most cases, only slightly—from those of ordinary quantum mechanics, and which can be tested experimentally. There are two type of tests: interferometric and non-interferometric ones.

Interferometric experiments are the most direct way of testing collapse models. Since their purpose is to forbid the existence of macroscopic superpositions, then the obvious way to test them is to take an object, which is as big as possible, create a large enough superposition and keep it there for as long a time as possible, and then detect the presence of quantum interference. Collapse models predict a negative outcome, contrary to ordinary quantum mechanics. Of course, performing such an experiment is not easy at all, for serval technical reasons.

The group of Markus Arndt in Vienna succeeded in superimposing and detecting quantum interference of the center of mass of a macro-molecule of 20,000 a.m.u. [21]. Currently this is the world record. Fig. 1 shows the bound such an experiment places on the CSL parameters. As we can see, neither Adler's values for λ and r_C, nor GRW's, are disproved by the experiment, meaning that the mass here involved, the superposition times and distances are large enough. Optomechanics promises to significantly go beyond the state of the art [22], though that requires a improvement of the technology.

Non-interferometric tests [23, 24, 25, 26] aim at measuring the Brownian motion induced by the noise, independently of the collapse. There are two effects, which have been investigated. The first one is the increase of the kinetic energy, discussed in Sec. 2. A system confined in a (harmonic) trap should tremble around the equilibrium position more than what is predicted by quantum mechanics. A recent experiment with a cantilever [27] was able to place a bound, which is strong enough to partly exclude Adler's value for λ. The second effect is the spontaneous emission of radiation from charged particles, due to the acceleration induced by the noise. Experimental data [28] have been taken in the X-ray region, which also exclude Adler's value for λ. Future experiment promise to set bounds, which are at least two orders of magnitude stronger.

One comment is in order. It has been shown that interferometric bounds coming from matter-wave interferometry are robust against changes in the collapse equation, e.g. by introducing dissipative effects and/or by changing the type of noise [29]. Therefore they serve the purpose of testing the whole class of collapse models (in space), not just a single

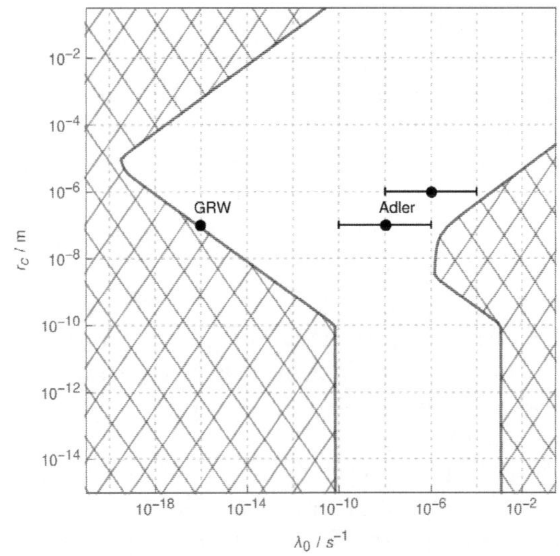

Figure 1. Left exclusion region: lower bound on the CSL parameters, coming from the requirement that macroscopic objects should be localized in space [29]. This is the fundamental motivation for introducing collapse models. Right exclusion region: upper bounds set by matter-wave interferometry [29]. The black dots/lines denote the numerical values for λ and r_C proposed by GRW and Adler.

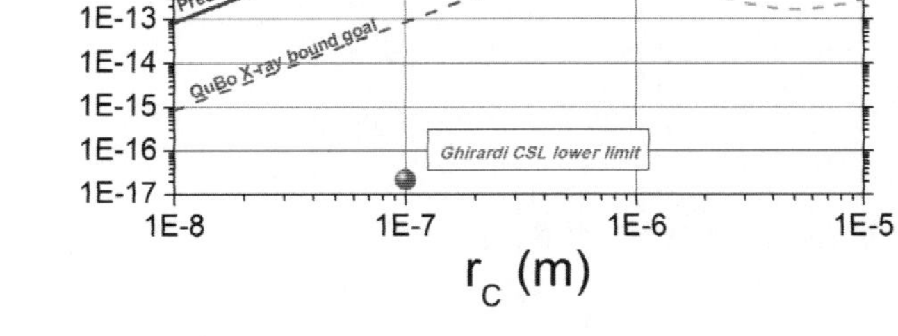

Figure 2. Upper bounds on the CSL parameters from non-interferometric experiments. The red solid line refers to the measurement of a violation of energy conservation with cantilevers [27], while the purple solid line refers to measurements of induced Brownian via spontaneous X-ray emission [28]. Adler's value for λ and r_C are excluded (but possibly rescued by suitable modifications of the CSL dynamics). Future upgrades of the experiments (dashed lines) promise to strengthen the bounds by two order of magnitude (QuBo project submitted to MIUR, Italy).

model—CSL, in our case. On the other hand, non-interferometric tests are more sensitive to the type of model. For example, predictions about spontaneous photon emission highly depend on the correlation function of the noise [20]. These bounds are strong, but probably are not able to span the whole class of collapse models. A combination of interferometric and non-interferometric experiments will probably be needed to test collapse models—therefore the quantum superposition principle—in a meaningful way.

6. Conclusion

Collapse models provide a consistent phenomenology for describing the collapse of the wave function as a dynamical effect, described by a suitable modification of the Schrödinger equation. One expects them to derive from an underlying theory, out of which both the quantum evolution and the collapse of the wave function emerge in a suitable coarse-grained description. This is a difficult problem, which has not been solved yet.

Meanwhile, it is interesting and relevant to test them experimentally, not least because any such test is a test of the superposition principle, the most important feature of quantum theory. Till recently, collapse models were confined to theoretical speculation. Now instead the experimental effort in testing them is increasing, and technological improvement allows (and will allow even more, in the near future) to explore significant regions of the parameter space. It will be interesting to see at the end, whether nature forbids nonlinear modification of quantum theory or, more exciting, whether the superposition principle will eventually break down.

Acknowledgments

The author wishes to thank the organisers of the Symposium *EmQM15 – Emergent Quantum Mechanics* for the kind invitation. Fig. 1 has been prepared by M. Toros (Trieste), and Fig. 2 by A. Vinante (Trento).

References

[1] Bassi A and Ghirardi G C 2003 *Phys. Rep.* **379** 257
[2] Bassi A, Lochan K, Satin S, Singh T P and Ulbricht H 2013 *Rev. Mod. Phys.* **85** 471
[3] Bassi A and Ippoliti E 2004 *Phys. Rev.* A **69** 012105
[4] Adler S L 2004 *Quantum Theory as an Emergent Phenomenon* (Cambridge University Press)
[5] Ghirardi G C, Rimini A and Weber T 1986 *Phys. Rev.* D **34** 47
[6] Ghirardi G C, Pearle P and Rimini A 1990 *Phys. Rev.* A **42** 78
[7] Diosi L 1989 *Phys. Rev.* A **40** 1165
[8] Bassi A, Drr D 2009 *J. Phys.* A **42** 485302
[9] Diosi L 1989 *Phys. Rev.* A **40** 1165
[10] Bassi A, Ippoliti E and Vacchini B 2005 *J. Phys.* A **38** 8017
[11] Smirne A, Vacchini B and Bassi A 2014 *Phys. Rev.* A **90** 062135
[12] Smirne A and Bassi A 2015 *Sci. Rep.* **5** 12518
[13] Bahrami M, Smirne A and Bassi A 2014 *Phys. Rev.* A **90** 062105
[14] Pearle P 1993 *Phys. Rev.* A **48** 913
[15] Adler S L and Bassi A 2007 *J. Phys.* A **40** 15083
[16] Adler S L and Bassi A 2008 *J. Phys.* A **41** 395308
[17] Bassi A and Ferialdi L 2009 *Phys. Rev. Lett.* **103** 050403
[18] Ferialdi L and Bassi A 2012 *Phys. Rev. Lett.* **108** 170404
[19] Ghirardi G C, Grassi R and Benatti F 1995 *Found. Phys.* **25** 5
[20] Adler S L 2007 *J. Phys.* A **40** 13501
[21] Eibenberger S, Gerlich S, Arndt M, Mayor M and Txen J 2013 *Phys. Chem. Chem. Phys.* **15** 14696
[22] Bateman J, Nimmrichter S, Hornberger K and Ulbricht H 2014 *Nat. Commun.* **5** 4788
[23] Bahrami M, Bassi A and Ulbricht H 2014 *Phys. Rev.* A **89** 032127
[24] Bahrami M, Paternostro M, Bassi A and Ulbricht H 2014 *Phys. Rev. Lett.* **112** 210404
[25] Nimmrichter S, Hornberger K and Hammerer K 2014 *Phys. Rev. Lett.* **113** 020405
[26] Diosi L 2015 *Phys. Rev. Lett.* **114** 050403

[27] Vinante A, Bahrami M, Bassi A, Usenko O, Wijts G and Oosterkamp T H 2015 *Upper bounds on spontaneous wave-function collapse models using millikelvin-cooled nanocantilevers Preprint* arXiv:1510.05791
[28] Curceanu C, Hiesmayr B C and Piscicchia K 2015 *J. Advanced Phys.* **4** 263
[29] Toros M and Bassi A 2016 *Bounds on collapse models from matter-wave interferometry Preprint* arXiv:1601.03672

Exploring the propagation of relativistic quantum wavepackets in the trajectory-based formulation

Hung-Ming Tsai and Bill Poirier

Department of Chemistry and Biochemistry, and Department of Physics,
Texas Tech University, Box 41061, Lubbock, Texas 79409-1061, USA

E-mail: Bill.Poirier@ttu.edu

Abstract. In the context of nonrelativistic quantum mechanics, Gaussian wavepacket solutions of the time-dependent Schrödinger equation provide useful physical insight. This is not the case for *relativistic* quantum mechanics, however, for which both the Klein–Gordon and Dirac wave equations result in strange and counterintuitive wavepacket behaviors, even for free-particle Gaussians. These behaviors include zitterbewegung and other interference effects. As a potential remedy, this paper explores a new trajectory-based formulation of quantum mechanics, in which the wavefunction plays no role [*Phys. Rev. X*, **4**, 040002 (2014)]. Quantum states are represented as ensembles of trajectories, whose mutual interaction is the source of all quantum effects observed in nature—suggesting a "many interacting worlds" interpretation. It is shown that the relativistic generalization of the trajectory-based formulation results in well-behaved free-particle Gaussian wavepacket solutions. In particular, probability density is positive and well-localized everywhere, and its spatial integral is conserved over time—in any inertial frame. Finally, the ensemble-averaged wavepacket motion is along a straight line path through spacetime. In this manner, the pathologies of the wave-based relativistic quantum theory, as applied to wavepacket propagation, are avoided.

1. Introduction

Quantum mechanics is generally regarded to be a theory based on wavefunctions [1–3]. The quantum wavefunction, Ψ, is thought to be the fundamental mathematical representation of the state of a quantum system. As such, Ψ has always enjoyed a hallowed status, despite much historical and ongoing disagreement about its precise interpretation or physical significance [4–11]. Even "alternative" interpretations of quantum mechanics such as Bohmian mechanics [5, 6], which attribute physical reality to a "quantum trajectory," nevertheless still adopt a hybrid ontology in practice, wherein it is a combination of both the wavefunction *and* the quantum trajectory, *together*, that specify the quantum state.

This article explores a fundamentally different kind of quantum theory, that makes no direct or indirect recourse to wavefunctions. It resembles Bohmian mechanics in that quantum trajectories are indeed employed. However, unlike Bohmian mechanics, *only* trajectories are used—the wave being replaced with a trajectory *ensemble*, which thereby represents the quantum state [12–20]. The trajectory ensemble is continuous, with each individual member trajectory labeled by the parameter C. The time evolution of the trajectory ensemble, $x(t, C)$, is governed by some partial differential equation (PDE) in (t, C) that replaces the usual time-independent Schrödinger equation governing the $\Psi(t, x)$ evolution. All quantum effects manifest as a dynamical interaction between neighboring trajectories in the ensemble—i.e., as partial derivatives in C. A new interpretation of quantum mechanics is thus also suggested by the new mathematics—what has come to be known as "many interacting worlds" [14, 19, 20].

Interpretations notwithstanding, in the nonrelativistic context described above, the wave-based and (continuous) trajectory-based mathematics are equivalent; thus, no new experimental predictions are proffered by the latter. On the other hand, a recent extension of the trajectory-based formulation to the *relativistic* quantum regime [21]—for single, massive, spin-zero, free particles, propagating on a flat Minkowski spacetime—does indeed provide new experimental predictions. However there are also other motivations for considering the relativistic extension of the trajectory theory. For example, the analogous wave-based relativistic quantum PDE for massive spin-zero particles—i.e. the Klein–Gordon equation—is riddled with foundational difficulties, when interpreted as a single-particle theory. Most problematically, the temporal component of the four-current is not positive over all spacetime, and therefore cannot be interpreted as a probability density [22, 23].

This "negative probability" issue is mostly resolved when one moves to spin-1/2 particles such as electrons, for which the Dirac equation is used to describe the relativistic quantum dynamics. On the other hand, both the Klein–Gordon and Dirac equations must contend with another difficulty, i.e. negative-energy solutions. Conceptually, one can address the negative-energy quantum states somewhat satisfactorily—in familiar terms involving positrons, the "Dirac sea," etc. In practice, however, negative-energy states lead to highly undesirable outcomes, that manifest even in that simplest and most pedagogically useful of examples, the free-particle Gaussian wavepacket [24, 25]. By far the most famous of these undesired outcomes is zitterbewegung—rapid oscillations of the position expectation value over time, about a straight-line path through spacetime. However, this is but one of several "strange" behaviors observed in relativistic free-particle wavepacket dynamics, associated with interference between positive- and negative-energy components. These can be so severe that even 85+ years after they were first theoretically predicted, the physical origin and interpretation of these effects is still controversial, leaving even the standard probabilistic interpretation of the Dirac theory in doubt [24, 25]. A trajectory-based description may help substantially, in at least two different ways. First, it may admit an *analytical* wavepacket solution; according to one author, much of the remaining controversy "is all because there is no known analytical expression for a Dirac packet" [25]. Second, the relativistic quantum trajectory-based PDE *has* no negative-energy solutions, and is thereby possibly able to side-step the above difficulties, at least in principle. In any event, it is perhaps telling that true zitterbewegung has yet to be observed experimentally [25].

In this paper, we analyze the trajectory-based dynamical PDEs for spin-zero relativistic quantum particles, as derived in Ref. [21]. We also propagate those PDEs numerically, for various (initially) Gaussian wavepacket examples. We begin in Sec. 2, by introducing the dynamical PDEs for a relativistic quantum particle in $(3 + 1)d$ spacetime. Then, in Sec. 3, we reduce the spacetime dimension to $(1 + 1)d$ and derive dynamical equations specific to Gaussian wavepackets. Numerical solutions are also presented. The symmetry properties of the dynamical PDEs, together with their numerical solutions, are then investigated in Sec. 4, with respect to a set of scale transformations that preserves the equations of motion. The predicted scale invariance is verified both analytically and numerically. In this section, also, we examine the spacetime dependence of the spatial metric, which provides important dynamical information—e.g., as pertains to the relativistic restriction on wavepacket broadening.

Whereas an analytical solution may someday be tenable for the relativistic free-particle Gaussian wavepacket, at present we must rely on numerical solutions. Sec. 5 therefore addresses numerical issues, especially the issue of how to systematically characterize and improve the convergence accuracy of the numerical trajectory ensemble solutions. This is more difficult than for many standard numerical analyses, owing to the fact that the true boundary conditions are

Journal of Physics: Conference Series **701** (2016) 012013 doi:10.1088/1742-6596/701/1/012013

unknown. As a consequence, numerical instabilities are sometimes encountered under certain conditions—although we provide one simple strategy for ameliorating this difficulty, involving nonuniform grids, which seems to work rather well.

The analysis up to this point in the paper pertains only to "stationary" Gaussian wavepackets, which remain centered at $x = 0$ throughout time. In Sec. 6, we generalize everything for arbitrary moving free-particle Gaussian wavepackets, by simply applying Lorentz boosts to the stationary wavepacket solutions. In doing so, we confirm the Lorentz invariance of the dynamical PDEs and their numerical solutions, as well as the associated flux four-vectors. We also examine the probability density in both inertial frames, confirming that this quantity—the temporal component of the flux four-vector—is indeed positive throughout spacetime, and in all inertial frames. Finally, we confirm probability conservation by explicitly integrating the probability density over all space, and showing that the result is conserved over time—in any inertial frame. Thus, the trajectory-based treatment of relativistic free-particle Gaussian wavepackets does indeed appear to avoid the undesirable features of the corresponding wave-based approaches.

2. Dynamical equations in $(3+1)d$ spacetime

The following is a very brief overview of the derivation provided in Ref. [21], which should be consulted for further details. In the trajectory-based formulation for a single spin-free relativistic quantum particle in flat $(3+1)d$ spacetime, a quantum state is represented as an ensemble of quantum trajectories, $x^\alpha(X^\mu)$. The *dependent* field variables are the inertial or extrinsic coordinates $x^\alpha = (ct, \boldsymbol{x})$, whereas the independent variables are the natural or intrinsic coordinates $X^\mu = (c\lambda, \boldsymbol{C})$. The quantity λ is a global time-like parameter called the "ensemble time," whereas the three space-like $\boldsymbol{C}$ coordinates label the individual quantum trajectories in the ensemble. Contours of the former, expressed in inertial coordinates, represent "simultaneity submanifolds"—i.e., global space-like-separated sets of events that occur simultaneously for the quantum particle. Contours of the latter—or rather, their mutual intersections—define the quantum trajectories themselves. These foliate spacetime, and may be interpreted as a continuous spectrum of "worlds," in each of which resides a single "copy" of the particle. Both λ and $\boldsymbol{C}$ may be separately reparametrized, in arbitrary fashion.

We denote the metric tensor of the natural coordinate system as $g^{\mu\nu}$, and that of the inertial coordinates as $\eta^{\alpha\beta}$. These are related via the transformation,

$$g_{\mu\nu} = \eta_{\alpha\beta} \frac{\partial x^\alpha}{\partial X^\mu} \frac{\partial x^\beta}{\partial X^\nu}. \tag{1}$$

In matrix form, the inertial metric tensor is the familiar $\tilde{\eta} = \text{diag}(-1, 1, 1, 1)$. For the natural coordinates, we impose local orthogonality between infinitesimal displacements in λ and $\boldsymbol{C}$. This results in a block-diagonal metric tensor, i.e.

$$\tilde{g} = \begin{pmatrix} g_{00} & 0 \\ 0 & \tilde{\gamma} \end{pmatrix}, \tag{2}$$

where the "spatial metric" $\tilde{\gamma}$ denotes the 3×3 spatial block of $\tilde{g}$. We further define

$$g = \det \tilde{g} \qquad ; \qquad \gamma = \det \tilde{\gamma}. \tag{3}$$

Finally, we note that $g_{00} = -(d\tau/d\lambda)^2$, where τ is the usual proper time, defined via

$$d\tau^2 = -\frac{1}{c^2} \eta_{\alpha\beta} dx^\alpha dx^\beta. \tag{4}$$

Probability conservation along individual quantum trajectories is presumed [21], as a result of which the spatial scalar probability density on $\boldsymbol{C}$-space, $f(\boldsymbol{C})$, must be independent of λ. In addition to this scalar quantity, there is also the *flux four-vector*,

$$j^\mu = f(\mathbf{C})\,\frac{dX^\mu}{d\lambda} = (cf(\mathbf{C}),0,0,0), \tag{5}$$

which transforms as a contravariant vector density of weight $W = -1$. Due to a fortuitous cancellation of Christoffel symbols, j^μ satisfies the following covariant continuity equation,

$$\partial_\mu\, j^\mu = 0, \tag{6}$$

in *all* coordinate frames. In an inertial frame, in particular, this leads to the familiar integrated form of probability conservation,

$$\int j^0(x^\alpha)\, d^3x = \text{const.} \tag{7}$$

It is noteworthy that j_0 is can be interpreted as a true *probability* density (apart from a scaling factor), as in nonrelativistic quantum mechanics. The analogous Klein–Gordon j_0 quantity, for example, can change sign, and must therefore be interpreted as a *charge* density [22, 23].

In addition to its usual probabilistic role, $f(\mathbf{C})$ also plays an indirect *dynamical* role in the quantum trajectory theory. The quantum potential Q and the quantum force f^α are given by

$$Q = -\frac{\hbar^2}{2m}\gamma^{-1/4} f^{-1/2}\frac{\partial}{\partial C^i}\left[\gamma^{1/2}\gamma^{ij}\frac{\partial}{\partial C^j} f^{1/2}\gamma^{-1/4}\right], \tag{8}$$

$$f^\alpha = -\eta^{\alpha\beta}\frac{\partial C^i}{\partial x^\beta}\frac{\partial Q}{\partial C^i} = -\frac{\partial x^\alpha}{\partial C^i}\gamma^{ij}\frac{\partial Q}{\partial C^j}, \tag{9}$$

where γ^{ij} is the matrix form of $\tilde{\gamma}^{-1}$. The above equations are valid for any parametrization of λ and $\mathbf{C}$. On the other hand, we can make a special choice for λ, known as the "ensemble proper time," $\mathcal{T}$, which reduces to τ when $Q = 0$. More generally [21], the relation is

$$\frac{d\tau}{d\mathcal{T}} = \exp\left(-\frac{Q}{mc^2}\right). \tag{10}$$

This leads, finally, to the dynamical equations of motion for a free particle:

$$\frac{\partial^2 x^\alpha}{\partial \mathcal{T}^2} = \exp\left(\frac{-2Q}{mc^2}\right)\frac{f^\alpha}{m} - \frac{1}{mc^2}\frac{\partial Q}{\partial \mathcal{T}}\frac{\partial x^\alpha}{\partial \mathcal{T}}. \tag{11}$$

Solving Eq. (11) requires specification of the initial conditions, $x_0^\alpha = x^\alpha(0,\mathbf{C})$ and $(\partial x^\alpha/\mathcal{T})(0,\mathbf{C})$.

Note that Eq. (11) is second order in $\mathcal{T}$ and fourth order in $\mathbf{C}$—in exact analogy with the time and space orders of the Schrödinger PDE (the order doubling is due to the fact that the field quantity is in this case real- rather than complex-valued). Thus, *natural* time and *natural* space are not treated on an equal footing in this relativistic picture—but then again, neither should they be. In contrast, the *inertial* coordinates, x^α *are* treated equally, in the sense that Eq. (11) is Lorentz-invariant. In this manner—i.e., by separating natural from inertial coordinates—we are able to avoid the relativistic quantum "paradox" that, for example, led Dirac to consider multiple wave components. Moreover, the negative energy solutions that one finds with Klein–Gordon [22, 26, 27] or Dirac [22, 26, 27] also go away, because one finds that positive-negative pairs of wavefunction solutions correspond to a single trajectory ensemble solution $x^\alpha(X^\mu)$ [21].

3. Gaussian wavepacket solutions in $(1+1)d$ spacetime

Some analytical and numerical solutions of Eq. (11) are provided in Ref. [21], for $(1+1)d$ spacetime. In this reduced-dimensional context, there is but a single inertial (natural) spatial coordinate, $x = x^1$ ($C = C^1$), and the dynamical equations simplify as a consequence of

$$\frac{1}{\gamma^{ij}} = \gamma_{ij} = \gamma = \left(\frac{\partial x}{\partial C}\right)^2 - c^2 \left(\frac{\partial t}{\partial C}\right)^2. \tag{12}$$

For the remainder of this article, we will focus on $(1+1)d$ spacetime, and specifically on the free particle propagation of the relativistic Gaussian wavepacket. In nonrelativistic quantum mechanics, the Gaussian wavepacket is well known to disperse while moving in a straight line—yet retains a perfect Gaussian shape at all times [2]. The wavepacket spreading is therefore linear. This behavior cannot also characterize the relativistic regime, for it would imply the existence of faster-than-light quantum trajectories in the Gaussian tails, which are precluded by the $(1+1)d$ Eq. (11). At best, the relativistic wavepacket can be truly Gaussian at a single time only. Moreover, one has to resolve which "time" is being referred to, i.e. $\mathcal{T}$ or t.

Both issues are resolved by considering coherent-state or minimum-uncertainty Gaussians, which exhibit uniform flux/velocity throughout x. All nonrelativistic free-particle Gaussian wavepackets evolve into a coherent state at exactly one point in time during their lifecycles. By setting this time equal to zero, and transforming to a frame in which the uniform velocity is zero, all quantum trajectories in the ensemble become instantaneously stationary. The relativistic and nonrelativistic descriptions must therefore agree in that initial instant. In particular, the $t = 0$ and $\mathcal{T} = 0$ initial conditions become identical. Our ansatz for a free-particle relativistic Gaussian wavepacket therefore becomes a solution of $(1+1)d$ Eq. (11) whose initial conditions satisfy:

$$t_0(C) = 0 \quad ; \quad x_0(C) = C \quad ; \quad \frac{\partial x}{\partial \mathcal{T}}(0, C) = 0, \qquad \text{with } f(C) = \exp\left(-aC^2\right). \tag{13}$$

Note that Eq. (13) implies a specific parametrization for C—i.e., for any given trajectory, C is the initial x value, x_0 [14, 21]. Likewise, our earlier specification of the natural time coordinate, $\lambda = \mathcal{T}$, implies the following as the one remaining initial condition:

$$\frac{\partial t}{\partial \mathcal{T}}(0, C) = \exp\left[-\frac{Q_0(C)}{mc^2}\right] = \exp\left[\frac{1}{2}\left(\frac{\hbar}{mc}\right)^2 \left(a^2C^2 - a\right)\right]. \tag{14}$$

In Eq. (14) above, $Q_0(C) = Q(0, C)$ is the initial quantum potential. More generally, Eqs. (8) and (9) reduce to the following, for $(1+1)d$ spacetime with the $f(\mathbf{C})$ choice of Eq. (13):

$$Q = -\frac{\hbar^2}{2m} \frac{e^{\frac{1}{2}aC^2}}{\gamma^{1/4}} \frac{\partial}{\partial C}\left[\frac{1}{\gamma^{1/2}} \frac{\partial}{\partial C}\left(e^{-\frac{1}{2}aC^2} \gamma^{-1/4}\right)\right]; \tag{15}$$

$$f^0 = -\frac{c}{\gamma} \frac{\partial t}{\partial C} \frac{\partial Q}{\partial C} \quad ; \quad f^1 = -\frac{1}{\gamma} \frac{\partial x}{\partial C} \frac{\partial Q}{\partial C}. \tag{16}$$

The resultant dynamical equations become

$$\begin{aligned} -\frac{1}{m} \frac{1}{\gamma} \frac{\partial t}{\partial C} \frac{\partial Q}{\partial C} &= \exp\left(\frac{Q}{mc^2}\right) \frac{\partial}{\partial \mathcal{T}}\left[\exp\left(\frac{Q}{mc^2}\right) \frac{\partial t}{\partial \mathcal{T}}\right], \\ -\frac{1}{m} \frac{1}{\gamma} \frac{\partial x}{\partial C} \frac{\partial Q}{\partial C} &= \exp\left(\frac{Q}{mc^2}\right) \frac{\partial}{\partial \mathcal{T}}\left[\exp\left(\frac{Q}{mc^2}\right) \frac{\partial x}{\partial \mathcal{T}}\right]. \end{aligned} \tag{17}$$

To date, analytical solutions of Eq. (17) for $t(\mathcal{T}, C)$ and $x(\mathcal{T}, C)$ have eluded us. However, we have obtained numerical solutions, for $a = 1/2$, $\hbar = m = 1$, and various values of the speed

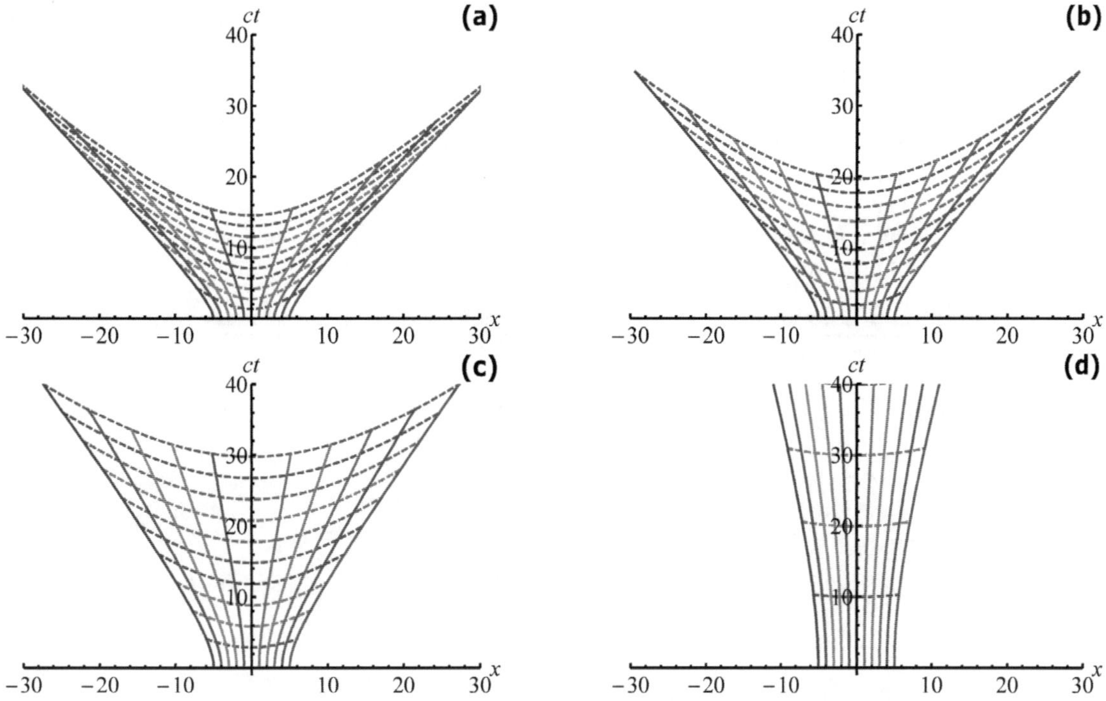

Figure 1. Quantum trajectories (solid curves) and simultaneity submanifolds (dashed curves) for the relativistic free-particle Gaussian wavepacket of Eq. (13), for $a = 1/2$, $\hbar = m = 1$, and several choices of the speed of light, c: (a) $c = 1.5$; (b) $c = 2.0$; (c) $c = 3.0$; (d) $c = 10$.

of light c, as presented in Fig. 1. Wavepacket spreading is evident, as in the nonrelativistic case. The quantum trajectories approach the light cone faster in the ultrarelativistic limit of small c, as expected. The simultaneity submanifolds also become increasingly distorted in this limit, whereas in the nonrelativistic (large c) limit, they approach horizontal lines (i.e., contours of t).

4. Scale invariance of the dynamical solutions, and the spatial metric

The PDE of Eq. (17) governs the time evolution of the quantum trajectory ensemble for the relativistic free-particle Gaussian wavepacket. Rescaling the system parameters of this PDE alters its solutions. However, it is possible to rescale the parameters and dynamical variables together, in such a way that the solutions remain invariant. Consider the double scale transformation defined in terms of the two parameters, ζ and η, with ζ associated with c, and η with the independent and dependent variables, as follows:

$$c \to \zeta c \quad ; \quad X^\mu \to \eta X^\mu \quad ; \quad x^\alpha \to \eta x^\alpha. \tag{18}$$

$$\mathcal{T} \to \frac{\eta}{\zeta}\mathcal{T} \quad ; \quad C \to \eta C \quad ; \quad t \to \frac{\eta}{\zeta} t \quad ; \quad x \to \eta x. \tag{19}$$

For the quantum trajectory ensemble solution to remain invariant, Eq. (13) implies that $f(C)$ must also remain invariant. Under the rescaling $C \to \eta C$, this requires $a \to (1/\eta^2)a$. More generally, the invariance of the solution under $C \to \eta C$ rescaling implies:

$$\gamma \to \gamma \text{ from Eqs. (12) and (19)} \quad ; \quad \frac{Q}{mc^2} \to \frac{Q}{mc^2} \text{ from Eq. (17)} \tag{20}$$

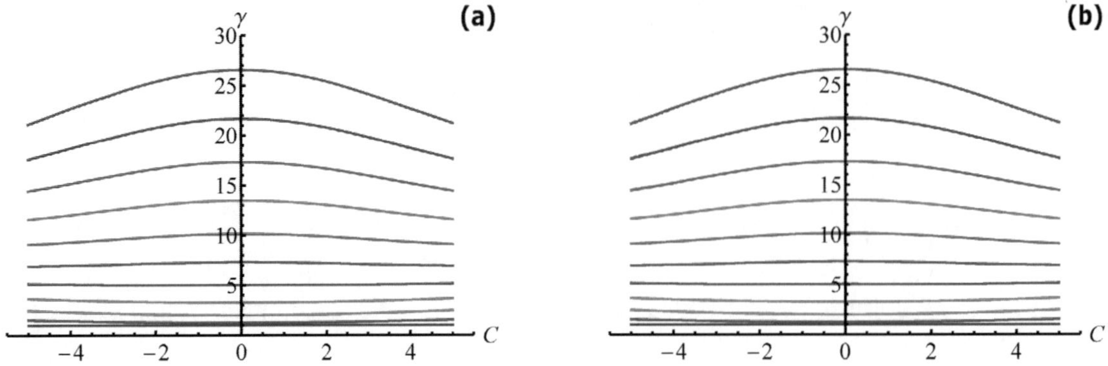

Figure 2. Spatial metric $\gamma(\mathcal{T}, C)$ for $(1+1)d$ relativistic free-particle Gaussian wavepackets. Each curve represents $\gamma(C)$ for fixed $\mathcal{T} = \{0, 1, 2\ldots, 10\}$, with increasing $\mathcal{T}$ values corresponding to increasing γ: (a) $(a = 1/2, \hbar = 1, c = 3)$ solution; (b) comparison between $(a = 1/2, \hbar = 1, c = 3)$ and $(a = 1/5, \hbar = 5\sqrt{10}/3, c = 10)$ solutions, which are numerically nearly indistinguishable.

$$\left(\frac{\hbar}{mc}\right)^2 \to \eta^2 \left(\frac{\hbar}{mc}\right)^2 \text{ from Eqs. (8) and (20).} \tag{21}$$

Now we consider the invariance of the solution under $c \to \zeta c$, which implies:

$$\frac{\hbar}{m} \to \eta\zeta\left(\frac{\hbar}{m}\right) \text{ from Eq. (21)} \quad ; \quad \frac{Q}{m} \to \zeta^2 \frac{Q}{m} \text{ from Eq. (20)} \tag{22}$$

$$\frac{f^\alpha}{m} \to \frac{\zeta^2}{\eta}\left(\frac{f^\alpha}{m}\right) \text{ from Eqs. (16), (19) and (22).} \tag{23}$$

Finally, we examine the rescaling of Eq. (17) directly. Under the simultaneous rescaling $C \to \eta C$ and $c \to \zeta c$, the left hand side of Eq. (17) is found to rescale as ζ^2/η. The right hand side also rescales as ζ^2/η—thus implying scale invariance for the free-particle Gaussian wavepacket evolution. More generally, similar arguments are also found to hold for Eq. (11). The scale invariance has also been verified numerically. In Fig. 2, we compare $\gamma(\mathcal{T}, C)$ for the $(a = 1/2, \hbar = m = 1, c = 3)$ Gaussian wavepacket solution vs. that for $(a = 1/5, \hbar = 5\sqrt{10}/3, m = 1, c = 10)$—corresponding to the $(\zeta = 10/3, \eta = \sqrt{5/2})$ rescaling. We observe that all of the fixed-$\mathcal{T}$ $\gamma(C)$ curves are numerically nearly indistinguishable.

Since γ in fact represents the spatial metric, Fig. 2 provides useful insight into the wavepacket dynamics. The primary feature evident in the plots is that the $\gamma(C)$ curves increase in magnitude with increasing $\mathcal{T}$—an indication of wavepacket broadening. Note, however, that this increase is not uniform across C—i.e., the curves are not horizontal lines. If they were, this would imply preservation of the Gaussian form over time, which only occurs in the nonrelativistic limit. Conversely, the curvature or bowing of these curves—which for this example, is seen to be rather pronounced—is an indication of relativistic quantum dynamical effects.

Another interesting feature may also be observed in Fig. 2 , which is that the direction of the curvature changes over time. The early curves bow upward, whereas at later times, the curves bow downward. This can be explained as a competition between the dynamical influences of Q and f^α. At early times, all trajectories are "moving" in unison; their acceleration has not yet had a chance to manifest as a fanning out of velocities. However, even at $\mathcal{T}=0$, the local proper time evolves at a higher rate towards the exterior fringes of the ensemble (i.e., towards

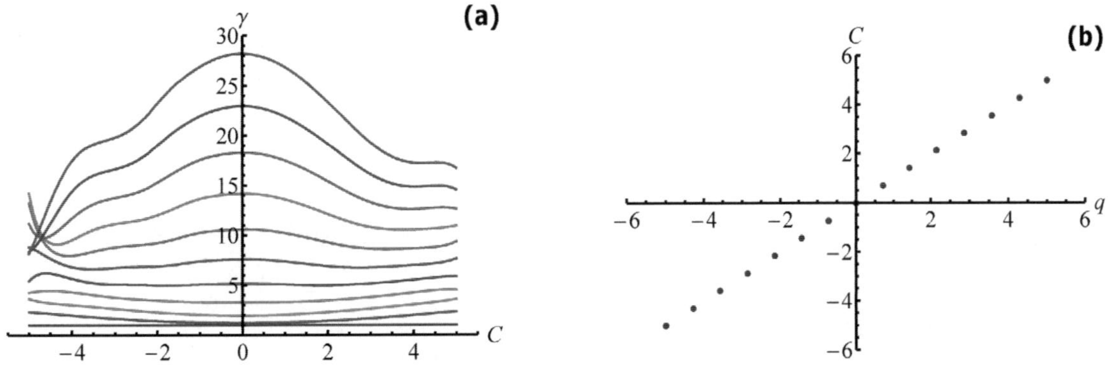

Figure 3. Calculations with a uniformly-spaced C grid: (a) numerically computed $\gamma(\mathcal{T}, C)$, as per Fig. 2, but for $c = 1.5$; (b) the $N_g = 15$ grid points themselves, uniformly spaced over $-5 \le C \le 5$.

larger $|C|$)—in accord with the relativistic quantum *time compression* effect [21]. The early simultaneity submanifolds thus curve away from the x axis, resulting in larger intervals between nearby trajectories that lie further from $C{=}0$—and thus, in upward-bowing $\gamma(C)$ curves. Over time, acceleration gives rise to trajectory fanout, as discussed. However, as the velocities of the exterior trajectories reach the order of the speed of light, further broadening of the wavepacket is relativistically hindered, and so the resultant $\gamma(C)$ curves bow downward, rather than upward.

5. Numerical details

From a numerical standpoint, the solution of Eq. (17) offers important advantages over both conventional, x-grid-based Crank–Nicholson propagation of Ψ, and traditional quantum trajectory methods [28]. Briefly, one has the simultaneous advantages of *both* a regular grid (in C) *and* probability-conserving trajectories (in x) [14]. There are nevertheless some nontrivial numerical issues, stemming from the fact that the true boundary conditions are unknown—unlike for Ψ propagation, for which Dirichlet boundary conditions are in effect. In practice, as a result of the small asymptotic density $f(C)$, the less significant boundary regions seem to have very little effect on the far more important interior. Consequently, one can choose any reasonable boundary condition—or none at all, as is the case with most of our *Mathematica* calculations (performed using **NDSolve**)—and still obtain excellent results. On the other hand, there are also situations when this boundary condition issue can cause numerical instabilities. This is an ongoing area of investigation, though we discuss one simple example and remedy here.

For the present Gaussian wavepacket calculations, errors and instabilities tend to increase when the numerical grid is extended over a larger C domain, and also in the ultrarelativistic limit. Figure 3 presents $\gamma(C)$ curves for the $c = 1.5$ example of Fig. 1(a), as computed using a simple uniformly-spaced C grid of $N_g = 15$ points, over the domain $-C_{\max} \le C \le C_{\max}$, with $C_{\max} = 5$. The lack of symmetry, curve-crossing, and sharp changes are all evidence of numerical inaccuracy in the boundary regions, although the interior trajectories where probability is significant are still fairly accurate. Simply increasing the number of grid points, however, without changing the C domain, does not lead to improved accuracy, and actually makes the boundary behavior worse. The same is true if $C_{\max}$ is increased, using the same grid density.

Based on these observations, we instead adopt a *non*uniform C grid, wherein a higher density of grid points is used in the more important interior region. This is achieved via the map,

$$C(q) = A \tanh^{-1}(\beta q), \tag{24}$$

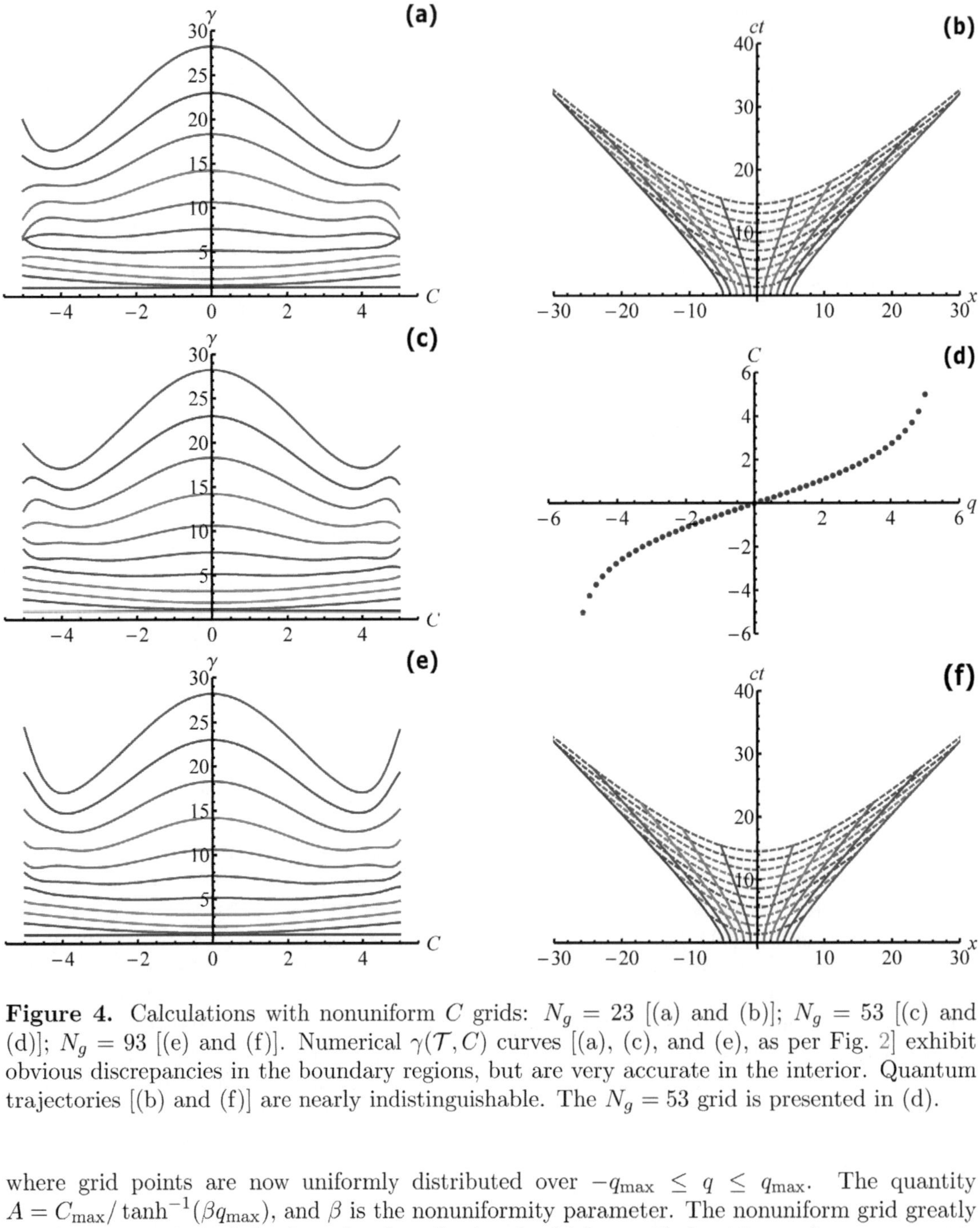

Figure 4. Calculations with nonuniform C grids: $N_g = 23$ [(a) and (b)]; $N_g = 53$ [(c) and (d)]; $N_g = 93$ [(e) and (f)]. Numerical $\gamma(\mathcal{T}, C)$ curves [(a), (c), and (e), as per Fig. 2] exhibit obvious discrepancies in the boundary regions, but are very accurate in the interior. Quantum trajectories [(b) and (f)] are nearly indistinguishable. The $N_g = 53$ grid is presented in (d).

where grid points are now uniformly distributed over $-q_{\rm max} \leq q \leq q_{\rm max}$. The quantity $A = C_{\rm max}/\tanh^{-1}(\beta q_{\rm max})$, and β is the nonuniformity parameter. The nonuniform grid greatly improves asymptotic behavior, thereby allowing for higher grid densities and more accurate calculations in the interior.

For the above $c = 1.5$ example, the choice ($q_{\rm max} = 5, C_{\rm max} = 5, \beta = 0.19$) enables stable, accurate calculations to be performed up to $N_g = 93$ grid points, as depicted in Fig. 4. Although numerical discrepancies in the boundary regions of the Fig. 4 curves are still evident, we emphasize that the wavepacket as a whole is *extremely* well-converged. This is evident in Fig. 5, which depicts the *probability-weighted* errors as a function of C and $\mathcal{T}$, for two different nonuniform grid calculations, of sizes $N_g = 53$ and $N_g = 83$, respectively. The larger grid size

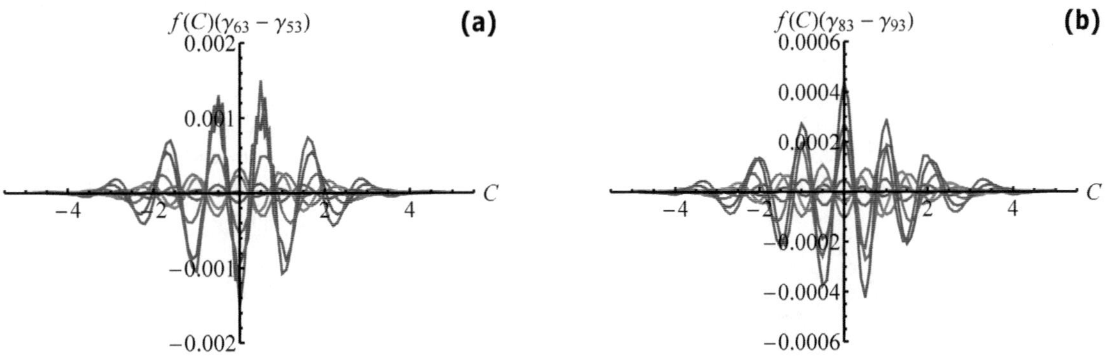

Figure 5. Probability-weighted convergence errors for nonuniform C grid calculations of $\gamma(\mathcal{T}, C)$, as per Fig. 2: (a) for $N_g = 53$; (b) for $N_g = 83$. In each case, the error is defined relative to a slightly larger calculation, as $f(C)(\gamma_{N_g+10} - \gamma_{N_g})$. Numerical errors are substantially smaller for the larger grid, but in both cases the largest probability-weighted errors are in the interior.

of the latter leads to substantial improvements in accuracy in the interior region. This is where the largest errors are to be found, in the probabilistic sense, although they are still quite small.

6. Probability conservation and Lorentz-boosted inertial frames

In Secs. 3 and 5, we considered only free-particle Gaussian wavepackets which, at the initial time $\mathcal{T} = 0$, comprise a stationary coherent state (i.e., $\partial x/\partial \mathcal{T}|_{\mathcal{T}=0} = 0$) that is centered at the origin [i.e., $x_0(C = 0) = 0$]. More generally, we wish to consider *non*stationary wavepackets, for which the wavepacket center moves along an arbitrary (time-like) straight line. The most general possible relativistic free-particle Gaussian wavepacket solutions may be obtained from the stationary solutions through an arbitrary combination of: (a) translation in t; (b) translation in x; (c) Lorentz boost. Whereas (a) and (b) are trivial and will not be considered further, (c) serves as an explicit check on the Lorentz-invariance of the dynamical equations (despite the different PDE orders in $\mathcal{T}$ and C), as well as the inertial coordinate probability conservation laws, as discussed in Sec. 2.

Let S denote the inertial frame of reference (ct, x), in which the free-particle Gaussian wavepacket adopts the "stationary" form, in the sense described above. Let $S'_{[v]}$ denote a new inertial frame, (ct', x'), obtained from S via application of the Lorentz boost corresponding to the positive velocity $v > 0$. In this manner, we obtain a new set of solutions, $\{x'^{\alpha}_{[v]}(X^\mu)\}$, each of which describes a quantum particle moving toward the negative x'-direction with ensemble-averaged velocity $-v$. For each boosted trajectory ensemble solution $x'^{\alpha}_{[v]}(X^\mu)$, the corresponding flux "four"-vector $j'^{\alpha}(x')$ can be easily obtained in the boosted frame $S'_{[v]}$.

It is of interest to verify the conservation of probability [Eq. (7)] and covariant continuity relation [Eq. (6)] of Sec. 2 numerically, in both stationary and boosted inertial frames. We do so for the stationary frame as follows. The stationary inertial flux four-vector is expressed as [21],

$$j^\alpha(x) = \frac{f(C)}{\sqrt{\gamma}} \frac{dx^\alpha}{d\tau} = \frac{e^{-aC^2}}{\sqrt{\gamma}} \frac{dx^\alpha}{d\tau}. \tag{25}$$

We obtain the probability density and flux, respectively, as follows:

$$j^0(ct, x) = c \frac{f(C)}{\sqrt{\gamma}} \frac{dt}{d\tau} \quad ; \quad j^1(ct, x) = \frac{f(C)}{\sqrt{\gamma}} \frac{dx}{d\tau}, \tag{26}$$

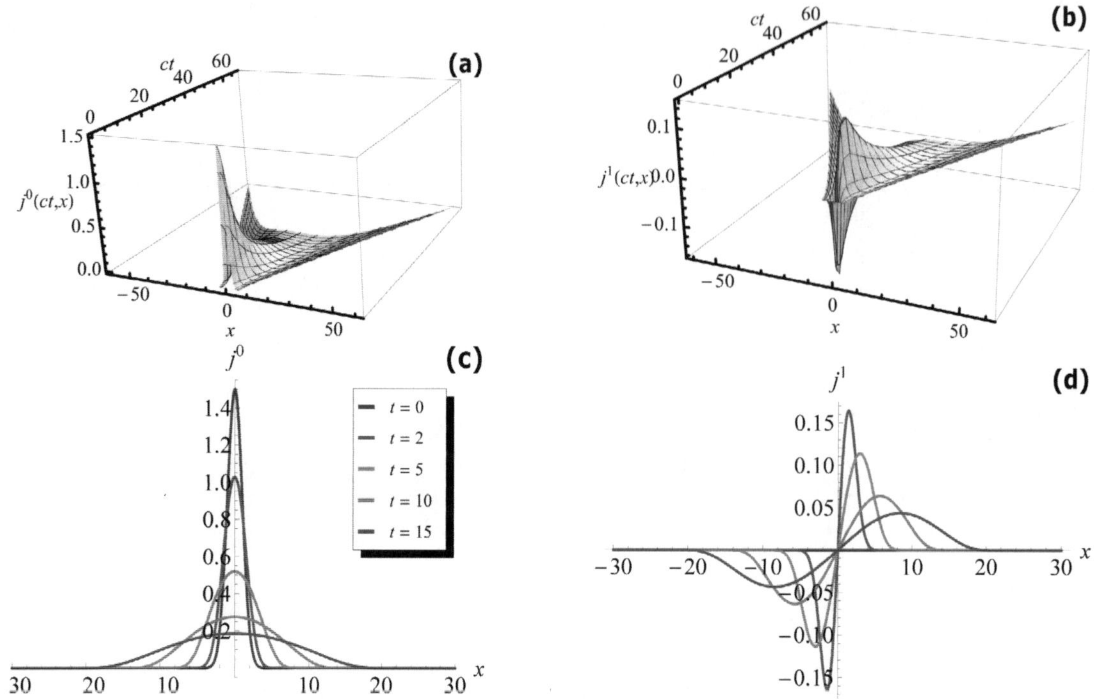

Figure 6. Flux four-vector for the relativistic free-particle Gaussian wavepacket for $c = 1.5$: (a) three-dimensional plot of $j^0(ct,x)$; (b) three-dimensional plot of $j^1(ct,x)$; (c) time slice plots of $j^0(ct,x)$; (d) time slice plots of $j^1(ct,x)$. The time slices are for $t = \{0, 2, 5, 10, 15\}$.

where

$$\frac{dt}{d\tau} = \frac{1}{\sqrt{1 - \frac{1}{c^2}\left(\frac{dx}{dt}\right)^2}} \quad ; \quad \frac{dx}{d\tau} = \frac{\frac{dx}{dt}}{\sqrt{1 - \frac{1}{c^2}\left(\frac{dx}{dt}\right)^2}}. \tag{27}$$

To obtain $j^\alpha(ct,x)$ numerically, we start with the numerical quantum trajectory ensemble solution, $x^\alpha(X^\mu) = \big(ct(\mathcal{T},C), x(\mathcal{T},C)\big)$. From the latter we obtain dx/dt everywhere via

$$\frac{dx}{dt} = \left.\frac{\partial x}{\partial \mathcal{T}}\right|_C \left(\left.\frac{\partial t}{\partial \mathcal{T}}\right|_C\right)^{-1}. \tag{28}$$

Substituting Eq. (28) into Eq. (27), and again into Eq. (26), we obtain numerical results for $j^0(ct,x)$ and $j^1(ct,x)$. These are presented in Fig. 6, for the $c = 1.5$ example.

In Fig. 6(a) and (c), we observe that $j^0\,(ct,x) > 0$ for all (ct,x)—i.e., the probability density remains positive always. We also observe substantial broadening of the wavepacket over time. However, the initial Gaussian shape is also substantially distorted over time, with the fringe trajectories "bunching up" as they approach the speed of light (Sec. 4). As for the flux, $j^1\,(ct,x)$, it is always odd in x, being positive for $x > 0$ and negative for $x < 0$. This also reflects wavepacket broadening—i.e., trajectories to the right of $x = 0$ spread to the right, whereas those to the left spread to the left. Although the trajectory speeds increase monotonically with increasing $|C|$, the flux reaches a maximum and then decays, owing to the probability weighting. Finally, according to the $(1+1)d$ Eq. (7), the areas under each of the $j^0(x)$ curves in Fig. 6(c) should be constant, if probability is conserved. Numerical integration for each of the sixteen t values, $t = \{0, 1, 2, ..., 15\}$, results in a near-constant value of 3.7615, with an RMS error of only 0.0011.

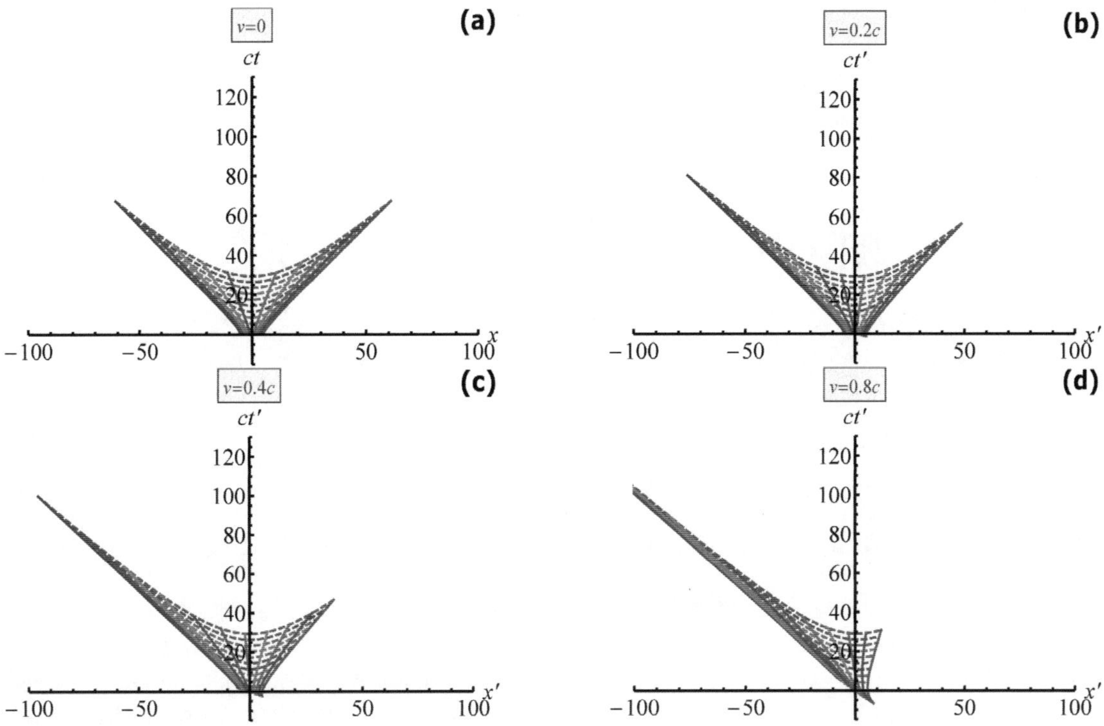

Figure 7. Quantum trajectories (solid curves) and simultaneity submanifolds (dashed curves) for the relativistic free-particle Gaussian wavepacket for $c = 1.5$, in various stationary and Lorentz-boosted frames: (a) $v = 0.0c$; (b) $v = 0.2c$, (c) $v = 0.4c$; (d) $v = 0.8c$.

In the Lorentz-boosted frame $S'_{[v]}$, we denote the flux four-vector as $j'^{\alpha}(ct', x')$. Being a true Lorentz-invariant vector, this quantity must transform as does x^{α}—i.e., as

$$x'^{\alpha} = \Lambda^{\alpha}{}_{\delta}\, x^{\delta} \qquad ; \qquad j'^{\alpha}(ct', x') = \Lambda^{\alpha}{}_{\delta}\, j^{\delta}(ct, x), \tag{29}$$

where the rank-2 boost tensor $\Lambda^{\alpha}{}_{\delta}$ can be expressed in terms of the parameter $\beta = v/c$ as

$$\Lambda^{\alpha}{}_{\delta} = \begin{pmatrix} \frac{1}{\sqrt{1-\beta^2}} & \frac{-\beta}{\sqrt{1-\beta^2}} \\ \frac{-\beta}{\sqrt{1-\beta^2}} & \frac{1}{\sqrt{1-\beta^2}} \end{pmatrix}, \tag{30}$$

in the usual manner.

In the Lorentz-boosted frame $S'_{[v]}$, a *different* probability conservation law than Eq. (7) must hold, namely,

$$\int j'^{0}(ct', x')\, dx' = \text{const.} \tag{31}$$

Once again, we will verify numerically that the Eq. (31) probability conservation law is indeed satisfied. First, however, we obtain the trajectory ensemble solution, $x'^{\alpha}(X^{\mu})$, by Lorentz-boosting $x^{\alpha}(X^{\mu})$ (dropping the '$[v]$' subscript). Then, we apply a second Lorentz boost to the dependent variables $j^{\alpha} = (j^0, j^1)$, to obtain the desired functions $j'^0(ct', x')$ and $j'^1(ct', x')$.

In Fig. 7, we depict the quantum trajectories and simultaneity submanifolds in the stationary frame S, and in several boosted frames $S'_{[v]}$ for $v = 0.2c$, $v = 0.4c$ and $v = 0.8c$. The boosted

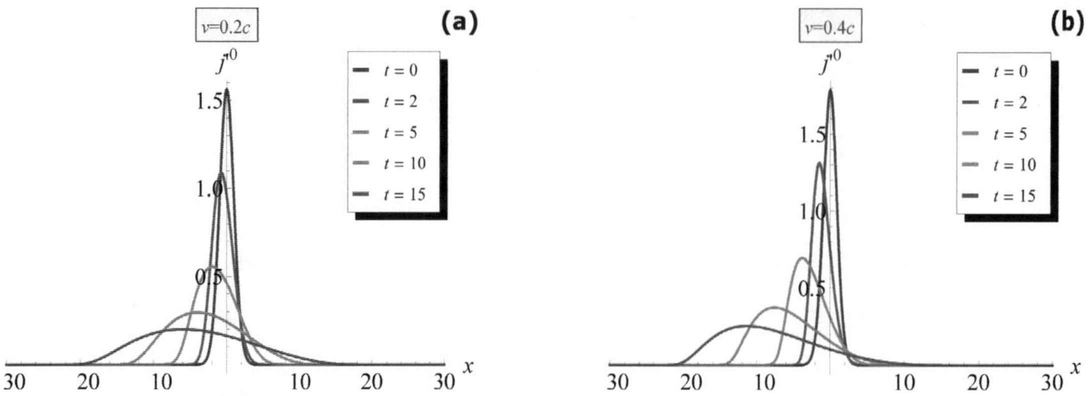

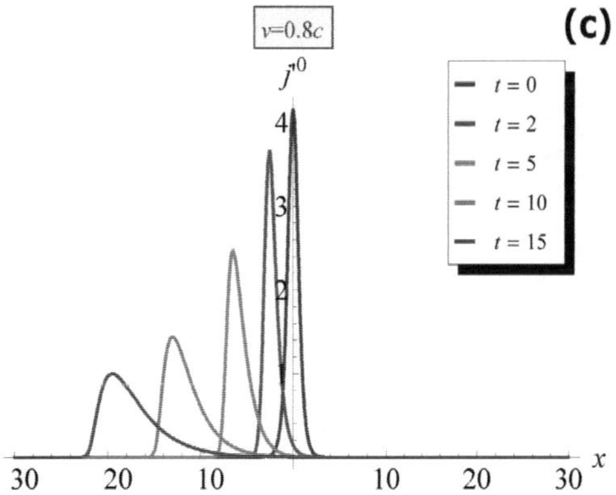

Figure 8. Time slice plots of the probability density $j'^0(ct', x')$, for the relativistic free-particle Gaussian wavepacket for $c = 1.5$, in various Lorentz-boosted frames: (a) $v = 0.2c$, (b) $v = 0.4c$; (c) $v = 0.8c$. The time slices are for $t' = \{0, 2, 5, 10, 15\}$.

quantum trajectories do indeed move generally to the left, increasingly so for increasing v. The trajectory ensemble is no longer symmetric in x', but it is symmetric under $(ct', x') \to (-ct', -x')$, as would be expected from the nonrelativistic case. Figure 8 depicts $j'^0(ct', x')$, plotted as a function of x' for various time slices t', for $v = 0.2c$, $v = 0.4c$ and $v = 0.8c$. In all cases, the motion of the wavepacket to the left over time is evident, as is wavepacket broadening. We also observe a "bunching" distortion of the Gaussian form in the wavepacket fringes, similar to that evident in Fig. 6(c) for the stationary case. Here, however, for all $t \neq 0$, the wavepacket is not even symmetric about its center—unlike both the travelling nonrelativistic Gaussian wavepacket, and the stationary relativistic example discussed previously. The reason is clear: due to the velocity bias in the ensemble, trajectories to the left of the wavepacket center experience much more pronounced bunching than those to the right—which may not even bunch at all, if the region of significant probability is always moving to the left over the time scale of interest.

Finally, we consider conservation of probability. First, note that according to Figure 8, $j'^0\,(ct', x') \geq 0$ for all (ct, x)—thus justifying our interpretation of this quantity as a probability density. Also, once again, for Eq. (31) to be satisfied, the areas under each of the curves in Fig. 8 must be the same, for a given v value. Via numerical integration, this is indeed found to be the case. Specifically: for $v = 0.2c$, the integration for each curve yields 3.8369, with an RMS error of 0.0008; for $v = 0.4c$, the integration for each curve yields 4.1029, with an RMS error of 0.0009; for $v = 0.8c$, the integration for each curve yields 6.2704, with an RMS error of 0.0097. Thus, we conclude that conservation of probability as per Eq. (31) holds in all Lorentz-boosted frames, $S'_{[v]}$, and therefore for all relativistic free-particle Gaussian wavepackets.

Journal of Physics: Conference Series **701** (2016) 012013 doi:10.1088/1742-6596/701/1/012013

7. Summary and Conclusions

In this paper, we analyze the dynamical equations for a single, spin-zero relativistic quantum particle, based on a recently-derived trajectory-based formulation [21]. The PDE [Eq. (11)] governing the trajectory ensemble solution $x^\alpha(X^\mu)$ is second order in *natural* time $\mathcal{T}$, and fourth-order in *natural* space $\mathbf{C}$, yet treats *inertial* t and x on an equal, Lorentz-invariant footing. As a result, the main issues plaguing the traditional wave-based equations of Klein–Gordon and Dirac (when viewed as single-particle theories) are resolved here.

This is especially true for wavepacket dynamics, which serves as the focus of this contribution. Note that even though the traditional wavefunction plays no role in the trajectory formulation, one can nevertheless extract a wavepacket description from the latter—in terms of the inertial probability density quantity, j^0, which in turn can be obtained from the $x^\alpha(X^\mu)$ solution itself, together with the natural spatial scalar probability density $f(\boldsymbol{C})$. Unlike wave-based relativistic quantum treatments, the wavepacket dynamics observed here for a free particle are always simple, straightforward, and well-behaved. In particular, j_0 is everywhere positive, localized, and interference-free, with an ensemble-averaged position that travels along a linear path through spacetime. Zitterbewegung and other complications—stemming ultimately from interference between positive- and negative-energy wave solutions—never arise here, because there are no negative-energy trajectory solutions.

We have also examined the symmetry properties of the $(3+1)d$ and $(1+1)d$ dynamical equations, and of their numerically computed Gaussian wavepacket solutions—with respect to parameter and dynamical variable rescaling, as well as Lorentz boosts. The invariance with respect to these symmetry operations has been verified both analytically and numerically. In particular, from scale invariance, we learn that narrowing the initial Gaussian width (by increasing a) is essentially equivalent to reducing the speed of light c (keeping both m and $\hbar$ fixed). This makes good physical sense; even in the nonrelativistic context, narrower initial coherent wavepackets are known to disperse more rapidly than broader ones—corresponding in the present context to significant-probability trajectories that approach the speed of light more rapidly. In the extreme ultrarelativistic limit $c \to 0$, one accordingly obtains a Dirac delta initial condition, whose corresponding singular solution has been obtained analytically [21], and may be relevant for cosmology. In the opposite scaling limit, one approaches the standard nonrelativistic Gaussian wavepacket solution—except for the $|C| \to \infty$ "fringe" trajectories that carry almost no probability.

In this work, only the stationary $x^\alpha(X^\mu)$ solutions were obtained explicitly numerically—although the term "stationary" is a bit of a misnomer, since quantum forces lead to accelerating trajectories that give rise to wavepacket spreading over time. To obtain completely general, moving Gaussian wavepacket solutions, Lorentz-boosting is applied to the stationary solutions. In the wave-based approach, Lorentz boosts in and of themselves can introduce undesirable interference effects [24]. This is not the case for the present trajectory-based approach—e.g., the transformed j'^0 is just as well behaved as j^0, and x^α and j^α exhibit perfect Lorentz invariance.

Numerical solution of the Eq. (17) PDE is complicated by the fact that the boundary conditions are not known *a priori*. The simple remedy proposed here, involving a nonuniform C grid, enables higher grid densities and accuracies to be obtained for the physically relevant trajectories that bear significant probability. On the other hand, experience with the nonrelativistic case strongly suggests that this approach will not be so effective in the presence of external fields. Thus, more robust strategies will have to be developed, going forward.

The present work and Ref. [21] serve to demonstrate that the trajectory-based formulation is a very promising approach in the fixed-particle-number relativistic quantum context, worthy of further development. One very convenient feature of this approach is that it derives from a trajectory-based Lagrangian. Thus, both the nonrelativistic and classical limits emerge straightforwardly. Moreover, the Euler–Lagrange and Noether procedures can be directly applied, to obtain the dynamical PDE and conservation laws, respectively. All of these aspects will be examined in future publications. Another, quite obvious generalization will also be considered: just as the present approach constitutes the trajectory-based analog of the Klein–Gordon equation, by incorporating spin into the theory, we hope to derive the trajectory version of the Dirac equation. Finally, one is not restricted to a trajectory-based approach. Recent work [21] suggests that a mathematically equivalent wave equation can also be developed, which may offer certain advantages, although it would necessarily have to be nonlinear.

Acknowledgments
This work was supported in part by a grant from the Robert A. Welch Foundation (D-1523). The authors also acknowledge useful discussions with Jeremy Schiff.

References
[1] von Neumann J 1932 *Mathematical Foundations of Quantum Mechanics* (New Jersey: Princeton University Press)
[2] Cohen-Tannoudji C, Diu B and Laloë F 1977 *Quantum Mechanics* (New York: Wiley)
[3] Bohm D 1979 *Quantum Theory* (New York: Dover)
[4] Styer D 2002 Nine formulations of quantum mechanics *Am. J. Phys.* **70** 288
[5] Bohm D 1952 A suggested interpretation of the quantum theory in terms of "hidden" variables. I *Phys. Rev.* **85**(2) 166–179
[6] Holland P R 1993 *The Quantum Theory of Motion: An Account of the De Broglie–Bohm Causal Interpretation of Quantum Mechanics* (Cambridge, England: Cambridge University Press)
[7] Einstein A, Podolsky B and Rosen N 1935 Can quantum-mechanical description of physical reality be considered complete? *Phys. Rev.* **47**(10) 777–780
[8] Ballentine L E 1970 The statistical interpretation of quantum mechanics *Rev. Mod. Phys.* **42** 358–381
[9] Home D and Whittaker M A B 1992 Ensemble interpretations of quantum mechanics. a modern perspective *Phys. Rep.* **210** 223–317
[10] Everett III H 1957 "Relative state" formulation of quantum mechanics *Rev. Mod. Phys.* **29** 454
[11] Wheeler J A and Zurek W H (eds) 1983 *Quantum Theory of Measurement* (Princeton: Princeton University Press)
[12] Bouda A 2003 From a mechanical lagrangian to the Schrödinger equation: A modified version of the quantum Newton law *Int. J. Mod. Phys.* A **18** 3347
[13] Holland P 2005 Computing the wavefunction from trajectories: particle and wave pictures in quantum mechanics and their relation *Ann. Phys.* **315** 505
[14] Poirier B 2010 Bohmian mechanics without pilot waves *Chem. Phys.* **370** 4–14
[15] Holland P 2010 Quantum field dynamics from trajectories, ed Chattaraj P *Quantum Trajectories* (Boca Raton: Taylor and Francis/CRC Press) pp 73–85; Chap. 5
[16] Poirier B 2011 Trajectory-based derivation of classical and quantum mechanics, ed Hughes K H and Parlant G *Quantum Trajectories* (Daresbury Laboratory: CCP6) p 6
[17] Schiff J and Poirier B 2012 Quantum mechanics without wavefunctions *J. Chem. Phys.* **136** 031102
[18] Parlant G, Ou Y C, Park K and Poirier B 2012 Classical-like trajectory simulations for accurate computation of quantum reactive scattering probabilities *Comput. Theoret. Chem.* **990** 3
[19] Hall M J W, Deckert D A and Wiseman H M 2014 Quantum phenomena modeled by interactions between many classical worlds *Phys. Rev.* X **4**(4) 041013 (*Preprint* `arXiv:1402.6144 [quant-ph]`)
[20] Poirier B 2014 Does research on foundations of quantum mechanics fit into PRX's scope? *Phys. Rev.* X **4** 040002
[21] Poirier B 2012 Trajectory-based theory of relativistic quantum particles (*Preprint* `arXiv:arXiv:1208.6260 [quant-ph]`)
[22] Ryder L H 1996 *Quantum Field Theory* 2nd ed (Cambridge, England: Cambridge University Press)
[23] Wachter A 2011 *Relativistic Quantum Mechanics* (Dordrecht: Springer)
[24] Thaller B 2005 *Advanced Visual Quantum Mechanics* (New York: Springer)
[25] Park S T 2012 Propagation of a relativistic electron wave packet in the Dirac equation *Phys. Rev.* A **86** 062105
[26] Schweber S S 1961 *An Introduction to Relativistic Quantum Field Theory* (New York: Harper & Row)
[27] Gross F 1999 *Relativistic Quantum Mechanics and Field Theory* (New York: Wiley)
[28] Wyatt R E 2005 *Quantum Dynamics with Trajectories: Introduction to Quantum Hydrodynamics* (New York: Springer)

How quantization of gravity leads to a discrete space-time

Gerard 't Hooft

Institute for Theoretical Physics, Utrecht University
and
Spinoza Institute, Postbox 80.195, 3508 TD Utrecht, the Netherlands

Abstract. The idea that the Planck length is the smallest unit of length, and the Planck time the smallest unit of time, is natural, and has been suggested many times. One can, however, also derive this more rigorously, using nothing more than the fact that black holes emit particles, according to Hawking's theory, and that these particles interact gravitationally. It is then observed that the particles, going in and out, form quantum states bouncing against the horizon. The dynamics of these microstates can be described in a partial wave expansion, but Hawking's expression for the entropy then requires a cut-off in the transverse momentum, in the form of a Brillouin zone, and this implies that these particles live on a lattice.

1. Introduction

During the second half of the 20th century, a quite detailed understanding was acquired of how to reconcile Einstein's theory of special relativity with quantum mechanics. Starting with various types of fields that obey Lorentz covariant field equations, one identifies the Poisson brackets and the Hamiltonian for such field systems, after which one can apply standard procedures to turn a classical system into a quantum system, where the Poisson brackets are replaced by commutators. The energy quanta of these fields then represent the elementary particles, which automatically obey Lorentz invariance if the fields do. Since many of the equations of this theory contain the speed of light c and Planck's constant $\hbar$, units are usually chosen in such a way that $c = \hbar = 1$, so that only one dimensionful unit remains: a unit of energy (eV), which now corresponds to a frequency or an inverse length scale.

This theory, which successfully unites *special* relativity with quantum mechanics, is called quantum field theory (QFT). *Second quantization* is an essential ingredient of the theory. Here, we shall argue that this will no longer be the case when *general* relativity is used as a starting point.

One first notices that there is one more dimensionful constant, Newton's constant for gravity, G, so that, when that is normalized to one, natural units for mass, length and time emerge. Since

$$\begin{aligned} c &= 2.99792458 \times 10^{-8}\ \text{m/sec}\ , \\ h/2\pi = \hbar &= 1.0546 \times 10^{-34}\ \text{kg m}^2\ \text{sec}^{-1}\ , \end{aligned}$$

Journal of Physics: Conference Series **701** (2016) 012014 doi:10.1088/1742-6596/701/1/012014

$$G = 6.674 \times 10^{-11} \text{ m}^3 \text{ kg}^{-1} \text{ sec}^{-2} , \tag{1.1}$$

these natural units are

$$\begin{aligned} M_{\text{Planck}} &= \sqrt{\frac{\hbar c}{G}} &= 21.76 \ \mu\text{g} , \\ E_{\text{Planck}} &= M_{\text{Planck}} c^2 &= 1.221 \times 10^{28} \text{ eV} , \\ L_{\text{Planck}} &= \sqrt{\frac{\hbar G}{c^3}} &= 1.616 \times 10^{-33} \text{ cm} , \\ T_{\text{Planck}} &= L_{\text{Planck}}/c &= 5.391 \times 10^{-44} \text{ sec} . \end{aligned} \tag{1.2}$$

However, a properly unified theory combining quantum mechanics with general relativity is notoriously difficult to formulate. Although superstring theory [1] is often acclaimed to be "the prime candidate" for being such a theory, it appears to leave quite a few gaps and mysteries,. It is widely suspected that more theoretical understanding will be needed. Ideally, a future all-embracing theory should be simple and straightforward, but we are still very much in the dark as for the fundamental axioms on which such a theory should be based.

Realising that the Planck length is some 19 orders of magnitude smaller than the length scale of the Standard Model of the sub-atomic particles, one may have reasons to doubt whether the ultimate theory should be based on the same principles at all. For one, we have quantum mechanics. Will the rather mysterious logic on which Standard Model quantum mechanics is based, also apply to this completely different domain of physics?

2. Hidden variables

The idea that quantum mechanics might find its roots in a classical theory – the "hidden variable theory", is old and well-known. It is also well-known that there are very powerful arguments against such an idea. It appears to be impossible to account for the fact that quantum mechanics violates the Bell inequalities [2], unless one makes two assumptions: the assumption of *superdeterminism* and that of *conspiracy*.

Superdeterminism [3] is here taken to mean that observers cannot decide out of 'free will' which property of a system they should measure. If there is a choice between two observable quantities that do no commute with one another, the decision which of the two (or more) should be determined, must be rooted in the distant past. By itself, such an assumption is obviously not crazy; of course, in a deterministic theory, everything we do is rooted in phenomena that happened in the distant past. A way to phrase this in physical terms is that there could be delicate *correlations* between the settings chosen by distantly separated observers, and the polarisations of the quantum-entangled particles that they might be observing. Several researchers pointed this out as being a 'loophole' around Bell's no-go theorem [4]. However, this loophole is often dismissed as being physically untenable.[5]

This is because there is a more disturbing aspect in this loophole, which is that we should also assume some form of 'conspiracy', which is that, somehow, the entangled particles that the two observers, Alice and Bob, decide to study, behave in a way that depends, in advance, on how Alice and Bob will choose their settings. The photons are correlated to these settings. How can such a thing ever happen? Any theory based on classical, that is, non-quantum mechanical probability distributions, must answer this question. The issue is often rejected rather emotionally: *Alice and Bob have free will, why should they choose settings that are correlated with the particles yet to be observed?*

The author's position in this dispute, which is admitted to be a minority's view, is extensively reviewed in Ref. [6]. Here, we give a very brief summary of the 'Cellular Automaton Interpretation of Quantum Mechanics' (CAI). First, the conspiracy objection should be rephrased in such a way that the emotional overtones are replaced by a purely mathematical definition of free will. Secondly, a physical mechanism is suggested that can accommodate for the kind of correlations needed. Correlations occur everywhere in physics, so what we need is a credible scenario explaining where the 'conspiracy' can come from. The basic idea worked out in [6], is that there exists a simple, physical conservation law that indeed applies to the ways Alice and Bob decide to choose their settings, simply because this conservation law affects the entire world, even if, under normal circumstances, it is hardly noticeable.

It was found, in addition, that the assumption of a classical hidden variable scenario can not only explain the quantum mechanical nature of today's experiences with the physical laws, it also takes away several of the mysteries quantum mechanics appears to confront us with:

- The CAI theory provides a very natural description and explanation of what is called the "collapse of the wave function", whenever a measurement or observation is done. This collapse does *not* require any modification in Schrödinger's equation.
- It gives a natural explanation of Max Born's probability interpretation of the absolute squares of the wave function coefficients, $|\langle\vec{x}|\psi\rangle|^2$, where $\vec{x}$ is a quantity, or set of quantities, that is observed or measured.
- The theory ensures the exact validity of quantum mechanical mathematical calculations over the full range between sub-microscopic and classical scales.

The conservation law associated with superdeterminism emerges when we make our distinction between quantum states that are *ontological*, or "ontic", and states that we use as "templates". Ontic states are states that describe things "really going on", called 'beables'. Whenever we consider superpositions of different ontic states, we have no ontic state anymore but a template. Today's quantum physics is all about templates. The novelty that is absolutely essential for our understanding is that ontic states continue to evolve into ontic states. So, the 'ontology' of a wave function is absolutely conserved in time. Alice and Bob can never replace the state of their detectors by templates; this would violate the ontology conservation law.

Having made all these statements, we admit that the only proof that these ideas are correct could come from models that reproduce non-trivial quantum systems from classical underlying physics, but this turns out to be extremely difficult. What needs to be understood better is not only how nature could behave as an information processing machine, but also how these principles can be implemented.

An important observation is that to reproduce today's quantum mechanics in a deterministic theory, one has to reproduce quantum field theory. QFT is based on second quantization. Second quantization works very well in practice, but only when we perform perturbation expansions for small couplings. The apparent contradictions between quantum mechanics and locality may be due to the fact that these expansions are formally non convergent, even though they work very well in practice [6].

3. Lorentz invariance

There are some very tough obstacles. One is that there is strong evidence that the number of distinct ontic states building up the fabric of space, time and matter, within a given volume, is finite, and involves the Planck scale. In modelling this, one has to realise that we do wish to maintain local Lorentz invariance: whenever there is a domain with a vacuum in it – to some

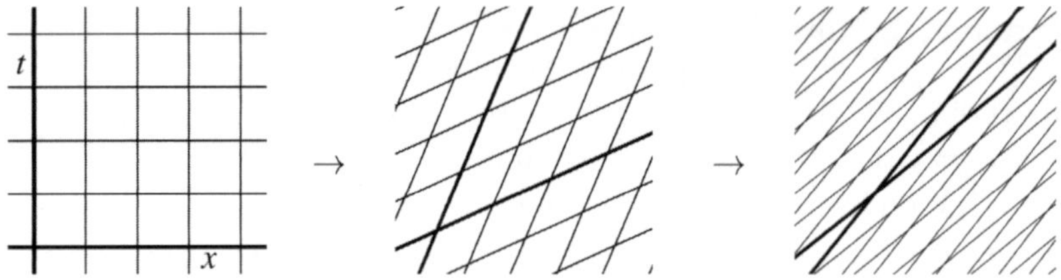

Figure 1. Lorentz transforming a space-time structure. The number of data in a space-time volume remains the same, but the concept of "neighbours" is affected.

reasonable approximation[1] – the Lorentz group should relate particles entering that domain, to particles entering at a different velocity, by means of local Lorentz transformations, see Fig. 1.

Many schemes of space-time lattice structures have been considered [7][8][9], but this demand, the existence of a local, Lorentz invariant vacuum, did not often receive the attention that it would deserve. There is a proposal that is formally correct: F. Dowker e.a. [10] observed that the only distribution of points in space-time that is Lorentz invariant is the *random* distribution (obtained by some random sprinkling procedure). One could add that these points are still allowed to have non-trivial correlations in their distributions, as long as these correlations are Lorentz invariant.

Consider some sprinkling of points $x_1, \ldots, x_n$. Define the space-time distribution function $\phi(x)$, a probability variable ϱ , and a 2 point distribution $F(x-y)$ as

$$\phi(x) = \sum_n \delta^4(x-x_n) \ , \qquad \langle\phi(x)\rangle = \varrho > 0 \ ,$$
$$\langle\phi(x)\,\phi(y)\rangle \ = \ F(x-y) \ = \ F((x-y)^2) \ > \ 0 \ ; \tag{3.1}$$

If $F(x-y) - \varrho^2 \neq 0$ then we have non-trivial correlations.

A proposal not yet fully explored is that the physical features in our world could all be reduced to aspects of these correlation functions. In any case, it will be very difficult to produce and understand the laws of physics that reproduce a quantum field theory resembling that of the Standard Model along such lines.

As stated before, discretized theories for quantum gravity should incorporate the condition mentioned above that the dominant distribution of lattice points should reproduce the metric of the vacuum state, allowing for local Lorentz transformations.

The easiest thing to introduce will then be the dilaton field, for which the density field $\varrho(x)$ in Eq. (3.1) is the obvious candidate:

$$g_{\mu\nu}(x) \equiv g(x)^{1/4}\,\hat g_{\mu\nu}(x) \ , \qquad g = \det(g_{\mu\nu}) = \varrho^8(x) \ . \tag{3.2}$$

An interesting feature of such a scheme would be that, by construction, there is no structure at all at distance scales tinier than one unit of volume in terms of the metric $g_{\mu\nu}$.

4. Conformal gravity

There are various reasons to suspect that, at distance scales comparable to the Planck scale, conformal invariance may be an important symmetry. Usually, conformal symmetry is thought of

[1] Remember: the exact vacuum state is not locally, but only globally defined.

Journal of Physics: Conference Series **701** (2016) 012014 doi:10.1088/1742-6596/701/1/012014

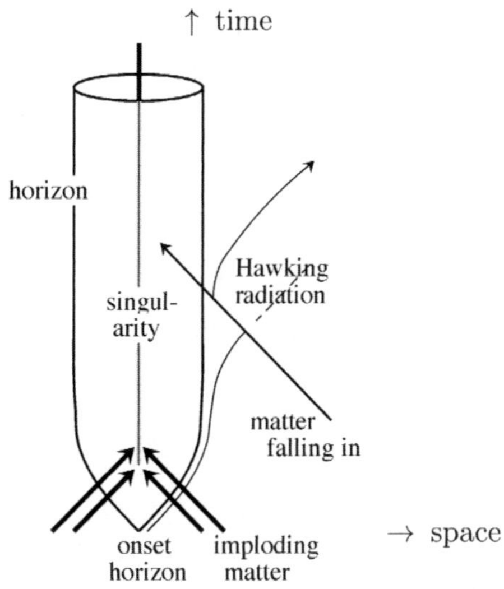

Figure 2. space-time near a black hole. matter falling in causes the pattern of Hawking particles to shift over a distance that varies over the angular coordinates θ and φ.

as an approximate, accidental symmetry that could be used as a simplifying approximation, while it is broken by various types of anomalies, particularly when we try to quantize a theory [11]. However, the dilaton could also be a much more elementary ingredient of a theory, if local conformal symmetry is merely *spontaneously* broken [12][13]. One might argue that the dilaton field, which could be introduced as a common factor for all components of the metric field $g_{\mu\nu}(x)$, settles for a non-vanishing expectation value in the vacuum. In this case, it could be the *only* field that defines the absolute length and time scales for the vacuum. By introducing such principles in a theory, it may be a lot easier to understand the huge scale differences in nature, while the Planck scale might emerge as the final distance scale.

For the quantization of gravity, it is crucial to understand the role of (real as well as virtual) black holes. The region in the immediate vicinity of the horizon, is characterised by the fact that time translations are substituted by Lorentz transformations. This means that observables near the horizon are undergoing unlimited processes of Lorentz contraction.

Light rays are essential for defining the exact location of a black hole horizon: it is the boundary that separates regions from which escape to the outside universe by light rays is still possible, from the domain where all light rays are trapped. We shall now show how to use light rays to describe the back reaction of Hawking particles upon the presence of matter entering the black hole.

The *density matrix*, that is, the probability distribution, of Hawking particles basically only depends on mass, charge and angular momentum of the black hole, but the *actual* configuration of the out-going particles characterises the microstate of the black hole, of which we have a large number ($\approx e^{4\pi M^2}$ in natural units) of distinct elements[2]. Whenever a particle enters the hole from outside, transitions to different microstates take place. This happens because a particle entering a black hole interacts with what comes out. The most disruptive interaction is the gravitational one, in spite of its apparent weakness. We calculate this as follows.

[2] The number $4\pi M^2$, corresponding to the black hole entropy, also represents the total number of Hawking particles emitted by a black hole in its life time [14].

The gravity field surrounding a particle causes a very slight *Shapiro effect*: a passing light beam is slowed down and dragged towards the particle [15]. For a particle at rest, this effect is very small, but when the particle is Lorentz boosted, its gravity field increases in strength. At the black hole horizon, this Lorentz boost enhances its energy exponentially in time.

Let $p^\mu_{\rm in}(\Omega)$ be the momentum distribution of the in-going particles at the spot $\Omega=(\theta,\varphi)$ on the horizon, and $\delta x^\mu_{\rm out}(\Omega)$ the Shapiro displacement for the Hawking particles going out. An elementary calculation shows that

$$\delta x^\mu_{\rm out}(\Omega)=8\pi G\int {\rm d}^2\Omega'\,f(\Omega,\Omega')\,p^\mu_{\rm in}(\Omega')\ , \tag{4.1}$$

where $f(\Omega,\Omega')$ is a Green function, defined by $\Delta_S f(\Omega,\Omega')=-\delta^2(\Omega,\Omega')$, and here, the angular operator Δ_S is defined by

$$\Delta_S=\Delta_\Omega-1=-\ell(\ell+1)-1\ . \tag{4.2}$$

Often, we ignore the -1. It was further assumed that the momentum $p^\mu_{\rm in}$ is dominated by the radial component $p^-_{\rm in}$, so that, also, its derivatives w.r.t. θ and φ are small.

Thus, whatever the positions $x^\mu_{\rm out}(\Omega)$ of the Hawking particles in the original microstate were, they are now replaced by the displaced positions, and since the particles emerge with a trajectory that separates itself from the horizon exponentially in Schwarzschild time, the significance of this displacement also increases exponentially in time. This is how information regarding in-going particles is imprinted in the out-going ones. In Ref [17], a formal expression for the ensuing unitary evolution matrix is derived, to be summarised in the next section. One important feature is seen to emerge: the effect is purely geometrical. Only the momentum distribution $p^\mu_{\rm in}(\Omega)$ (mainly the minus component) enters. Consequently, unitarity of this evolution (scattering) matrix implies that these in-going particles are to be *entirely* characterised by their momenta.

We do use approximations, however. Other long-range forces may affect this result, in particular electro-magnetism. Thus, besides the momentum distribution, also the electric charge distribution (and eventually also other local gauge charges) are allowed for the characterisation of the in-going particles. In turn, also the out-going particles are characterised geometrically, by their positions $x^\mu_{\rm out}(\Omega)$, and their electro-magnetic charge distributions.

5. Formal expression of the black hole scattering matrix

The following sections of this contribution had been sent to the Archive earlier [16]. The question is now, how to deduce the black hole microstates from this evolution matrix. Let us simplify the situation a bit by replacing the Schwarzschild black hole metric by Rindler space. This implies that the angular coordinates $\Omega=(\theta,\varphi)$ are replaced by two transverse, flat coordinates $\tilde x=(x,y)$, and the angular Laplacian operator Δ_S by $\tilde\Delta=\tilde\partial^2$. The Green function $f(\Omega,\Omega')$ becomes

$$f(\tilde x,\tilde x')=-\tfrac{1}{2\pi}\log(|\tilde x-\tilde x'|)\ . \tag{5.1}$$

Reserving the symbol z^- for the radial component of the positions, we write the displacement (4.1) due to the Shapiro shift as

$$\delta z^-_{\rm out}(\tilde x)=-4G\,\delta p^-(\tilde x')\,\log(|\tilde x-\tilde x'|/C)\ . \tag{5.2}$$

The constant C will soon drop out. Assume now that a black hole produced by one given initial state $|\,{\rm in}_0\rangle$, upon its final explosion leads to a given final state $|{\rm out}_0\rangle$. We then calculate the final state when a *slight modification*[3] is brought about to the state $|\,\mathrm{in}\rangle$. Let the modification consist of adding one light particle with momentum δp^- entering the Rindler horizon at the transverse position $\tilde x$. All particles at the transverse position $\tilde x\,'$ in the final state $|\mathrm{out}\rangle$ are then dragged along such that their out-coordinate z^- is modified by an amount given by Eq. (5.2).

We write this modification as a property of the black hole scattering matrix:

$$S\,|\,\mathrm{in}_0\rangle=|\mathrm{out}_0\rangle\quad\rightarrow\quad S|\,\mathrm{in}_0+\delta p^-(\tilde x)\rangle=e^{-i\int\mathrm{d}^2\tilde x\,p^+_{\mathrm{out}}(\tilde x\,')\delta z^-}|\mathrm{out}_0\rangle\ , \tag{5.3}$$

where we used the displacement operator $e^{-ip^+_{\mathrm{out}}(\tilde x\,')}$ to describe a displacement of the out-going particles at the transverse position $\tilde x\,'$.

The modification (5.3) can be repeated as many times as we wish, either by adding, or by removing amounts of momenta $\delta p^-(\tilde x)$ at positions $\tilde x$, and this means that now we can reach any other initial state $|\,\mathrm{in}\rangle$, when described by the distribution of the total momentum going in, $p^-_{\mathrm{tot}}(\tilde x)$ (as compared to the original initial state $|\,\mathrm{in}_0\rangle$), to find the new final state $|\mathrm{out}\rangle$ as a displacement of the original finite state $|\mathrm{out}_0\rangle$:

$$\langle\mathrm{out}|S|\,\mathrm{in}\rangle=\langle\mathrm{out}_0|S|\,\mathrm{in}_0\rangle e^{4iG\int\mathrm{d}^2\tilde x\,'\log(|\tilde x\,'-\tilde x|/C)\,p^+_{\mathrm{out}}(\tilde x\,')\,p^-_{\mathrm{in}}(\tilde x)}\ . \tag{5.4}$$

Note that, here, the operators $p^-_{\mathrm{in}}(\tilde x)$ and $p^+_{\mathrm{out}}(\tilde x\,')$ both describe the *total* momenta of all in- and out going particles as distributions on the Rindler horizon. The important step to be taken now is to postulate that the entire Hilbert space of the in-particles is spanned by the function $p^-_{\mathrm{in}}(\tilde x)$, and the black hole scattering matrix maps that Hilbert space onto the space of all particles going out, spanned by the function $p^+_{\mathrm{out}}(\tilde x\,')$. We arrive at the unitary scattering matrix S:

$$\langle p^+_{\mathrm{out}}|S|p^-_{\mathrm{in}}\rangle=\mathcal N\,e^{4iG\int\mathrm{d}^2\tilde x\,'\log(|\tilde x\,'-\tilde x|/C)\,p^+_{\mathrm{out}}(\tilde x\,')\,p^-_{\mathrm{in}}(\tilde x)}\ . \tag{5.5}$$

This procedure leads to one common factor $\mathcal N$, which now can be fixed (apart from an insignificant overall phase) by demanding unitarity.

Since we only take their gravitational interactions into account in describing the back reaction, particles can *only* be distinguished by their mass distributions. If other interactions are added, such as electro-magnetism, we will also be able to differentiate particles further, for instance by their electric charges. Therefore, Eq. (5.5) must be regarded as an approximation. Concerns were expressed in Ref. [19] that some particles are also bent sideways so that these fall back into the horizon. In principle, this does not affect our analysis, since these particles do not occur in the final state at all, so that they are excluded in our picture. In any case this happens only at small transverse distance scales. We do expect that the transverse components of the gravitational force will invalidate our calculations at very high values of the transverse momenta.

6. Algebra

From Eq. (5.3) and the expression (5.5) for the scattering matrix, we can now deduce the relations and commutation rules between the operators $p^-_{\mathrm{in}}(\tilde x)$, $p^+_{\mathrm{out}}(\tilde x)$, $z^+_{\mathrm{in}}(\tilde x)$ and $z^-_{\mathrm{out}}(\tilde x)$, where the latter can be regarded as the coordinates of in- and out-going particles relative to the Rindler horizon:

$$z^-_{\mathrm{out}}(\tilde x)\ =\ -4G{\textstyle\int}\mathrm{d}^2\tilde x\,'\,p^-_{\mathrm{in}}(\tilde x\,')\,\log(|\tilde x-\tilde x\,'|/C)\ , \tag{6.1}$$

[3] I registered a complaint that this absolutely crucial step was not sufficiently emphasised in previous papers.

$$\begin{aligned} z^+_{\rm in}(\tilde x) &= 4G\textstyle\int {\rm d}^2\tilde x'\, p^+_{\rm out}(\tilde x')\,\log(|\tilde x-\tilde x'|/C)\ , & (6.2)\\ [z^-_{\rm out}(\tilde x),\,p^+_{\rm out}(\tilde x')] &= [z^+_{\rm in}(\tilde x),\,p^-_{\rm in}(\tilde x')]\ =\ i\delta^2(\tilde x-\tilde x')\ , & (6.3)\\ [p^-_{\rm in}(\tilde x),\,p^+_{\rm out}(\tilde x')] &= -\tfrac{i}{8\pi G}\,\tilde\partial^2\delta^2(\tilde x-\tilde x')\ , & (6.4)\\ [z^+_{\rm in}(\tilde x),\,z^-_{\rm out}(\tilde x')] &= -4iG\log|\tilde x-\tilde x'|\ . & (6.5)\end{aligned}$$

In Eq. (6.4), we used the fact that the logarithm obeys a Laplace equation,

$$\tilde\partial^2\log|\tilde x| = 2\pi\delta^2(\tilde x)\ . \qquad (6.6)$$

It would have been more accurate to write the momentum distributions $p^\pm(\tilde x)$ as energy-momentum tensors $T^{\pm,r}(\tilde x,0,0)$, but we wished to emphasise that these are momenta, when integrated over the horizon.

So-far, we just reproduced the results of Refs. [17] and [18]. The algebra (6.1)—(6.5) is quite different from the usual Fock space algebra for the elementary particles. In fact, it resembles a bit more the algebra of excited states of a closed string theory, but even that is not the same. It is therefore instructive to ask how one can decompose the new physical degrees of freedom into eigen modes.

7. Eigen modes

The operators $z^\pm(\tilde x)$ and $p^\pm(\tilde x)$ form a linear set. Let us therefore consider plane waves in transverse space:

$$p^\pm(\tilde x) = \tfrac{1}{2\pi}\int {\rm d}^2\tilde k\,\hat p^\pm(\tilde k)\,e^{i\tilde k\cdot\tilde x}\ ,\qquad z^\pm(\tilde x) = \tfrac{1}{2\pi}\int {\rm d}^2\tilde k\,\hat z^\pm(\tilde k)\,e^{i\tilde k\cdot\tilde x}\ , \qquad (7.1)$$

to find that the $\tilde k$-waves indeed decouple:

$$\tilde k^2\,\hat z^-_{\rm out}(\tilde k) = 8\pi G\,\hat p^-_{\rm in}(\tilde k)\ ,\qquad \tilde k^2\,\hat z^+_{\rm in}(\tilde k) = -8\pi G\,\hat p^+_{\rm out}(\tilde k)\ , \qquad (7.2)$$

$$[\hat z^-_{\rm out}(\tilde k),\,\hat p^+_{\rm out}(\tilde k')] = [\hat z^+_{\rm in}(\tilde k),\,\hat p^-_{\rm in}(\tilde k')] = i\delta^2(\tilde k-\tilde k')\ , \qquad (7.3)$$

etc. Eqs. (7.2) can be seen as boundary conditions on the horizon; the waves moving in are transformed into waves going out.

Note that the waves (7.1) on Rindler space should not be interpreted as particles; they each may consist of many particles, and we should only consider their real parts as physical objects. A single physical particle is typically represented by a Dirac delta distribution, see Eq. (5.2), where we have $p^-_{\rm in}(\tilde x') = \delta p^-\,\delta^2(\tilde x'-\tilde x)$. Thus, at this stage of our presentation, the momenta $p^\pm$ are to be interpreted quite differently from the transverse wave number variables $\tilde k$ on Rindler space. The waves $\hat z^\pm(\tilde k)$ should be split into sines and cosines, which each describe a wavy displacement operator on the horizon.

The above boundary equations are not very intuitive. This is because the time evolution of the free waves are in the form of scaling equations. In terms of the Rindler time coordinate τ we have: $p^-_{\rm in}(\tau) = e^\tau\,p^-_{\rm in}(0)$, $p^+_{\rm out}(\tau) = e^{-\tau}\,p^+_{\rm out}(0)$, etc. Also, $p^\pm$ and $z^\pm$ are operators, not amplitudes.

What happens becomes more transparent if we replace the Rindler space coordinates z and the momenta p by tortoise (Eddington-Finkelstein) coordinates ϱ and ω, Then, however, we must take into account that both the z and the p coordinates may be positive or negative. Therefore, we also introduce signs $\alpha=\pm 1$ and $\beta=\pm 1$:

$$z^+_{\rm in}\equiv\alpha\,e^{\varrho}\ ,\quad p^-_{\rm in}\equiv\beta\,e^{\omega}\ . \tag{7.4}$$

How are the operators β and ω related to α and ϱ? This should be easy. The transformations (7.4) imply for the wave functions,

$$\psi(\alpha\,e^{\varrho})=e^{-\frac12\varrho}\,\varphi(\alpha,\,\varrho)\ , \tag{7.5}$$

where ψ is the amplitude in the variable $z^+_{\rm in}(\tilde k)$ for some fixed $\tilde k$, while φ is a wave function in the new tortoise coordinate ϱ; and ψ is normalized in the z variable while φ is normalized in the ϱ variable, which required the factor $e^{-\varrho/2}$. Similarly, we write in momentum space,

$$\hat\psi(\beta\,e^{\omega})=e^{-\frac12\omega}\,\hat\varphi(\beta,\,\omega) \tag{7.6}$$

(Note, that the caret ($\hat{}$) is now used for Fourier transformation in z^+ space, as the caret in Eqs. (7.2) and (7.3) is no longer needed; from now on we look at just one value for the transverse wave number $\tilde k$).

Now, the relation between the wave function ψ and its Fourier transform $\hat\psi$ is

$$\begin{aligned}\varphi(\alpha,\,\varrho)&=\tfrac{1}{\sqrt{2\pi}}e^{\frac12\varrho}\int_{-\infty}^{\infty}{\rm d}p^-_{\rm in}e^{ip^-_{\rm in}z^+_{\rm in}}\,\hat\psi(p^-_{\rm in})\ =\\ &\tfrac{1}{\sqrt{2\pi}}\sum_{\beta=\pm}\int_{-\infty}^{\infty}e^{\frac12(\varrho\,+\,\omega)}{\rm d}\omega\,e^{\alpha\beta\,ie^{\varrho+\omega}}\,\hat\varphi(\beta,\omega)\ =\\ &\tfrac{1}{\sqrt{2\pi}}\sum_{\beta=\pm}\int_{-\infty}^{\infty}{\rm d}u\,A(\alpha\beta,\,u)\hat\varphi(\beta,\,u-\varrho)\ ;\quad A(\sigma,\,u)=e^{\frac12u\,+\,i\sigma e^u}\ .\end{aligned} \tag{7.7}$$

We see the emergence of a matrix $A(\alpha,\beta)=\left(\begin{smallmatrix}A(+)&A(-)\\A(-)&A(+)\end{smallmatrix}\right)$, which is easy to diagonalise; simply write for φ,

$$\varphi(+,\omega)+\varphi(-,\omega)=\varphi_1(\omega)\ ,\quad\varphi(+,\omega)-\varphi(-,\omega)=\varphi_2(\omega)\ , \tag{7.8}$$

and similarly for $\hat\varphi(\beta,\,\omega)$ and $A(\sigma,\,u)$, to find the diagonalised expressions

$$\varphi_i(\varrho)=\int_{-\infty}^{\infty}{\rm d}u\,A_i(u)\hat\varphi_i(u-\varrho)\qquad(i=1,2)\ , \tag{7.9}$$

$$A_1(u)=\sqrt{\tfrac{2}{\pi}}\,e^{\frac12u}\cos(e^u)\ ,\quad A_2(u)=i\sqrt{\tfrac{2}{\pi}}\,e^{\frac12u}\sin(e^u)\ . \tag{7.10}$$

Similarly,

$$\hat\varphi_i(\omega)=\int_{-\infty}^{\infty}{\rm d}u\,A_i^*(u)\varphi_i(u-\omega)\ . \tag{7.11}$$

We can write the boundary conditions (7.2) as

$$z^-_{\rm out}=\lambda\,p^-_{\rm in}\ ,\quad p^+_{\rm out}=-\tfrac{1}{\lambda}z^+_{\rm in}\ , \tag{7.12}$$

where we take for granted that we are dealing with the Fourier coefficient with Rindler wave vector $\tilde{k}$, and define $\lambda = 8\pi G/\tilde{k}^2$. And now consider p^- as the Fourier variable to z^+. Applying Eqs. (7.9) and (7.11) then gives

$$\varphi_i^{\text{out}}(\varrho) = \int_{-\infty}^{\infty} \mathrm{d}u\, A_i^*(u)\, \varphi_i^{\text{in}}(u + \log\lambda - \varrho)\ . \tag{7.13}$$

In the tortoise coordinates ϱ and ω, we see that the waves move in and out with velocity one. This means that an in-going wave Ψ_i^{in} and an out-going wave Ψ_i^{out} can be written in terms of plane waves as

$$\Psi_i^{\text{in}}(\varrho,\,\tau) = \int \mathrm{d}\kappa \tilde{\Psi}_i^{\text{in}}(\kappa)\, e^{i\kappa(-\varrho\,-\,\tau)}\ , \tag{7.14}$$

$$\Psi_i^{\text{out}}(\varrho,\,\tau) = \int \mathrm{d}\kappa \tilde{\Psi}_i^{\text{out}}(\kappa)\, e^{i\kappa(\varrho\,-\,\tau)}\ , \tag{7.15}$$

where τ is the Rindler time variable, κ the Fourier parameter in the tortoise coordinates, and $\tilde{\Psi}$ now denotes the Fourier coefficient in the tortoise coordinates (we use the tilde ($\tilde{}$) rather than the caret because, again, this is a different Fourier transformation than the one in Eqs. (7.6)—(7.11)).

This then yields the final part of our diagonalization process, since these plane waves also diagonalise Eq. (7.13):

$$\tilde{\Psi}_i^{\text{out}}(\kappa) = \tilde{A}_i^*(\kappa)\, e^{-i\kappa\log\lambda}\, \tilde{\Psi}_i^{\text{in}}(\kappa)\ . \tag{7.16}$$

The coefficients $\tilde{A}_i(\kappa)$ are the Fourier coefficients of $A_i(u)$ and can be given in closed form. By contour integration, one derives:

$$\tilde{A}_1(\kappa) = \tfrac{1}{\sqrt{\pi}}\Gamma(\tfrac{1}{2} + i\kappa)(\cosh\frac{\pi\kappa}{2} + i\sinh\frac{\pi\kappa}{2})\ , \tag{7.17}$$

$$\tilde{A}_2(\kappa) = \tfrac{1}{\sqrt{\pi}}\Gamma(\tfrac{1}{2} + i\kappa)(\sinh\frac{\pi\kappa}{2} + i\cosh\frac{\pi\kappa}{2})\ . \tag{7.18}$$

These coefficients have norm one. This is verified using

$$\cosh^2(x) + \sinh^2(x) = \cosh(2x) \quad \text{and} \quad \Gamma(\tfrac{1}{2} + i\kappa)\Gamma(\tfrac{1}{2} - i\kappa) = \frac{\pi}{\cosh\pi\kappa}\ . \tag{7.19}$$

This completes our diagonalization process. We see that Eqs. (7.13) and (7.16) can be seen as a real bounce against the horizon. The information is passed on from the in-going to the out-going particles. We do emphasise that in- and out-going particles were not assumed to affect the metric of the horizon, which is fine as long as they do not pass by one another at distances comparable to the Planck length or shorter; in that case, the gravitational effect of the transverse momenta must be taken into account. For the rest, no other assumptions have been made than that the longitudinal components of the gravitational fields of in- and out-going particles should not be ignored. This must be accurate as long as we keep the transverse distances on the horizon large compared to the Planck length.

It is also important to emphasise that, even though we describe modes of in-going matter that "bounce back against the horizon", these bounces only refer to the information our particles are carrying, while the particles will continue their way falling inwards as seen by a co-moving observer. In accordance with the notion of Black Hole Complementarity [20], an observer falling in only sees matter going in all the way, and nothing of the Hawking matter being re-emitted, since that is seen as pure vacuum by this observer. Rather than stating that this would violate no-cloning theorems, we believe that this situation is asking for a more delicate quantum formalism.

8. The black hole entropy

One could try to compute the black hole entropy from the contributions of these reflecting modes. For each mode, the result is finite. The entropy is found from the free energy F, which is defined by

$$e^{-\beta F}=\sum_{\text{states}}e^{-\beta E}=\sum_{\kappa_n}e^{-\beta\kappa_n}\ , \tag{8.1}$$

where β is now taken to be the inverse of the Hawking temperature, later to be substituted by its (presumable) value 2π.

The energy modes κ_n are derived by assuming that Eq. (7.16) provides for the proper boundary condition near the horizon; the wave bounces at the value ϱ_0 of ϱ where $\Psi_i^{\rm out}(\varrho,\,\tau)=\Psi_i^{\rm in}(\varrho,\,\tau)$, which is where

$$\begin{aligned} e^{i\kappa(\varrho-\tau)}\,\tilde A_i^*(\kappa)\,e^{-i\kappa\log\lambda} &= e^{i\kappa(-\varrho-\tau)}\ , \quad \text{so that}\\ i\kappa(2\varrho_0-\log\lambda)-i\alpha_i(\kappa) &= 0\ , \end{aligned} \tag{8.2}$$

where the angles $\alpha_i(\kappa)$ are the arguments of the coefficients $\tilde A_i(\kappa)$:

$$\tilde A_i(\kappa)\equiv e^{i\alpha_i(\kappa)}\ . \tag{8.3}$$

Assuming a box with outer edge $\varrho=\varrho_1=\log\Lambda$, one finds that the values for κ_n must obey

$$\pi n\ =\ \kappa_n(\varrho_1-\varrho_0)\ =\ \kappa_n(\log\Lambda-\tfrac12\log\lambda)-\tfrac12\alpha_i(\kappa_n)\ . \tag{8.4}$$

A rough estimate for $\alpha_i(\kappa)$ is obtained by applying Stirling for large κ:

$$\tilde A_i(\kappa)=e^{i\alpha_i(\kappa)}\ \to\ e^{i\kappa(\log\kappa\,-\,1)\,+\,\pi i/4}\ . \tag{8.5}$$

Taking L sufficiently large, we elaborate Eq. (8.1):

$$\begin{aligned} e^{-\beta F} &= \textstyle\sum_{i=1}^2\displaystyle\int_0^\infty {\rm d}n\,e^{-\beta\kappa_n} = \textstyle\sum_{i=1}^2\displaystyle\int_0^\infty e^{-\beta\kappa}\,{\rm d}\kappa\frac{{\rm d}n}{{\rm d}\kappa}\ \approx\\ &\approx \tfrac{2}{\pi}\int_0^\infty e^{-\beta\kappa}\,{\rm d}\kappa\,(\log\Lambda-\tfrac12\log\lambda-\tfrac12\log\kappa)\ =\\ &= \tfrac{1}{\pi\beta}(2\log\Lambda+\log\beta+\gamma-\log\lambda)\ , \end{aligned} \tag{8.6}$$

where γ is Euler's constant. The cut-off Λ refers to the edges of the box in which we keep the black hole, so $\log\Lambda$ in Eq. (8.6) merely refers to the contribution of Hawking radiation in the empty space far from the black hole.

Using the thermodynamical equations

$$U=\tfrac{\partial}{\partial\beta}(\beta F)\ ,\quad S=\beta(U-F)\ , \tag{8.7}$$

one can derive the contribution of each mode with transverse wave number $\tilde k$ to the total entropy.

Journal of Physics: Conference Series **701** (2016) 012014 doi:10.1088/1742-6596/701/1/012014

9. Discreteness

The expression we obtained must now be summed over the values $\tilde{k}$. If we take these to describe a finite part of the black hole horizon area, we see that the summed expression will be proportional to the area, as expected, but the sum diverges quadratically for large $\tilde{k}$. Since $\lambda = 8\pi G/\tilde{k}^2$, Eq. (8.6) does depend on $\tilde{k}$, but too weakly, even slowly increasing for large $\tilde{k}$.

The explanation for this divergence is that, as noted at the end of section 7, our expressions are inaccurate at very large $\tilde{k}$, where transverse gravitational forces should be taken into account.

It is not easy to correct for this shortcoming, but we can guess how one ought to proceed. It was remarked already in Refs. [18], that the algebraic expressions we obtain on the 2-dimensional horizon, take the form of functional integrals very much resembling those of string theory. We did treat the transverse position variables $\tilde{x}$ and wave number variables $\tilde{k}$ very differently from the longitudinal variables $z^\pm$ and $p^\pm$, but it is clear that we are dealing with the full expressions of an S_2 sphere. This sphere should be given two arbitrary coordinates $\tilde{\sigma} = (\sigma^1, \sigma^2)$, after which these should be fixed by a gauge condition relating them to the transverse coordinates $\tilde{x}$. We took $\tilde{\sigma} = \tilde{x}$, but apparently this fails when the longitudinal variables fluctuate too wildly.

As long as a more precise procedure has not been found, we can simply insert a cut-off for the transverse wave numbers $\tilde{k}$, or equivalently, the angular momentum quantum numbers ℓ for a finite-size black hole. The divergence is quadratic. The expected expression, Hawking's entropy $S = 4\pi M^2$, comes about of we introduce a sharp cut-off at $|\tilde{k}| \approx M_{\text{Planck}}$. Such wave number cut-offs imply that the conjugate variables, $\tilde{x}$, or equivalently, $\Omega - (\theta, \varphi)$, form a discrete lattice [21].

Finally, we add a note in proof, not discussed in the conference: the careful reader may have been wondering about the signs α and β in our expressions (7.4)—(7.18). They refer to the *two* asymptotic regions at both sides of the horizon, conventionally labelled as universe I and III in the Penrose diagram. Our bounce would not be unitary if we omitted the region III. What does this mean physically?

This is not totally obvious. Evidently, we have to be dealing here with two asymptotic regions, both referring to observable states of the black hole, while one is a CPT image of the other. The only suggestion that we can make here is, that these are two different spots on the horizon, related by a $\mathbb{Z}_2$ symmetry. The most natural implementation is that these refer to antipodal points on the horizon:

> *In 3-space, the regions I and III form a wormhole connecting antipodal points on the horizon.*

The micro-states at the antipodal points are correlated! This is by far a more natural and a more likely interpretation than suggestions made long ago that regions I and III refer to different universes, or to different space-time locations in the same universe.[4]

10. Discussion

Note, that we did not apply second quantization, such as in Ref. [17], since now we are not dealing with a quantum field theory. At every value of $\tilde{k}$, there are exactly two wave functions $\Psi(\pm, \varrho, \tau)$ (one at each side of the horizon, which mix). Second quantization would fail here, since the microstates appear to be determined by a single function such as $p^-(\tilde{x})$.

[4] Stories about interstellar travel using the wormholes inside black holes abound in the popular press, but scientific references supporting such ideas are difficult to find. Even the idea that individual particles could make such a trip appears not to be strongly supported by serious physicists.

It was observed that the in-going and out-going particles with which we started, produce vertex insertions as in a string world sheet, as if all particles considered should be regarded as closed string loops. It all takes the form of a string theory. Strings were not put in, however, rather, they come out as inevitable objects! But beware, these are not "ordinary" strings. The black hole horizon is the string world sheet [18]. If ordinary strings were to be Wick rotated to form space-like string world sheets, all factors i would disappear from the action, whereas our expressions are still in the complex plane, *as if the string slope parameter* α' *should have the purely imaginary value* $4Gi$. In most string treatments of black holes, the string world sheets are assumed to be in the longitudinal direction, that is, the world sheets are taken to be orthogonal or dual to the horizon.

Our analysis appears to be closely related to ideas using the BMS approach [22], although there, the emphasis seems to be specially on the in- and out-going gravitational waves, while we focus on all particle types entering or leaving the black hole. Secondly, although it is clearly of importance to consider measurements made at $\mathfrak{I}^+$ and $\mathfrak{I}^-$, we attribute the black hole properties to the immediate surroundings of the future and past event horizon. Also, one may note that both approaches now focus on light-like geodesics, which justifies attempts to employ conformally invariant (or covariant) descriptions of quantum gravity.

References

[1] Green M B, Schwarz J H and Witten E 1987 *Superstring Theory* (Cambridge Univ. Press, ISBN 0-521-32384-3); Polchinski J G 1998 *Introduction to the Bosonic String*, String Theory Vol. **I** (Cambridge Univ. Press, ISBN 0-521-63303-6); id. 1998 *Superstring Theory and Beyond*, String Theory Vol. **II** (Cambridge Univ. Press, ISBN 0-521-63304-4)

[2] Bell J S 1964 On the Einstein Podolsky Rosen paradox, *Physica* **1** 195;
Clauser J F, Horne M A, Shimony A and Holt R A 1969 *Proposed experiment to test local hidden-variable theories Phys. Rev. Lett.* **23**(15) 880–4, `doi:10.1103/PhysRevLett.23.880`

[3] Hossenfelder S 2011 Testing super-deterministic hidden variables theories *Found. Phys.* **41** 1521 (*Preprint,* `arXiv:1105.4326 [quant-ph]`); id. 2014 Testing superdeterministic conspiracy (*Preprint,* `arXiv:1401.0286 [1401.0286]`)

[4] Vervoort L 2013 Bell's theorem: Two neglected solutions *Found. Phys.* **43**(6) 769–791 *Preprint,* `arXiv:1203.6587v2 [quant-ph]`, `doi:10.1007/s10701-013-9715-7`

[5] Zeilinger A 2010 *Dance of the Photons* (New York, Farrar, Straus and Giroux, p. 266)

[6] 't Hooft G, The Cellular Automaton Interpretation of Quantum Mechanics (*Preprint,* `arXiv:1405.1548v3 [quant-ph]`), to be publ.

[7] Sorkin R 2002 *Causal Sets: Discrete Gravity*, Notes for the Valdivia Summer School (*Preprint,* `arXiv:gr-qc/0309009, and references therein`)

[8] Rovelli C 1998 Loop quantum gravity *Living Reviews in Relativity* **1** 1; Smolin L 2001 *Three Roads to Quantum Gravity* (Basic Books, ISBN 0-465-07835-4)

[9] Ambjorn J et al 2013 *Quantum Gravity via Causal Dynamical Triangulations* (*Preprint,* `arXiv:1302.2173v1 [hep-th]`), to appear in *Handbook of Spacetime* (Springer)

[10] Benincasa D M T, Dowker F and Schmitzer B 2011 The random discrete action for 2-dimensional spacetime *Class. Quant. Grav.* **28** 105018 (*Preprint,* `arXiv:1011.5191v2 [gr-qc]`)

[11] Fradkin E S and Tseytlin A A 1982 Renormalizable asymptotically free quantum theory of gravity *Nuclear Phys.* **B201** 469; Duff M, private communication

[12] Mannheim P D 2007 Solution to the ghost problem in fourth order derivative theories *Found. Phys.* **37** 532–571 (*Preprint,* `arXiv:hep-th/0608154`)

[13] 't Hooft G 2014 Local conformal symmetry: The missing symmetry component for space and time (*Preprint,* `arXiv:1410.6675v2 [gr-qc]`); id. 2015 Singularities, horizons, firewalls, and local conformal symmetry, presented at the 2nd Karl Schwarzschild Meeting on Gravitational Physics, Frankfurt, July 23, 2015 (*Preprint,* `arXiv:1511.04427[gr-qc]`)

[14] Dvali G, Gomez C and Lüst D 2016 Classical limit of black hole quantum N-portrait and BMS symmetry *Phys. Lett.* **B753** 173–177 (*Preprint,* `arXiv:1509.02114 [hep-th]`); Dvali G, personal communication

[15] Aichelburg P C and Sexl R U 1971 On the gravitational field of a massless particle *Gen. Rel. and Gravitation* **2** 303; Bonner W B 1969 *Commun. Math. Phys.* **13** 163; Dray T and 't Hooft G 1985 Gravitational shock

wave of a massless particle *Nucl. Phys.* **B253** 173

[16] 't Hooft G 2015 Diagonalizing the black hole information retrieval process (*Preprint,* arXiv:1509.01695 [gr-qc])

[17] 't Hooft G 1996 The scattering matrix approach for the quantum black hole: an overview *J. Mod. Phys.* **A11** 4623–4688 (*Preprint,* arXiv:gr-qc/9607022)

[18] 't Hooft G 1987 Strings from gravity, in: *Unification of Fundamental Interactions*, Proceedings of Nobel Symposium 67, Marstrand, Sweden, June 2-7, 1986. Eds. Brink L et al. *Physica Scripta* Vol. **T15** 143–150; id. 1995 Black holes, Hawking radiation, and the information paradox *Nucl. Phys.* **B43** (Proc. Suppl.) 1–11;

[19] Itzhaki N 1996 Some remarks on 't Hooft's S matrix for black holes (*Preprint,* arXiv:hep-th/9603067)

[20] Susskind L, Thorlacius and Uglum 1993 The stretched horizon and black hole complementarity (*Preprint,* arXiv:hep-th/9306069); 't Hooft G 1985 On the quantum structure of a black hole *Nuclear Physics* **B 256** 727–745

[21] Kempf A 2000 A generalized Shannon sampling theorem, fields at the Planck scale as bandlimited signals *Phys. Rev. Lett.* **85** 2873

[22] Hawking S W 2015 *The Information Paradox for Black Holes*, talk delivered at the Conference on Hawking Radiation, Stockholm, Aug. 25, 2015 (*Preprint,* arXiv:1509.01147 [hep-th])

Journal of Physics: Conference Series **701** (2016) 012015 doi:10.1088/1742-6596/701/1/012015

Probing a gravitational cat state: Experimental Possibilities

M. Derakhshani
New York, New York.
E-mail: maaneliD@yahoo.com

C. Anastopoulos
Department of Physics, University of Patras, 26500 Patras, Greece.
E-mail: anastop@physics.upatras.gr

B. L. Hu
Maryland Center for Fundamental Physics and Joint Quantum Institute,
University of Maryland, College Park, Maryland 20742-4111 U.S.A.
E-mail: blhu@umd.edu

Abstract. This is a progress report on a preliminary feasibility study of experimental setups for preparing and probing a gravitational cat state [1].

1. Introduction

This is a follow-up to the theoretical studies of [1], exploring the best suited schemes for preparing and probing a gravitational cat state, within the currently available experimental proposals. In the nature of a progress report, we aim to share our thoughts for further discussions, leaving plenty of room for improvements and collective wisdom.

2. Experimental proposals

One of us (MD) considered several possible ways of experimentally implementing the setup suggested by Anastopoulos and Hu in [1], that is, the cat state gravitationally interacting with a classical or quantum probe (Figure 1). Let us consider these sequentially.

3. Preparing a GravCat state

The most promising experimental proposal for the preparation of a gravitational cat state appears to be Romero-Isart *et al.*'s superconducting microsphere trapped in a harmonic potential created by a magnetic quadrupole field [2]. They suggest that this trapping method should make it possible to isolate a lead (Pb) microsphere of mass $M \sim 10^{14}$ amu and radius $R = 2$ microns from various sources of environmental decoherence, to a degree sufficient that it should be possible to place the microsphere in a coherent superposition of two position eigenstates. The

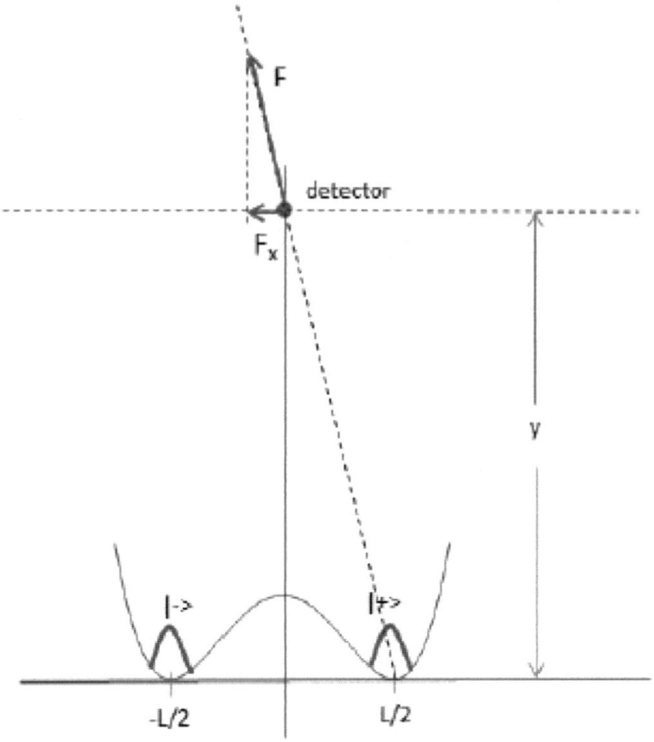

Figure 1. Force on a probe/detector exerted by a massive particle in a gravitational cat state, $c_+|+> + c_-|->$, between locations $\pm L/2$.

protocol they propose for creating this microsphere 'cat state' is parametric coupling to a qubit state.

How do Romero-Isart *et al.* obtain $M \sim 10^{14}$ amu and $R = 2$ microns? They suppose that the (superconducting) microsphere is trapped in a 3-D harmonic oscillator potential of the form (Figures 2-b and 2-c)

$$V_{trap} = \frac{M}{2}\left[\omega_t^2 \hat{x}^2 + \omega_\perp^2 (\hat{y}^2 + \hat{z}^2)\right], \tag{1}$$

where the trapping frequency

$$\omega_t \simeq (1.05)\,\frac{\left(\sqrt{\mu_0/\rho}\right) I}{l^2}, \tag{2}$$

and $\omega_\perp = \frac{\omega_t}{2}$, because the potential is created by a quadrupole magnetic field that traps the microsphere via the Meissner effect.

Here M is the mass of the microsphere and ρ is its mass density (assuming a homogeneous microsphere). The parameter l is the radius of and separation between the anti-Helmholtz coils surrounding the microsphere in the setup illustrated in Figure 2-a, and I is the current through the coils. In order for the microsphere to be trapped, it is required that the magnetic field at any point of the sphere be smaller than the critical field, B_{crit}, in order to allow superconductivity.

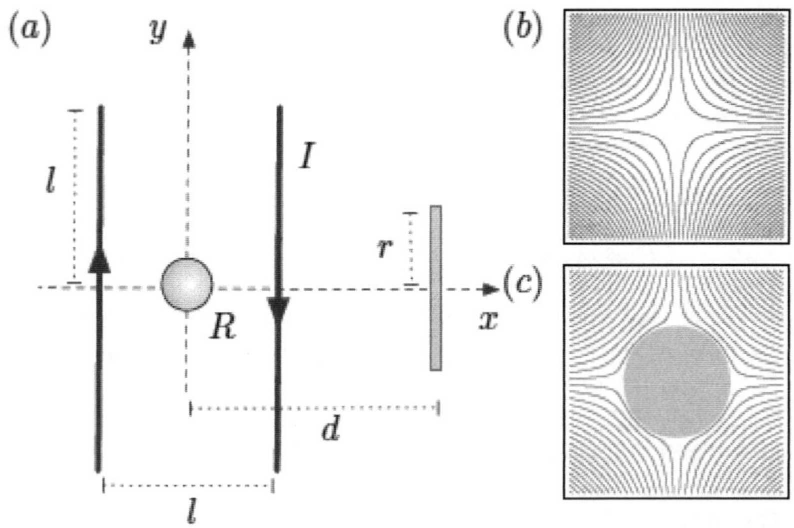

Figure 2. Microsphere trap setup in [2]: (a) Scheme of anti-Hemholtz coils (AHC) with a pickup coil. (b) Quadrupole field created by AHC. (c) Field expelled by microsphere in Meissner state.

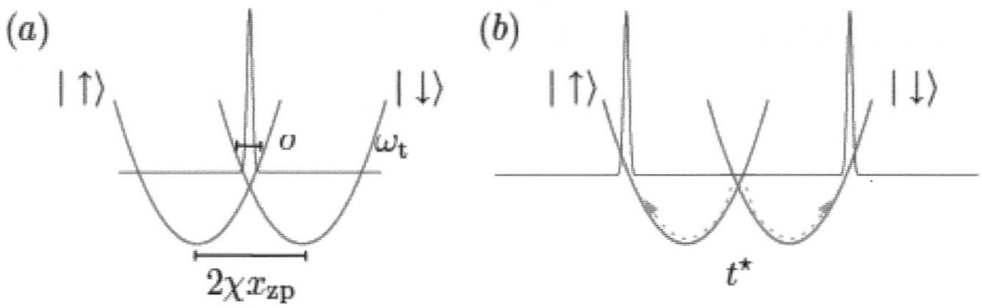

Figure 3. Scheme for parametric coupling of microsphere to qubit in [2]: (a) The state $|+>|0>$ prepped at $t=0$ and recovered at $t=2t^*$. (b) The superposition state Eq. (4) created at $t=t^*$.

This yields an upper bound on the radius of the sphere as

$$R < R_{max} \simeq (0.98)\,\frac{B_{crit}}{\omega_t\sqrt{\mu_0\rho}}. \tag{3}$$

They also require that $R \gg \lambda, \xi$, where λ is the sphere's penetration depth and ξ is its coherence length (at temperature $\mathrm{T}=0$). For Figure 2-a, they assume $l \gg R$, although this does not appear to be a mandatory condition. To determine the size of the sphere, they choose the parameters $\rho = 11,360\frac{kg}{m^3}$, $\lambda = 30.5nm$, $\xi = 96nm$ (at $\mathrm{T}=0$), $B_{crit} = 0.08T$, $l = 25\mu m$, $I = 10A$ (assuming materials with a critical current density of $J_{crit} = 7\times10^{11}\frac{A}{m^2}$), $\omega_t \simeq 2\pi\times 28kHz$, and $R_{max} = 3.7\mu m$. Thus they consider a microsphere of $R = 2\mu m$, which implies a microsphere mass of $M = \rho\frac{4}{3}\pi R^3 \simeq 10^{14}amu \simeq 10^{-13}kg$.

In a detailed decoherence analysis, they argue that trapping a microsphere this way makes negligible common sources of environmental decoherence (clamping losses, scattering of photons, damping created by background gas, blackbody radiation, and internal vibrational modes). They also characterize other sources of decoherence specific to their setup (damping due to hystereses losses in the superconducting anti-Helmholtz coils, fluctuations in the trap frequency, and fluctuations in the trap center) and argue that they do not undermine their proposal.

Then, they show that parametric coupling to a qubit state puts the microsphere in a spatial superposition (Figure 3) described by the wave function

$$|\Psi_s>= \frac{1}{\sqrt{2}}\left[\hat{T}(-2\chi)|\uparrow,0> +\hat{T}(2\chi)|\downarrow,0>\right], \tag{4}$$

after a time $t^* = \pi/\omega_t$, where χ is a dimensionless parameter that characterizes the parametric coupling, and $\hat{T}(...)$ is the usual translation operator. The two superposed wave packets are separated by a distance of $l_s = 4\chi x_{zp}$, where x_{zp} is the zero-point motion of the microsphere in the trap, with overlap given by $< 0|\hat{T}^\dagger(-2\chi)\hat{T}(2\chi)|0 >= exp[-l_s^2/(8\sigma^2)]$ and the requirement that $8\sigma^2 < l_s^2$. This allows for an l_s of the order of 100 nm or more [3].

4. Probing a GravCat state

One of us (MD) has examined the leading state-of-the-art proposals in the past five years for ultrasensitive force measurement. A survey of which is contained in the Appendix.

We separate the consideration of a classical versus a quantum probe.

4.1. Classical Probe

For the role of the classical probe, the most promising experimental proposal appears to be Reinhardt *et al.*'s trampoline resonator [4], with effective mass 4.5 ng, diameter 100 microns (Figure 4-i), and projected force sensitivity of $\sim$ 14 zN at cryogenic temperatures (14 mK).

Assuming an experimental configuration between the resonator and the microsphere like that in Figure 1, and using Eq. (62) of [1] to calculate the resonator-microsphere Newtonian gravitational force we have:

$$f_0 = \frac{GM_{res}M^{Pb}_{sphere}L}{2\left(y^2+\frac{L^2}{4}\right)^{3/2}} = \frac{(6.67\times10^{-11}units)(4.5\times10^{-12}kg)(3.8\times10^{-13}kg)}{[(10^{-6}m)^2+(5\times10^{-7}m)^2]^{3/2}}(5\times10^{-7}m) = 4.1\times10^{-23}N, \tag{5}$$

which is only around 3 orders of magnitude away from the projected force sensitivity range of the resonator. The specific values of L and y used in Eq. (5) were imported from Bahrami *et al.* [5], who proposed a cat state experiment very similar to that suggested by Anastopolous and Hu. Bahrami *et al.* indicated that for micromechanical systems (which the microsphere would be), reasonable values are those used in Eq. (5).

Two ways we might be able to boost the resonator-microsphere gravitational interaction to cover this three-orders-of-magnitude gap are:

(1) We might be able to change one or more of the non-mass parameters in the force calculation (within plausibility). For example, rather than a resonator-microsphere spatial separation y of one micron, we might use half a micron. This already increases the gravitational force to 1.6×10^{-22} N, which is only 88 times smaller than 14 zN.

(2) We can try increasing the mass of the microsphere, consistent with the constraints of the Romero-Isart *et al.* setup.

Recall that a necessary condition for the microsphere to be trapped is that the magnetic field at any point on the sphere be smaller than the critical field, B_{crit}, in order to keep the Meissner

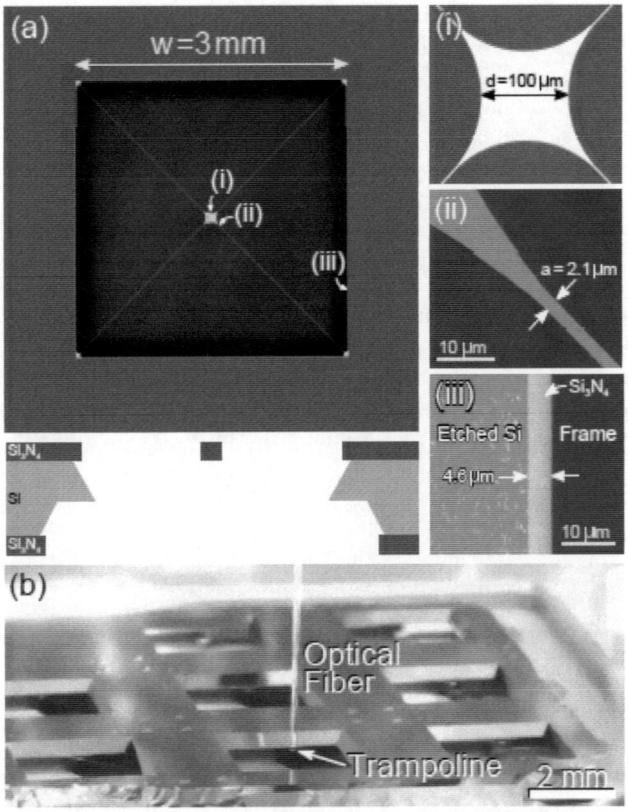

Figure 4. Fabricated Si_3N_4 trampoline resonators in [4].

state. This condition limits the size of the microsphere, as can be seen from the interaction Hamiltonian between the microsphere and the magnetic field gradient [3]

$$H_{int} = -\frac{1}{2}\mathbf{m} \cdot \mathbf{B} = \frac{3V}{4\mu_0}|\mathbf{B}|^2 = \frac{\rho V}{2}\omega_t^2 x^2, \tag{6}$$

which implies that the trapping frequency is given by

$$\omega_t = b\sqrt{\frac{3}{2\rho\mu_0}} \sim 2\pi \times \sqrt{2}b[T/m]Hz, \tag{7}$$

where b is the magnetic field gradient, and $|\mathbf{B}| = B = bx$. Now, if we assume that the center of mass of the sphere is at the position of $B = 0$, the field at the surface will just be bR. This implies that the maximum field gradient that can be used is

$$b_{max} = \frac{B_{crit}}{R}, \tag{8}$$

and a maximum trapping frequency of

$$\omega_t^{max} = b_{max}\sqrt{\frac{3}{2\rho\mu_0}}. \tag{9}$$

Now, in order to cool the center of mass to the ground state, we need trap frequencies of at least tens of kHz, or, equivalently, a field gradient of $10^4\frac{T}{m}$. For a Pb microsphere, we have $B_{crit}^{Pb} = 0.08T$, $\rho_{Pb} = 11,360\frac{kg}{m^3}$, and $T_{crit}^{Pb} = 7.2K$. For $R_{max}^{Pb}[\mu m] = 3.7$, these values give

$$b_{max}^{Pb} = \frac{B_{crit}^{Pb}}{R_{max}^{Pb}} = \frac{8}{R_{max}^{Pb}[\mu m]} \times 10^4\frac{T}{m} = 2.16 \times 10^4\frac{T}{m}, \tag{10}$$

and

$$\omega_{t-max}^{Pb} \sim 2\pi \times \frac{131}{R_{max}^{Pb}[\mu m]} kHz = 2\pi \times 35.4kHz, \tag{11}$$

as required. This is why the Pb microsphere size is bounded to a few microns.

What about other superconducting elements? To investigate this, one of us (MD) examined the transition temperatures and critical fields of all known superconducting elements [6, 7]. He found that the most promising element for increasing the microsphere size and mass, consistent with the above constraints, is Technetium (Tc), a Type-II superconductor. A Tc microsphere would have $B_{crit}^{Tc} = 0.141T$, $\rho_{Tc} = 11,000\frac{kg}{m^3}$, and $T_{crit}^{Tc} = 7.8K$. If in Eqs. (10)-(11) we use B_{crit}^{Tc} and $R_{max}^{Tc}[\mu m] = 10$, we get

$$b_{max}^{Tc} = \frac{B_{crit}^{Tc}}{R^{Tc}} = \frac{14.1}{R^{Tc}[\mu m]} \times 10^4\frac{T}{m} = 1.41 \times 10^4\frac{T}{m}, \tag{12}$$

and

$$\omega_{t-max}^{Tc} \sim 2\pi \times \frac{234}{R^{Tc}[\mu m]} kHz = 2\pi \times 23.4kHz, \tag{13}$$

as required. This allows a Tc microsphere of mass

$$M_{max}^{Tc} = \rho_{Tc}\frac{4}{3}\pi(R_{max}^{Tc})^3 = 4.6 \times 10^{-11}kg = 2.77 \times 10^{16}amu. \tag{14}$$

Even if we use a sphere radius slightly less than R_{max}^{Tc}, for example $R^{Tc}[\mu m] = 8$, we would still obtain

$$M^{Tc} = \rho_{Tc}\frac{4}{3}\pi(R^{Tc})^3 = 2.36 \times 10^{-11}kg = 1.42 \times 10^{16}amu. \tag{15}$$

So Tc seems to afford us a microsphere mass two orders of magnitude greater than Pb. Using a Tc microsphere of $R_{max}^{Tc} = 10$ microns in Eq. (5), we then obtain

$$f_{01} = \frac{GM_{res}M_{sphere}^{Pb}L}{2\left(y^2 + \frac{L^2}{4}\right)^{3/2}} = \frac{(6.67 \times 10^{-11}units)(4.5 \times 10^{-12}kg)(4.6 \times 10^{-11}kg)}{[(10^{-6}m)^2 + (5 \times 10^{-7}m)^2]^{3/2}}(5\times10^{-7}m) = 4.94\times10^{-21}N, \tag{16}$$

which is very close to the 14 zN projected force sensitivity of the resonator. Furthermore, if we allow $y = 5 \times 10^{-7}m$, we get

$$f_{02} = \frac{GM_{res}M_{sphere}^{Pb}L}{2\left(y^2 + \frac{L^2}{4}\right)^{3/2}} = \frac{(6.67 \times 10^{-11}units)(4.5 \times 10^{-12}kg)(4.6 \times 10^{-11}kg)}{[(5 \times 10^{-7}m)^2 + (5 \times 10^{-7}m)^2]^{3/2}}(5\times10^{-7}m) = 1.95\times10^{-20}N, \tag{17}$$

which *exceeds* 14 zN.

It therefore appears that superconducting microspheres made of Tc are what's needed to make our proposal feasible. However, since Tc is a Type-II superconductor, this complicates the theoretical feasibility of Romero-Isart *et al.*'s scheme for fashioning coherent microsphere cat states. Also, increasing the microsphere mass makes fluctuations in the trap center more appreciable as a source of decoherence. These issues are currently under study.

Journal of Physics: Conference Series **701** (2016) 012015 doi:10.1088/1742-6596/701/1/012015

4.2. Quantum Probe

For the role of a quantum probe, Anastopoulos and Hu [1] propose to use a von-Neumann probe, i.e., a quantum harmonic oscillator coupled to a macroscopic detector. To model this, they specialize the Hamiltonian description of the quantum harmonic oscillator to a single-mode Jaynes-Cummings model and consider the deep strong-coupling limit. By treating the oscillator in the ground state and entangling it with the two-state spatial superposition that characterizes the gravitational cat state, they find that the reduced density matrix for the oscillator degrees of freedom has distinguishable off-diagonal terms only when

$$\omega_{osc}^3 \ll \left(\frac{f_0}{M_{osc}}\right)^2 M_{osc}. \tag{18}$$

Hence, a position measurement of the oscillator by a macroscopic detector will collapse the wave function of the joint system and find the oscillator in a neighborhood of phase space associated with the cat state collapsing to either $-L/2$ or $L/2$, with Born-rule relative frequencies. They also find that a continuous position measurement will lead to Rabi-type oscillations between the two neighborhoods of phase space, with transition probability $p(t) = sin^2\nu t$, where ν is the transition frequency.

In considering possible experimental implementations of this von-Neumann probe, one of us (MD) found that the most promising candidate seems to be the state-of-the-art optomechanical harmonic oscillator described in [8, 5]. Such an oscillator has a mass of 100 ng, and would experience a Newtonian gravitational force of $f_{01} = 9.07 \times 10^{-22}$ Newtons from a Pb microsphere cat state, using Eq. (5). And for a Tc microsphere, it would experience $f_{02} = 1.10 \times 10^{-19}$ Newtons. However, for a Tc microsphere, Eq. (18) gives

$$2.4 \times 10^{23} \frac{rad^3}{sec^3} \gg 1.21 \times 10^{-28} \frac{N^2}{kg}, \tag{19}$$

while for a Pb microsphere, we get

$$2.4 \times 10^{23} \frac{rad^3}{sec^3} \gg 8.23 \times 10^{-33} \frac{N^2}{kg}. \tag{20}$$

This means that, even with state-of-the-art optomechanical oscillators, we are still a long way off from experimentally realizing the quantum probe of a gravitational cat state.

There are many additonal theoretical issues related to both the gravitational cat state and the probes, e.g., 1) how intact a gravcat state could remain, how long it can exist, in the presence of massive objects (such as the Earth); 2) why the gravitational force interaction with a classical probe should necessarily collapse the cat state wave function (in contrast to other massive bodies in nature, such as the Earth); 3) whether the collapse postulate is actually compatible with the hypothesis of a semiclassical gravitational source; 4) in the experimental schemes discussed above, the impact of motional decoherence and decoherence by vortices in Type II superconductors, to name just two. They are currently under investigation.

Appendix: Experimental schemes for classical probes of a GravCat State

One of us (MD) examined all the state-of-the-art proposals in the past five years for ultra-sensitive force measurements. He then identified a handful of proposals that seemed initially promising for our project, and did some back-of-the-envelope estimates on each. Among these, he was able to identify only one ultra-sensitive force measurement scheme that can play the role of a classical probe of sufficiently large mass and sufficiently high force-measurement sensitivity that, in combination with Romero-Isart *et al.*'s superconducting microsphere cat state proposal [2], might yield a feasible experimental scheme for detecting the gravitational field of a cat state.

Those handful of proposals were:

(i) Schreppler *et al.*'s scheme involving an ultra-cold atom cloud in a high finesse cavity [9] which, to date, produced the smallest externally applied force measured of 42 yN (yoctonewtons).
(ii) Moser *et al.*'s scheme involving carbon nanotube mechanical resonators with quality factors greater than a million [10, 11], which yields force measurements on the zN (zeptonewton) scale.
(iii) Tao *et al.*'s scheme using single-crystal diamond nanomechanical resonators with quality factors exceeding one million [12], which produces force sensitivities of a few hundred zN.
(iv) Ranjit *et al.*'s scheme involving laser-cooled silica microspheres as force sensors in a dual beam optical dipole trap in high vacuum [13], which yields force measurement sensitivity at the aN (attonewton) scale.
(v) Kleckner *et al.*'s [14] and Reinhardt *et al.*'s [4] schemes involving the use of optomechanical trampoline resonators, which yield projected maximum force sensitivities on the aN and zN scales, respectively.
(vi) Wagner *et al.*'s use of state-of-the-art torsion balance pendulums to test for WEP violations with a precision of one part in 10^{13} [15].

Of all these, only proposal 5 seems promising for our purpose. To justify that, it will first be explained why the other proposals were deemed not viable.

Proposal 1 yields the greatest force sensitivity, but the atom cloud used has a miniscule mass of only 1.8×10^{-22} kg. Using the center of mass of this atom cloud in place of the resonator in Eq. (5) (keeping all other parameters the same), one obtains a gravitational force of only $\sim 10^{-34}$ N, which is eleven orders of magnitude smaller than the force sensitivity of Schreppler *et al.*'s scheme. In addition, the scheme requires that the measured force be an externally applied force that oscillates at the natural frequency of the center of mass motion of the atom cloud in the cavity ($\sim$ 12 kHz). It is unclear how this could be done with the gravitational force from a cat state, even if it were somehow possible to fashion an atom cloud with a center of mass of, let's say, 4.5 ng (which would yield a force magnitude that falls just within the sensitivity of Schreppler *et al.*'s scheme).

Proposal 2 has a similar obstacle in that the mass of these carbon nanotube mechanical resonators is only $\sim 10^{-20}$ kg, which when used in place of the resonator in Eq. (5) yields a gravitational force of only $\sim 10^{-32}$ N, or eleven orders of magnitude smaller than the force sensitivity of the Moser *et al.* scheme.

Proposal 3 used several different kinds of single-crystal diamond nanomechanical resonators, the largest of which has a mass of $\sim 10^{-12}$ kg (the mass isn't given in the paper, but the estimated volume is given, and this was used in combination with the estimated mass density of single-crystal diamond given in [16]). When the largest resonator is used in place of the resonator in Eq. (5), it yields a gravitational force of $\sim 10^{-24}$ N. Since this resonator has a force sensitivity of only 540 zN, the expected gravitational force is around five orders of magnitude smaller.

Proposal 4 uses silica microspheres of 3 micron diameter, with an estimated mass of at most $\sim 10^{-12}$ kg (the mass isn't given in the paper, but Barker [17] uses 10 micron diameter silica microspheres and gives this mass value). Thus these microspheres yield a gravitational force six to seven orders of magnitude smaller than the force sensitivity of Ranjit *et al.*'s scheme.

Proposal 6 is, of course, a very different force measurement scheme than all the others in that the goal of a torsion balance is to detect a difference in the directions of the external force vectors applied to the test bodies, rather than the absolute magnitudes of the forces. The scheme described in Wagner *et al.* uses test bodies of masses 10 g each and has a force sensitivity of one part in 10^{13} (for the Eötvös parameter). If we supposed a gravitational force interaction between one of these test bodies and the Romero-Isart *et al.* microsphere cat state, keeping all the other parameters the same in Eq. (5), the resulting force, $\sim 10^{-14}$ N, yields a differential acceleration of $\sim 10^{-12} m/s^2$ for one test body, which would be within this sensitivity range (the smallest differential acceleration their torsion balance was able to detect was $\sim 10^{-15} m/s^2$). However, for such macroscopic test bodies, it seems highly implausible that experimentalists could keep the other parameters the same; in particular, it seems doubtful that the spatial separation between a test body and the microsphere can be only one micron in practice; the experimental setup in Romero-Isart *et al.* [2] (Figure 2-a) involves surrounding the microsphere by an anti-Helmholtz coil configuration only 25 microns in width. Much more experimentally feasible, it seems, is a separation on the order of a centimeter or a millimeter. Using a one millimeter separation, the resulting gravitational force ($\sim 10^{-23}$ N) and differential acceleration ($\sim 10^{-21} m/s^2$) for a test body is significantly below the maximum sensitivity of the torsion balance.

Why then does proposal 5 seem promising? First, the largest trampoline resonator used in the Kleckner *et al.* proposal has a mass of 110 ng, diameter of 80 microns (see Figure 2 of [14]), and a projected force sensitivity on the aN scale at cryogenic temperatures. To be more precise about this last feature, Kleckner *et al.* write:

> Trampoline resonators are also suitable for use as ultra-high resolution force sensors. Assuming the quality factor increase is also seen for the lowest frequency devices [i.e., the resonator with $m = 110$ ng], it should be possible to obtain a thermal force noise in the aN/Hz regime at demonstrated [cryogenic] temperatures. This is comparable to or better than the single crystal Si resonators currently used in magnetic resonance force microscopy (MRFM) experiments [36, 37]. Furthermore, the rear side optical access can be used to provide extremely precise position sensitivity while leaving the front side free for surface modifications required for use as sensors. (Last page)

Now, the gravitational force between the 110 ng trampoline resonator and the microsphere, keeping all other parameters the same in Eq. (5), is equal to 1×10^{-21} N. This is only around three orders of magnitude away from the projected force sensitivity range of their resonator. Second, the largest trampoline resonator used in the Reinhardt *et al.* proposal [4] has a mass of 4.5 ng, diameter of 100 microns (Figure 4-i), and a projected force sensitivity of ~ 14 zN at cryogenic temperatures (14 mK). The gravitational force between the 4.5 ng resonator and the microsphere, keeping all other parameters the same in Eq. (5), is equal to 4.1×10^{-23} N, which is also only around three orders of magnitude away from the projected force sensitivity range of their resonator. But because of the greater specificity of the Reinhardt *et al.* proposal (in regards to the magnitude and conditions of maximum force sensitivity), it was chosen as the most promising to combine with Romero-Isart *et al.*'s proposal.

References

[1] Anastopoulos C and Hu B L 2015 *Class. Quant. Grav.* **32** 165022 (*Preprint* `arXiv:1504.03103 [cond-mat.mes-hall]`)

[2] Romero-Isart O, Clemente L, Navau C, Sanchez A and Cirac J I 2012 *Phys. Rev. Lett.* **109** 147205 (*Preprint* `arXiv:1112.5609 [quant-ph]`)

[3] Romero-Isart 2015 private communication

[4] Reinhardt C, Mueller T, Bourassa A and Sankey J C 2015 Ultralow-noise SiN trampoline MEMS for sensing and optomechanics (*Preprint* `arXiv:1511.01769 [cond-mat.mes-hall]`)

[5] Bahrami M, Bassi A, McMillen S, Paternostro M and Ulbricht H 2015 Is gravity quantum? (*Preprint* `arXiv:1507.05733 [quant-ph]`)

[6] Kittel C 1996 *Introduction to Solid State Physics* (Wiley) chap 12, 7th ed

[7] Myers H P 1997 *Introductory Solid State Physics* (Taylor & Francis) chap 13, 2nd ed

[8] Aspelmeyer M, Kippenberg T J and Marquardt F 2014 *Rev. Mod. Phys.* 1391 (*Preprint* `arXiv:1303.0733 [cond-mat.mes-hall]`)

[9] Schreppler S, Spethmann N, Brahms N, Botter T, Barrios M and Stamper-Kurn D M 2014 *Science* **344** 1486 (*Preprint* `arXiv:1312.4896 [quant-ph]`)

[10] Moser J, Guettinger J, Eichler A, Esplandiu M J, Liu D E, Dykman M I and Bachtold A 2013 *Nature Nanotechnol.* **8** 493 (*Preprint* arXiv:1305.6887 [cond-mat.mes-hall])
[11] Moser J, Eichler A, Guettinger J, Dykman M I and Bachtold A 2014 *Nature Nanotechnol.* **9** 1007 (*Preprint* arXiv:1503.09155 [cond-mat.mes-hall])
[12] Tao Y, Boss J M, Moores B A and Degen C L 2005 *Nature Commun.* **5** 3638 (*Preprint* arXiv:1212.1347 [cond-mat.mes-hall])
[13] Ranjit G, Atherton D P, Stutz J H, Cunningham M and Geraci A A 2015 Attonewton force detection using microspheres in a dual-beam optical trap in high vacuum *Phys. Rev.* A **91** (*Preprint* arXiv:1503.08799 [quant-ph])
[14] Kleckner D, Pepper B, Jeffrey E, Sonin P, Thon S M and Bouwmeester D 2011 Optomechanical trampoline resonators *Opt. Express* **19** 19708
[15] Wagner T A, Schlamminger S, Gundlach J H and Adelberger E 2012 *Class. Quant. Grav.* **29** (*Preprint* arXiv:1207.2442 [gr-qc])
[16] Pomorski M 2008 Ph.D. thesis University of Frankfurt
[17] Barker P F 2010 *Phys. Rev. Lett.* **105** 073002 (*Preprint* arXiv:1004.1443 [quant-ph])

Journal of Physics: Conference Series **701** (2016) 012016 doi:10.1088/1742-6596/701/1/012016

Uncertainty relations and precession of perihelion

Fabio Scardigli[a,b] **and Roberto Casadio**[c,d]

[a] Department of Mathematics, College of Engineering,
American University of the Middle East, P.O.Box 220 Dasman, 15453 Kuwait
[b] Dipartimento di Matematica, Politecnico di Milano,
Piazza Leonardo da Vinci 32, 20133 Milano, Italy
[c] Dipartimento di Fisica e Astronomia, Alma Mater Università di Bologna
via Irnerio 46, 40126 Bologna, Italy
[d] I.N.F.N., Sezione di Bologna, viale Bert Pichat 6/2, 40127 Bologna, Italy

E-mail: `fabio@phys.ntu.edu.tw`, `casadio@bo.infn.it`

Abstract. We compute the corrections to the Schwarzschild metric necessary to reproduce the Hawking temperature derived from a Generalized Uncertainty Principle (GUP), so that the GUP deformation parameter is directly linked to the deformation of the metric. Using this modified Schwarzschild metric, we compute corrections to the standard General Relativistic predictions for the perihelion precession for planets in the solar system, and for binary pulsars. This analysis allows us to set bounds for the GUP deformation parameter from well-known astronomical measurements.

1. Introduction

Research on generalizations of the uncertainty principle of quantum mechanics has nowadays a long history [1]. One of the main lines of investigation focuses on understanding how the Heisenberg Uncertainty Principle (HUP) should be modified once gravity is taken into account. Given the pivotal rôle played by gravitation in these arguments, it is not surprising that the most relevant modifications to the HUP have been proposed in string theory, loop quantum gravity, deformed special relativity, and studies of black hole physics [2, 3, 4, 5, 6, 7], just to mention some of the most notable frameworks.

An interesting novelty, emerged during the last decade or so, is a lively debate on the *measurable* features of various kinds of Generalized Uncertainty Principles (GUPs). From more theoretical shores, the discussion has therefore landed on the ground of experimental predictions about the *size* of these modifications, and several experiments have been proposed to test GUPs in the laboratory. Among the more elaborated proposals are those, for example, of the groups of Brukner, Cerdonio, Bonaldi [8, 9, 10].

Studies that aim at putting bounds on the dimensionless deforming parameter of the GUP, heretofore denoted by β, date back at least to Brau [11], and can be roughly divided into three different categories (actually, only two, as we will see). In the first group one finds papers such as those of Brau [11], Vagenas [29], Nozari [13], which use a specific (in general, non linear) representation of the operators in the deformed fundamental commutator [1]

$$[\hat{X},\hat{P}] = i\,\hbar(1+\beta\,\hat{P}^2/m_{\rm p}^2)\ , \tag{1}$$

in order to compute corrections to quantum mechanical predictions, such as energy shifts in the spectrum of the hydrogen atom, or to the Lamb shift, the Landau levels, Scanning Tunneling Microscope, charmonium levels, etc. The bounds so obtained on β are quite stringent, but the drawback of this approach is a potentially strong dependence of the expected shifts on the specific (non linear) representation chosen for the operators $\hat{X}$ and $\hat{P}$ in the fundamental commutator.

In the second group, we can find the works of, e.g., Chang [14], Nozari and Pedram [15], where a deformation of classical Newtonian mechanics is introduced by modifying the standard Poisson brackets in a way that resembles the quantum commutator,

$$[\hat{x},\hat{p}] = i\,\hbar\left(1+\beta_0\,\hat{p}^2\right) \quad\Rightarrow\quad \{X,P\} = \left(1+\beta_0\,P^2\right)\ , \tag{2}$$

where $\beta_0 = \beta/m_{\rm p}^2$. In particular, Chang in Ref. [14] computes the precession of the perihelion of Mercury directly from this GUP-deformed Newtonian mechanics, and interprets it as an extra contribution to the well known precession of 43"/century due to General Relativity (GR). He then compares this global result with the observational data, and the very accurate agreement between the GR prediction and observations leaves Chang not much room for possible extra contributions to the precession. In fact, he obtains the tremendously small bound $\beta \lesssim 10^{-66}$. A problem with this approach is that a GUP-deformed Newtonian mechanics is simply superposed linearly to the usual GR theory. One may argue that a modification of GR at order β should likewise be considered, but this is however omitted in Ref. [14]. In other words, it is not clear why the two structures, GR and GUP-modified Newtonian mechanics, should coexist independently, and why the two different precession errors add into a final single precession angle. Most important, as a matter of fact, in the limit $\beta \to 0$, Ref. [14] recovers *only* the Newtonian mechanics but not GR, and GR corrections must be added as an extra structure. Clearly, the physical relevance of this approach and the bound that follows for β, remain therefore questionable.

Finally, a third group of works on the evaluation of β contains, for example, papers by Ghosh [16] and Pramanik [17]. They use a covariant formalism, first defined in Minkowski space, with the metric $\eta_{\mu\nu} = {\rm diag}(1,-1,-1,-1)$, which can be easily generalized to curved space-times via the standard procedure $\eta_{\mu\nu} \to g_{\mu\nu}$. These papers should however be considered as belonging to the second group. In fact, a closer look reveals that they also start from a deformation of classical Poisson brackets, although posited in covariant form. From the deformed covariant Poisson brackets, they obtain interesting consequences, like a β-deformed geodesic equation, which leads to a violation of the Equivalence Principle. They do not deform the field equations or the metric. In Ref.[18], however, we show that this violation of the Equivalence Principle is completely due to the postulate of deformed Poisson brackets, and has nothing to do with the covariant formalism, or with a deformation of the GR field equations or solutions, or of the geodesic equation. Nonetheless, the Ghosh–Pramanik formalism remains covariant when $\beta \to 0$ and reproduces standard GR results in the limit $\beta \to 0$ (this differs, in general, from the results obtained by papers in the second group).

The novelties of our approach, when compared with the previous ones, are many and various. The main point is to start directly from a quantum mechanical effect, the Hawking evaporation, for which the GUP is necessarily relevant, rather than postulating specific representations of canonical operators or modifications of the classical equations of motion. We connect the

[1] We shall work with $c = k_B = 1$, but explicitly show the Newton constant $G_{\rm N}$ and Planck constant $\hbar$. We also recall that the Planck length is defined as $\ell_{\rm p}^2 = G_{\rm N}\,\hbar/c^3$, the Planck energy as $\mathcal{E}_{\rm p}\,\ell_{\rm p} = \hbar\,c/2$, and the Planck mass as $m_{\rm p} = \mathcal{E}_{\rm p}/c^2$, so that $G_{\rm N} = \ell_{\rm p}/2\,m_{\rm p}$ and $\hbar = 2\,\ell_{\rm p}\,m_{\rm p}$.

deformation of the Schwarzschild metric directly to the uncertainty relation, without relying on a specific representation of commutators. We leave the Poisson brackets and classical Newtonian mechanics untouched, and recover GR, and standard quantum mechanics, in the limit $\beta \to 0$. In particular, we preserve the Equivalence Principle, and the equation of motion of a test particle is still given by the standard geodesic equation. In the present work, this is obtained by deforming a specific solution of the standard GR field equations, namely the Schwarzschild metric.

2. Deforming the Schwarzschild metric

In this section, we start from a known way of deriving the Hawking temperature directly from the metric of a black hole, and then show how the GUP modifies the Hawking temperature. These two steps will pave the road to a deformation of the Schwarzschild metric, constructed so as to reproduce the GUP-modified Hawking temperature. We consider here a space-time with a metric that locally has the form

$$\mathrm{d}s^2 = g_{\mu\nu} dx^\mu dx^\nu = F(r)\,\mathrm{d}t^2 - F(r)^{-1}\,\mathrm{d}r^2 - r^2\,\mathrm{d}\Omega^2, \tag{3}$$

where $\mathrm{d}\Omega^2 = \mathrm{d}\theta^2 + \sin^2\theta\,\mathrm{d}\phi^2$. The horizons (if any), are located at the positive zeros of the function $F(r)$ (see, for example, Ref. [19]).

We loosely follow a standard derivation, as for example that in Ref. [20]. Suppose $r = r_{\mathrm{H}}$ is an horizon, so that $F(r_{\mathrm{H}}) = 0$, and consider $r \geq r_{\mathrm{H}}$. Then, a quantized scalar field outside the horizon lives in a heat bath with temperature

$$T = \hbar\,\frac{F'(r_{\mathrm{H}})}{4\pi}\,. \tag{4}$$

Therefore the temperature of the black hole horizon as seen by a distant observer is in general given by formula (4). In particular, for a Schwarzschild black hole the function $F(r)$ is given by $(1 - 2G_{\mathrm{N}}M/r)$, the horizon is at $r_H = 2\,G_{\mathrm{N}}\,M$, and we get $T_{\mathrm{H}} = \hbar/(8\pi\,G_{\mathrm{N}}\,M)$, which is the well-known Hawking temperature.

We now give here a derivation of the mass-temperature relation starting directly from the uncertainty relations. The most common form of deformation of the Heisenberg uncertainty relation (and the form of GUP that we are going to study in this paper) is without doubt the following

$$\Delta x\,\Delta p \;\geq\; \frac{\hbar}{2}\left(1 \,+\, \beta\,\frac{4\,\ell_{\mathrm{p}}^2}{\hbar^2}\,\Delta p^{\,2}\right) = \frac{\hbar}{2}\left[1 + \beta\left(\frac{\Delta p}{m_{\mathrm{p}}}\right)^2\right]\,. \tag{5}$$

The dimensionless parameter β is usually assumed to be of order one, in the most common quantum gravity formulations. Following the arguments of Refs. [21, 22, 23, 24, 25, 26, 27], we promptly arrive to translate relation (5) into a mass-temperature relation for a Schwarzschild black hole

$$M = \frac{\hbar}{8\pi\,G_{\mathrm{N}}\,T} + \beta\,\frac{T}{2\pi}\,. \tag{6}$$

To zero order in β, we recover the usual Hawking formula. Let us note that in this work we assume that the correction induced by the GUP has a thermal character, and therefore it can be cast in the form of a shift of the Hawking temperature. Of course, there are also different approaches (see e.g. Ref. [28]), where the corrections do not respect the exact thermality of the spectrum, and thus need not be reducible to a simple shift of the temperature.

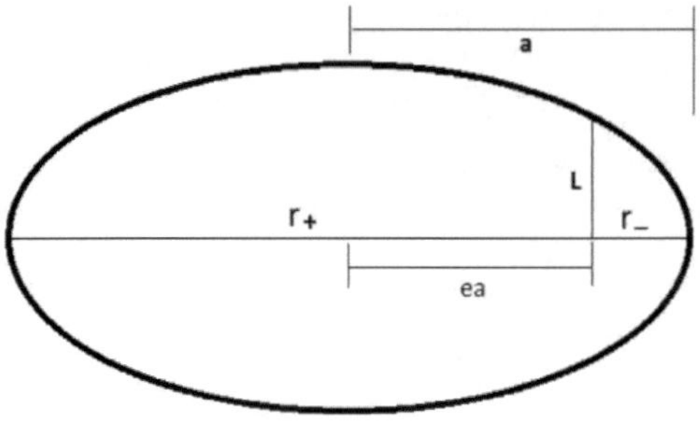

Figure 1. Elements of the ellipse used in the text for the calculations of precession of planetary orbits.

We can legitimately wonder what kind of (deformed) metric would predict a Hawking temperature like the one inferred from the GUP relation (6), for a given β. Since we are interested only in small corrections to the Hawking formula, we can consider a deformation of the Schwarzschild metric of the kind

$$F(r) = 1 - \frac{2\,G_{\rm N}\,M}{r} + \varepsilon\,\frac{G_{\rm N}^2\,M^2}{r^2}\,, \tag{7}$$

and we shall look for the lowest order correction in ε (for a different ansatz on $F(r)$ recently introduced, see Ref. [29]). We see that Eq.(7) is actually the simplest mathematical form, if one supposes that the metric can be expanded in powers of $1/r$. This is nothing else than the well known Eddington–Robertson expansion of a spherically symmetric metric. Note however that, since $R_{\rm H}/r \sim 10^{-5}$ on the surface of the Sun (where $R_{\rm H} = 2\,G_{\rm N}\,M$), the term proportional to ε can still be considered small even if ε is relatively large. The temperature predicted by this deformed Schwarzschild metric is

$$T(\varepsilon) = \hbar\,\frac{F'(r_{\rm H})}{4\pi} = \frac{\hbar}{2\pi\,G_{\rm N}\,M}\,\frac{\sqrt{1-\varepsilon}}{\left(\,1+\sqrt{1-\varepsilon}\,\right)^2}\,, \tag{8}$$

which must coincide with the temperature $T(\beta)$ predicted by Eq. (6), for any given β. This yields the following relation between β and ε,

$$\beta(\varepsilon) = -\pi^2\,\frac{G_{\rm N}\,M^2}{\hbar}\,\frac{\varepsilon^2}{1-\varepsilon}\,. \tag{9}$$

For $|\varepsilon| \ll 1$, to the lowest order in ε, we thus get $\beta = -\pi^2 M^2 \varepsilon^2/(4\,m_{\rm p}^2)$, where we notice that both β and ε are dimensionless. It is now of great interest to observe that Eq. (9) forces us to admit that $\beta < 0$, since $\varepsilon \leq 1$. Although quite unexpected, this might be a suggestion of fundamental importance. It seems that a metric is able to reproduce the GUP-deformed Hawking temperature *only if* the deforming parameter β is *negative*. We already encountered a situation like this when we studied the uncertainty relation formulated on a crystal lattice [30]. This could be a further hint that the physical space-time has actually a lattice or granular structure at the level of the Planck scale.

3. Perihelion precession by deformed Schwarzschild metric

Having established a connection between the GUP parameter β and the deformation ε of the Schwarzschild metric, we are now in a position to compute the physical (possible observable) consequences of such a deformed metric. Here, we consider a particle bound in a orbit around a massive body, typically a planet around the Sun. Again, we roughly follow the treatment of Ref. [31]. The relevant geometrical parameters for an elliptic orbit in a polar coordinates system, with the radial coordinate r which at aphelia and perihelia takes, respectively, the maximum value r_+ and minimum value r_-, are the eccentricity e, the semi-major axis a, and the *semilatus rectum* L (see Fig.1). These geometrical parameters are related by

$$r_\pm = (1 \pm e)\, a \tag{10}$$
$$L = (1 - e^2)\, a \tag{11}$$
$$\frac{2}{L} = \frac{1}{r_+} + \frac{1}{r_-} \, . \tag{12}$$

The angle swept out by the position vector when it increases from r_- to r is then given by the integral

$$\phi(r) - \phi(r_-) = \int_{r_-}^{r} \left[\frac{r_-^2 \left(\frac{1}{F(r)} - \frac{1}{F(r_-)} \right) - r_+^2 \left(\frac{1}{F(r)} - \frac{1}{F(r_+)} \right)}{r_-^2 r_+^2 \left(\frac{1}{F(r_+)} - \frac{1}{F(r_-)} \right)} - \frac{1}{r^2} \right]^{-1/2} \frac{dr}{r^2 \sqrt{F(r)}} \, . \tag{13}$$

The total change in ϕ at every lap is just twice the change as r increases from r_- to r_+. This would equal 2π if the orbit were a closed ellipse, so the total orbital precession in each revolution is given by

$$\Delta\phi = 2\,|\phi(r_+) - \phi(r_-)| - 2\pi \, . \tag{14}$$

We expand the integrand before integrating, and the small parameter is given by $R_{\rm H}/r_-$, or better $R_{\rm H}/L$. Finally the total precession after a single lap, to first order in $R_{\rm H}/L$, is given by

$$\Delta\phi \simeq \frac{6\,\pi\, G_{\rm N}\, M}{L} \left(1 - \frac{\varepsilon}{6} \right) , \tag{15}$$

which, of course, reproduces the usual GR prediction in the limit $\varepsilon \to 0$. This relation should now be compared with known observational data.

4. Solar system data

The perihelion precession for Mercury is by far the best known and measured GR precession in the Solar system. Referring to Ref. [32] for the latest most accurate and comprehensive data, we can report the relation

$$\langle \dot\omega \rangle = \frac{6\,\pi\, G_{\rm N}\, M}{L} \left[\frac{1}{3}(2 + 2\gamma - \bar\beta) + 3 \cdot 10^{-4} \frac{J_2}{10^{-7}}) \right] , \tag{16}$$

where $\langle \dot\omega \rangle$ is the measured perihelion shift, J_2 a dimensionless measure of the quadrupole moment of the Sun, and γ and $\bar\beta$ are the usual Eddington-Robertson expansion parameters. The latest data from helioseismology give $J_2 = (2.2 \pm 0.1) \cdot 10^{-7}$. The measured perihelion shift of Mercury is known to about 0.1% from radar observations of Mercury between 1966 and 1990 [33]. The solar oblateness effect due to the quadrupole moment is then smaller than the observational error, so it can be neglected. Substituting standard orbital elements and physical constants for Mercury and the Sun, we obtain

$$\langle \dot\omega \rangle = \left(1 + \frac{2\gamma - \bar\beta - 1}{3} \right) 42.98\text{"}/\text{century} \, , \tag{17}$$

where we can place a bound of $|2\gamma - \bar{\beta} - 1| \lesssim 3 \cdot 10^{-3}$. Comparing with $\Delta\phi$ from Eq. (15), we get $|\varepsilon| \lesssim 6 \cdot 10^{-3}$, which, replaced in Eq. (9), yields the lower bound

$$|\beta| = \frac{M^2}{4\,m_{\rm p}^2}\,\frac{\pi^2\,\varepsilon^2}{1-\varepsilon} \lesssim 3 \cdot 10^{72}\ . \tag{18}$$

We can also consider the most recent data from the Messenger spacecraft [34], which orbited Mercury in 2011-2013, and improved very much the knowledge of its orbit. Then we can push this bound even lower, to $|2\gamma - \bar{\beta} - 1| \lesssim 7.8 \cdot 10^{-5}$, although the knowledge of J_2 would have to improve simultaneously. If just the error in $|2\gamma - \bar{\beta} - 1|$ were taken into account, this would imply $|\varepsilon| = 2\,\left|2\gamma - \bar{\beta} - 1\right| \lesssim 1.56 \cdot 10^{-4}$ and therefore

$$|\beta| \lesssim 2 \cdot 10^{69}\ . \tag{19}$$

But of course this limit should not be considered completely reliable in this contest, since the less accurate bound on J_2 cannot be brutally neglected, at least in principle. Once again the perihelion shift appears to be one of the most precise tests of GR, a true GR effect not present at all in Newtonian gravity (as it is well known).
We can try to put this limits on a firmer ground by looking for even larger effects of this kind in Binary Pulsars.

5. Pulsar PRS B 1913+16 data

Clearly, binary pulsars are very good candidates for measurements of periastron shifts. Among the known pulsar systems, the best tested pair is the Pulsar PRS B 1913+16. Discovered in 1974 by Hulse and Taylor, this system has become, after 40 years of observations, one of the most reliable celestial laboratories for precise GR measurements. For example, prediction of GR for the period decay rate due to emission of gravitational waves coincides with the measured value up to an error on the 14th decimal figure.

The state of the art on this system is described in Ref. [35]. In Table 1, we report the orbital parameters of interest for us. The parameters e and P_b are called Keplerian parameters, since they are well defined quantities also in the Newtonian theory. On the contrary, $\langle\dot{\omega}\rangle$, γ, $\dot{P}_b$ are known as Damour–Deruelle post-Keplerian parameters [36], quantities typically well defined in GR only. In Ref. [37], Taylor and Weisberg have shown that each GR post-Keplerian parameter can be expressed in terms of the Keplerian parameters and of the unknown masses of the pulsar and its companion, m_1, m_2. The GR theoretical prediction of the periastron shift, $\langle\dot{\omega}\rangle^{\rm GR}$, can be compared with the observed value of $\langle\dot{\omega}\rangle^{\rm Obs}$ again given in Table 1. The relative error can then be defined as $\tilde{\varepsilon}$ by the relation

$$\langle\dot{\omega}\rangle^{\rm GR}(1+\tilde{\varepsilon}) = \langle\dot{\omega}\rangle^{\rm Obs}\,. \tag{20}$$

Parameter	Value
e (Eccentricity)	0.6171334(5)
P_b (days) (Orbital period)	0.322997448911(4)
$\langle\dot{\omega}\rangle$ (deg/year) (Periastron shift)	4.226598(5)
γ (s) (Time dilation-gravitational redshift)	$4.2992(8) \times 10^{-3}$
$\dot{P}_b$ (s/s) (Orbital period decay)	$-2.423(1) \times 10^{-12}$

Table 1. Orbital parameters of PRS B 1913+16 [35]. Figures in parentheses represent estimated uncertainties in the last quoted digit.

On comparing with $\Delta\phi$ in Eq. (15), we get $|\varepsilon| = 6\,|\tilde{\varepsilon}|$. There is one further issue we should care of: measurements are now so precise that the observed value of $\dot{P}_b$ in Table 1 should be corrected for the relative acceleration between the pulsar reference frame and the solar system centre-of-mass frame (see Ref. [38]). Such relative acceleration is mainly due to the fact that the pulsar and our solar system are located in different arms of our Galaxy, at different distances from the galactic centre. We then get $\tilde{\varepsilon} = 8.9 \cdot 10^{-5}$, which means $|\varepsilon| \simeq 5.4 \cdot 10^{-4}$, and this translates into the bound

$$|\beta| \lesssim 2 \cdot 10^{71} \ , \tag{21}$$

which is tighter than the bound (18) coming from "standard" Mercury observations, but weaker than the "Messenger bound" (19) of the previous section. However, note that we do not have here the caveat of the error bounds on J_2.

6. Conclusions

We have shown that a suitable deformation of the Schwarzschild metric can reproduce the Hawking temperature for a black hole, when this is computed from a Generalized Uncertainty Principle. We have found in this way an analytic relation between the deformation parameter of the metric ε and the usual GUP deformation parameter β. In particular, when $\beta \to 0$, we correctly recover GR, and standard quantum mechanics. Neither the geodesic equation, nor the equivalence principle are violated, for any value of β or ε. Well-known astronomical measurements, in the Solar system as well as in binary pulsar systems, allowed us to put constraints on the parameter β. This direction seems to point towards promising research: at present we just deformed the Schwarzschild solution, but a future possibility is to deform the full field equations of GR, in order to get, among other things, a more stringent bound on the GUP parameter β. We would like to conclude by emphasizing once again that, although in the existing literature one can find bounds on β much tighter than those obtained in this paper, they seem to depend, at least partially, either on a specific (non linear) representation of the deformed commutator, or on the hypothesis of a deformation of Poisson brackets, which implies a violation of the equivalence principle. The line of reasoning presented in this paper avoids these possible difficulties.

References

[1] Snyder H S 1947 *Phys. Rev.* **71** 38; Yang C N 1947 *Phys. Rev.* **72** 874; Mead C A 1964 *Phys. Rev.* **135** B 849; Karolyhazy F 1966 *Nuovo Cim. A* **42** 390
[2] Amati D, Ciafaloni M and Veneziano G 1987 *Phys. Lett. B* **197** 81; Gross D J and Mende P F 1987 *Phys. Lett. B* **197** 129
[3] Maggiore M 1993 *Phys. Lett. B* **304** 65
[4] Kempf A, Mangano G and Mann R B 1995 *Phys. Rev. D* **52** 1108
[5] Scardigli F 1999 *Phys. Lett. B* **452** 39
[6] Adler R J and Santiago D I 1999 *Mod. Phys. Lett.* **A14** 1371
[7] Scardigli F and Casadio R 2003 *Class. Quantum Grav.* **20** 3915
[8] Brukner C *et al.* 2012 *Nature Phys.* **8** 393
[9] Marin F, Cerdonio M *et al.* 2013 *Nature Phys.* **9** 71
[10] Bawaj M, Biancofiore C, Marin F *et al.* 2015 *Nature Commun.* **6** 7503 (*Preprint*, arXiv:1411.6410)
[11] Brau F 1999 *J. Phys. A* **32** 7691
[12] Das S and Vagenas E 2008 *Phys. Rev. Lett.* **101** 221301
[13] Pedram P, Nozari K and Taheri S H 2011 *JHEP* **1103** 093
[14] Chang L N, Minic D, Okamura N and Takeuchi T 2002 *Phys. Rev. D* **66** 026003
[15] Nozari K and Akhshabi S 2008 *Chaos Solitons Fractals* **37** 324
[16] Ghosh S 2014 *Class. Quant. Grav.* **31** 025025
[17] Pramanik S and Ghosh S 2013 *Int. J. Mod. Phys. A* **28**(27) 1350131
[18] Scardigli F and Casadio R 2015 *Eur. Phys. J. C* **75** 425
[19] Walker M 1970 *J. Math. Phys. (N.Y.)* **11** 2280

[20] Zee A 2003 *Quantum Field Theory in a Nutshell* (Princeton University Press)
[21] Heisenberg W 1927 *Zeitschr. Physik* **43** 172
[22] Scardigli F 1995 *Nuovo Cim. B* **110** 1029
[23] Adler R J, Chen P and Santiago D I 2001 *Gen. Rel. Grav.* **33** 2101
[24] Cavaglia M, Das S and Maartens R 2003 *Class. Quant. Grav.* **20** L205
[25] Susskind L and Lindesay J 2005 *An Introduction to Black Holes, Information, and the String Theory Revolution* (Singapore: World Scientific); See chapter 10
[26] Nouicer K 2007 *Class. Quant. Grav.* **24** 5917
[27] Scardigli F 2008 Glimpses on the micro black hole Planck phase, (*Preprint,* `arXiv:0809.1832`)
[28] Dvali G and Gomez C 2011 Black hole's quantum N-portrait, (*Preprint,* `arXiv:1112.3359`)
[29] Ali F, Khalil M and Vagenas E 2015 *Europhys. Lett.* **112**(2) 20005
[30] Jizba P, Kleinert H and Scardigli F 2010 *Phys. Rev. D* **81** 084030
[31] Weinberg S 1972 *Gravitation and Cosmology* (New York: John Wiley and Sons)
[32] Will C M 2006 *Living Rev. Relativity* **9**, 3rd Update in `arXiv:1403.7377v1`
[33] Shapiro I I 1990 Solar system tests of general relativity: Recent results and present plans, ed. Ashby N, Bartlett D F and Wyss W *Proceed. 12th Int. Conf. on Gen. Rel. and Grav., Boulder, 1989* (Cabridge, UK, New York, NY: Cambridge University Press)
[34] Verma A K, Laskar J *et al.* 2014 *Astron. Astrophys.* **561** A115
[35] Weisberg J M, Nice D J and Taylor J H 2010 *Astrophys. J.* **722** 1030
[36] Damour T and Deruelle N 1986 *Ann. Inst. H. Poincare (Phys. Theorique)* **44** 263
[37] Taylor J H and Weisberg J M 1989 *Astrophys. J.* **345** 434
[38] Damour T and Taylor J H 1991 *Astrophys. J.* **366** 501

Journal of Physics: Conference Series **701** (2016) 012017 doi:10.1088/1742-6596/701/1/012017

Quantum features of natural cellular automata

Hans-Thomas Elze

Dipartimento di Fisica "Enrico Fermi", Università di Pisa, Largo Pontecorvo 3, I-56127 Pisa, Italia

E-mail: elze@df.unipi.it

Abstract. Cellular automata can show well known features of quantum mechanics, such as a linear rule according to which they evolve and which resembles a discretized version of the Schrödinger equation. This includes corresponding conservation laws. The class of "natural" Hamiltonian cellular automata is based exclusively on integer-valued variables and couplings and their dynamics derives from an Action Principle. They can be mapped reversibly to continuum models by applying Sampling Theory. Thus, "deformed" quantum mechanical models with a finite discreteness scale l are obtained, which for $l \to 0$ reproduce familiar continuum results. We have recently demonstrated that such automata can form "multipartite" systems consistently with the tensor product structures of nonrelativistic many-body quantum mechanics, while interacting and maintaining the linear evolution. Consequently, the Superposition Principle fully applies for such primitive discrete deterministic automata and their composites and can produce the essential quantum effects of interference and entanglement.

1. Why cellular automata?

The *Cellular Automaton Interpretation* of Quantum Mechanics (QM) has recently been outlined by G. 't Hooft [1]. This presents an attempt to redesign the foundations of quantum theory in accordance with essentially classical concepts, such as determinism and existence of ontological states of reality. Which entails a surprising and intuitive explanation of the Born rule and the apparent collapse of quantum mechanical states in measurement processes.

Such an approach may be founded on the observation that quantum mechanical features arise in a large variety of deterministic and, loosely speaking, "classical" models. – So far, however, practically all of these models have been exceptional in that they cannot easily be generalized to cover real phenomena, incorporating interactions and relativity. Yet Cellular Automata (CA) may provide the necessary versatility, as we shall presently continue to discuss [2, 3]. For a large variety of earlier attempts in this field, see, for example, Refs. [4, 5, 6, 7, 8, 9, 10, 11, 12, 13, 14, 15] and further references there.

To begin with, let us recall the linearity of quantum mechanics (QM) as a crucial aspect of the unitary dynamics embodied in the Schrödinger equation. This linearity does not depend on the particular object under study, provided it is sufficiently isolated from anything else. It is reflected in the Superposition Principle and implies the possibilities of interference effects and of non-factorizable states of composite objects, *i.e.* entanglement in multipartite systems.

The linearity of QM has been questioned and nonlinear modifications have been proposed before – notwithstanding various 'no-go' arguments – especially in order to test experimentally the robustness of QM against such nonlinear deformations. Until now no deviations from the predictions of QM have been observed, in particular no indications of nonlinearity.

We have studied a seemingly unrelated *discrete* dynamical theory, *i.e.*, which appears to deviate substantially from quantum theory. Nevertheless, it has been shown with the help of Sampling Theory that the deterministic mechanics of the class of discrete *Hamiltonian CA* can be mapped one-to-one to continuum models pertaining to nonrelativistic QM, which are modified by the presence of a fundamental time scale [2, 3].

Perhaps surprisingly, this construction of a linear relation between CA and QM with a nonzero discreteness scale, compatible with the consistency of the Action Principle underlying the discrete dynamics on one side and the locality of the continuum description on the other, requires that both theories are linear [16].

This result has led us to consider composite objects formed from CA subsystems [17]. – Clearly, QM is special in that it is characterized not only by interference effects, like any classical wave theory would be, but also by the tensor product structures of states and observables applying for composite systems, which allow for entanglement. – It is not obvious that CA can constitute composites which conform with QM in this aspect, in the limit of negligible discreteness scale. A main obstruction could have been the fact that the state space of Hamiltonian CA is not a complex projective space, since the norm of the analogue of state vectors is not conserved by the dynamics; instead there is a conserved two-time correlation function, as we shall recall, which becomes the familiar norm only in the continuum limit.

In Section 2., we will review the earlier results concerning individual CA and briefly indicate the construction of multipartite CA – in such a way that the Superposition Principle is respected and the composition rules of CA are compatible with those of QM. This 'outlandish' perspective based on CA may lead to additional insight in regard to interference and entanglement and eventually to an understanding of nonrelativistic QM in accordance with Ref. [1]. Concluding remarks are presented in Section 3.

2. Natural Hamiltonian CA – action, evolution, conservation laws and composites

We describe classical Hamiltonian CA with countably many degrees of freedom presently in terms of *complex integer-valued* state variables ψ_n^α (also known as *Gaussian integers*), where $\alpha \in \mathbf{N}_0$ denote different degrees of freedom and $n \in \mathbf{Z}$ different states labelled by this discrete *clock variable*. Various equivalent forms of the action for such CA exist, as indicated earlier [2]. We will employ a particularly compact form here, which is useful for the following construction of composite CA in analogy with multipartite QM systems.

Let $\hat{H} := \{H^{\alpha\beta}\}$ denote a self-adjoint matrix of Gaussian integers that will play the role of the Hamilton operator. Furthermore, we introduce the suggestive notation $\dot{O}_n := O_{n+1} - O_{n-1}$, for any quantity O_n depending on the clock variable n. The summation convention for Greek indices, $r^\alpha s^\alpha \equiv \sum_\alpha r^\alpha s^\alpha$, will often allow us to simplify notation further by suppressing them altogether, for example, writing $\psi_n^{*\alpha} H^{\alpha\beta} \psi_n^\beta \equiv \psi_n^* \hat{H} \psi_n$.

Then, with ψ_n^α and $\psi_n^{*\alpha}$ as independent variables, the CA action $\mathcal{S}$ is defined by:

$$\mathcal{S}[\psi, \psi^*] := \sum_n [\frac{1}{2i}(\psi_n^* \dot{\psi}_n - \dot{\psi}_n^* \psi_n) + \psi_n^* \hat{H} \psi_n] \equiv \psi^* \hat{\mathcal{S}} \psi \ , \tag{1}$$

where the operator $\hat{\mathcal{S}}$ is a useful abbreviation, *cf.* below. In order to set up the variational principle, we introduce *integer-valued* variations δf to be applied to a polynomial g as follows:

$$\delta_f g(f) := [g(f + \delta f) - g(f - \delta f)]/2\delta f \ , \tag{2}$$

and $\delta_f g \equiv 0$, if $\delta f = 0$. – Note that variations of terms that are *constant, linear, or quadratic* in integer-valued variables yield analogous results as standard infinitesimal variations of corresponding expressions in the continuum. – Making use of these ingredients, we postulate the variational principle:

[*CA Action Principle*] The discrete evolution of a CA is determined by stationarity of the action under arbitrary integer-valued variations of all dynamical variables, $\delta\mathcal{S} = 0$. •

Let us point out several characteristics of this *CA Action Principle*:
i) While infinitesimal variations do not conform with integer valuedness, there is *a priori* no restriction of integer variations. Hence *arbitrary* integer-valued variations must be admitted.
ii) One could imagine contributions to the action (1) which are of higher than second order in ψ_n or ψ_n^*. However, in view of arbitrary variations $\delta\psi_n^\alpha$ and $\delta\psi_n^{*\alpha}$, such additional contributions to the action must be absent for consistency. Otherwise the number of equations of motion generated by variation of the action, according to Eq. (2), would exceed the number of variables. (A limited number of such remainder terms, which are nonzero only for some fixed values of n, could serve to encode the *initial conditions* for the evolution.)

These features of the *CA Action Principle* are essential in constructing a map between Hamiltonian CA and equivalent quantum mechanical continuum models [2]. – For curiosity, generalizations of the variations defined in Eq. (2) have also been considered, which allow higher than second order polynomial terms in the action. While leading to consistent discrete equations of motion, however, these nonlinear equations generally are beset with undesirable nonlocal features in the corresponding continuum description [16].

2.1. Equations of motion

The equations of motion are obtained by applying the *CA Action Principle* to the action $\mathcal{S}$ of Eq. (1) with the variations as defined in Eq. (2). Thus, variations $\delta\psi_n^*$ and $\delta\psi_n$, respectively, yield discrete analogues of the Schrödinger equation and its adjoint:

$$\dot{\psi}_n = \frac{1}{i}\hat{H}\psi_n \ , \tag{3}$$

$$\dot{\psi}_n^* = -\frac{1}{i}(\hat{H}\psi_n)^* \ , \tag{4}$$

recalling that $\hat{H} = \hat{H}^\dagger$ and $\dot{\psi}_n = \psi_{n+1} - \psi_{n-1}$, *etc.* We remark that the action $\mathcal{S}$ vanishes when evaluated for solutions of these finite difference equations.

Note that by setting $\psi_n^\alpha =: x_n^\alpha + ip_n^\alpha$, with real integer-valued variables x_n^α and p_n^α, and suitably separating real and imaginary parts of Eqs. (3)–(4), the discrete equations assume a form that resembles Hamilton's equations for a network of coupled classical oscillators [18, 19]:

$$\dot{x}_n^\alpha = h_S^{\alpha\beta} p_n^\beta + h_A^{\alpha\beta} x_n^\beta \ , \quad \dot{p}_n^\alpha = -h_S^{\alpha\beta} x_n^\beta + h_A^{\alpha\beta} p_n^\beta \ , \tag{5}$$

where we also split the self-adjoint matrix $\hat{H}$ into real integer-valued symmetric and antisymmetric parts, respectively, $H^{\alpha\beta} =: h_S^{\alpha\beta} + ih_A^{\alpha\beta}$. These equations led us to the name *Hamiltonian CA*; this is furthermore justified by the fact that analogues of Poisson brackets and classical like observables can be introduced here [20].

2.2. Conservation laws

The time-reversal invariant equations of motion of Section 2.1. give rise to conservation laws which are in *one-to-one correspondence* with those of the related Schrödinger equation in the continuum. It is straightforward to demonstrate the following theorem.

[*Theorem A*] For any matrix $\hat{G}$ that commutes with $\hat{H}$, $[\hat{G}, \hat{H}] = 0$, there is a *discrete conservation law*:

$$\psi_n^{*\alpha} G^{\alpha\beta} \dot{\psi}_n^\beta + \dot{\psi}_n^{*\alpha} G^{\alpha\beta} \psi_n^\beta = 0 \ . \tag{6}$$

For self-adjoint $\hat{G}$, defined by Gaussian integers, this relation is about real integer quantities. •

A rearrangement of Eq. (6) yields the corresponding conserved quantity $q_{\hat{G}}$ (using matrix notation, as before):

$$q_{\hat{G}} := \psi_n^* \hat{G} \psi_{n-1} + \psi_{n-1}^* \hat{G} \psi_n = \psi_{n+1}^* \hat{G} \psi_n + \psi_n^* \hat{G} \psi_{n+1} \ , \tag{7}$$

i.e. a real integer-valued two-point correlation function which is invariant under a shift $n \to n+m$, $m \in \mathbf{Z}$. – In particular, for $\hat{G} := \hat{1}$, the corresponding conservation law amounts to a constraint on the state variables:

$$q_{\hat{1}} = 2\mathrm{Re}\ \psi_n^* \psi_{n-1} = 2\mathrm{Re}\ \psi_{n+1}^* \psi_n = \mathrm{const} \ . \tag{8}$$

This can be anticipated to play a similar role for discrete CA as the familiar normalization of state vectors in continuum QM. – We may also define the following symmetrized quantity:

$$\psi_n^* \hat{\mathcal{Q}} \psi_n := \frac{1}{2}\mathrm{Re}\ \psi_n^* (\psi_{n+1} + \psi_{n-1}) \equiv \frac{1}{2}\mathrm{Re}\ \psi_n^{*\alpha} (\psi_{n+1}^\alpha + \psi_{n-1}^\alpha) \ , \tag{9}$$

which, by Eq. (8), is conserved as well.

2.3. Continuum representation

We have shown earlier how to construct an one-to-one invertible map between the dynamics of discrete Hamiltonian CA and continuum QM in presence of a fundamental time scale [2, 3, 16]. Due to this discreteness scale l, continuous time wave functions are *bandlimited*, *i.e.*, their Fourier transforms have only finite support in frequency space, $\omega \in [-\pi/l, \pi/l]$. Hence, we can apply Sampling Theory, in order to reconstruct continuous time signals, wave functions $\psi^\alpha(t)$, from their discrete samples, the CA state variables ψ_n^α, and *vice versa* [21, 22, 23].

Let us represent here the resulting mapping rules obtained through the reconstruction formula provided by *Shannon's Theorem* [21, 22]:

$$\psi_n^\alpha \longmapsto \psi^\alpha(t) \ , \tag{10}$$

$$\psi_{n\pm 1}^\alpha \longmapsto \exp[\mp l \frac{\mathrm{d}}{\mathrm{d}t}] \psi^\alpha(t) = \psi^\alpha(t \mp l) \ , \tag{11}$$

$$\psi^\alpha(nl) \longmapsto \psi_n^\alpha \ , \tag{12}$$

keeping in mind that the continuum wave function is bandlimited.

These results allow to map CA equations of motion, in particular Eqs. (3)–(4) to appropriate continuum versions. Corresponding to Eqs. (6)–(9), there exist analogous conservation laws and conserved quantities, which are found by applying the mapping rules separately to all wave function factors that appear. Thus, for example, we obtain from Eq. (9) the conserved quantity:

$$\mathrm{const} = \psi_n^* \hat{\mathcal{Q}} \psi_n \longmapsto \psi^*(t) \hat{\mathcal{Q}} \psi(t) = \mathrm{Re}\ \psi^*(t) \cosh[l \frac{\mathrm{d}}{\mathrm{d}t}] \psi(t) \tag{13}$$

$$= \psi^{*\alpha}(t)\psi^\alpha(t) + \frac{l^2}{2}\mathrm{Re}\ \psi^{*\alpha}(t) \frac{\mathrm{d}^2}{\mathrm{d}t^2} \psi^\alpha(t) + \mathrm{O}(l^4) \ , \tag{14}$$

which shows l-dependent corrections to the continuum limit, which here amounts to the usual conserved normalization $\psi^{*\alpha}\psi^\alpha = \mathrm{const}$. Similarly, the Schrödinger equation with finite-l correction terms are obtained [2].

Up to this point, our considerations dealt with individual Hamiltonian CA. Based on these results, we will briefly indicate in the following the construction of multipartite systems.

2.4. From single to multipartite Hamiltonian CA [17]

Can discrete CA combine to form composite multipartite systems? – A positive answer to this question is crucial, if we would like the quantum features of CA that we have found, so far, to also include the full extent of the Superposition Principle. It is responsible for interference effects that we can find in single CA. However, a more dramatic consequence in QM is the possibility of entanglement in states or observables.

When scrutinizing possible constructions of multipartite CA, several requirements appear naturally. – One may wonder whether not only the *linearity* of the evolution law but also the *tensor product structure* of composite wave functions finds its analogue here. These are the fundamental ingredients of the usual continuum theory reflected in interference and entanglement. Which should be recovered in the continuum limit ($l \to 0$) of the CA picture, at least. Furthermore, when the discreteness scale l is finite, the dynamics of composites of CA which do not interact among each other should lead to *no spurious correlations* among them. This principle of "no correlations without interactions" holds in all of known physics.

However, we are facing some serious obstacles which seem to prevent satisfying these requirements, when trying to form composites of Hamiltonian CA.

The want-to-be discrete time derivative introduced before, $\dot{O}_n := O_{n+1} - O_{n-1}$, for any quantity O_n depending on the clock variable n, which appears all over in the CA equations of motion and conservation laws, does not obey the *Leibniz rule*:

$$[A_n B_n]^{\cdot} = \dot{A}_n \tfrac{B_{n+1}+B_{n-1}}{2} + \tfrac{A_{n+1}+A_{n-1}}{2}\dot{B}_n \neq \dot{A}_n B_n + A_n \dot{B}_n \ . \tag{15}$$

Similar observations can be expected for other definitions one might come up with. Ignoring this for a moment, consider a trial multi-CA equation of motion analogous to the single-CA Eq. (3):

$$\dot{\Psi}_n = \frac{1}{i}\hat{H}_0 \Psi_n \ , \tag{16}$$

where $\hat{H}_0$ may describe a block-diagonal Hamiltonian in the absence of interactions among the CA. Then, through Eq. (15), the required *factorization* of Eq. (16) is hindered on the left-hand side, since unphysical correlations will be produced among the components of a factorized wave function, such as

$$\Psi_n^{\alpha\beta\gamma\cdots} = \psi_n^\alpha \phi_n^\beta \kappa_n^\gamma \cdots \ , \tag{17}$$

or for superpositions of such factorized terms.

For a bipartite system, for example, we obtain immediately: $\dot{\Psi}_n^{\alpha\beta} = \dot{\psi}_n^\alpha(\phi_{n+1}^\beta + \phi_{n-1}^\beta)/2 + \psi \leftrightarrow \phi \neq \dot{\psi}_n^\alpha \phi_n^\beta + \psi_n^\alpha \dot{\phi}_n^\beta$. – Furthermore, applying the mapping rules of Section 2.3, before taking the limit $l \to 0$, we find that the bilinear terms here do not converge to the appropriate QM expression. Which should be $\partial_t(\psi^\alpha\phi^\beta) = (\partial_t\psi^\alpha)\phi^\beta + \psi^\alpha\partial_t\phi^\beta$, in order to allow the decoupling of two subsystems that do not interact.

This latter problem is a general one of nonlinear terms in the equations of motion of discrete CA, which we discussed before [16]: *The linear map provided by Shannon's Theorem does not commute with the multiplication implied by the nonlinearities.* This follows from the explicit reconstruction formula (or any variant thereof that is linear) [2, 21, 22]. Also on the right-hand side of Eq. (16) we find this obstruction, when trying to map the equation with a factorized wave function to its continuous time description.

2.4.1. The many-time formulation It can be observed that the difficulties we pointed out arise from the implicit assumption that the components of a multipartite CA are *synchronized* to the extent that they share a common clock variable n. In Ref. [17], we have considered a radical way out of the impasse encountered by resorting to a *many-time* formalism. This means giving up synchronization among parts of the composite CA by introducing a set of clock variables, $\{n(1),\ \ldots,\ n(m)\}$, one for each one out of m components.

It may be surprising to encounter this in the present nonrelativistic context, since the many-time formalism has been introduced by Dirac, Tomonaga, and Schwinger in their respective formulations of relativistically covariant many-particle QM or quantum field theory, where a global synchronization cannot be maintained [24, 25, 26].

We replace here the single-CA action of Eq. (1) by an integer-valued multipartite-CA action:

$$\mathcal{S}[\Psi,\Psi^*] := \Psi^*(\sum_{k=1}^{m}\hat{\mathcal{S}}_{(k)}\ +\ \hat{\mathcal{I}})\Psi\ , \tag{18}$$

with $\Psi := \Psi^{\alpha_1\ldots\alpha_m}_{n_1\ldots n_m}$ and, correspondingly, Ψ^* as independent *Gaussian integer* variables; the self-adjoint operator $\hat{\mathcal{I}}$ incorporates interactions between different CA; whereas $\hat{\mathcal{S}}_{(k)}$ is as introduced in Eq. (1), with the subscript $_{(k)}$ indicating that it acts *exclusively* on the pair of indices pertaining to the k-th single-CA subsystem:

$$\Psi^*\hat{\mathcal{S}}_{(k)}\Psi := \sum_{\{n_k\}}[(\mathrm{Im}\ \Psi^{*\ldots\alpha_k\ldots}_{\ldots n_k\ldots}\ \dot{\Psi}^{\ldots\alpha_k\ldots}_{\ldots n_k\ldots}\ +\ \Psi^{*\ldots\alpha_k,\ldots}_{\ldots n_k\ldots}\ H^{\alpha_k\beta_k}_{(k)}\Psi^{\ldots\beta_k\ldots}_{\ldots n_k\ldots}]\ , \tag{19}$$

with summation over *all* clock variables (summation over twice appearing Greek indices remains understood); the $\dot{}$-operation, however, acts only with respect to the explicitly indicated n_k, $\dot{f}(n_k) := f(n_k+1) - f(n_k-1)$, while the single-CA Hamiltonian, $\hat{H}_{(k)}$, requires a matrix multiplication, as before.

Obviously, we can apply the *CA Action Principle* also to the present situation with the generalized action of Eq. (18). This results in the following discrete equations of motion:

$$\sum_{k=1}^{m}\dot{\Psi}^{\ldots\alpha_k\ldots}_{\ldots n_k\ldots}\ =\ \frac{1}{i}(\sum_{k=1}^{m}H^{\alpha_k\beta_k}_{(k)}\Psi^{\ldots\beta_k\ldots}_{\ldots n_k\ldots}\ +\ \mathcal{I}^{\ldots\alpha_k\ldots\ \beta_1\ldots\beta_m}\Psi^{\beta_1\ldots\beta_m}_{\ldots n_k\ldots})\ , \tag{20}$$

together with the adjoint equations; here the interaction $\hat{\mathcal{I}}$, like $\hat{H}_{(k)}$, is assumed to be independent of the clock variables and the $\dot{}$-operation acts only with respect to n_k in the k-th term on the left-hand side.

We have verified in Ref. [17] that the many-time formulation avoids the problems of a single-time multi-CA equation, such as Eq. (16), which we discussed.

In particular, in the absence of interactions with each other, between CA subsystems, $\hat{\mathcal{I}} \equiv 0$, *no unphysical correlations* are introduced among independent CA subsystems.

Furthermore, *continuous multi-time equations* corresponding to Eqs. (20) are obtained by applying the mapping rules given in Section 2.3. to the discrete equations. We find no problem of incompatibility between multiplication according to nonlinear terms *vs.* linear mapping according to *Shannon's Theorem*, since a separate mapping is applied for each one of the clock variables. This effectively replaces $n_k \to t_k$, $k = 1,\ldots,m$, where t_k is a continuous real time variable. In this way, a *modified multi-time Schrödinger equation* is obtained:

$$\sum_{k=1}^{m}\sinh[l\frac{\mathrm{d}}{\mathrm{d}t_k}]\Psi^{\ldots\alpha_k\ldots}_{\ldots t_k\ldots}\ =\ \frac{1}{i}(\sum_{k=1}^{m}H^{\alpha_k\beta_k}_{(k)}\Psi^{\ldots\beta_k\ldots}_{\ldots t_k\ldots}\ +\ \mathcal{I}^{\ldots\alpha_k\ldots\ \beta_1\ldots\beta_m}\Psi^{\beta_1\ldots\beta_m}_{\ldots t_k\ldots})\ , \tag{21}$$

where an overall factor of two from the left-hand side has been absorbed into the matrices on the right. By construction, here Ψ is bandlimited with respect to each variable t_k.

Performing the continuum limit, $l \to 0$, we arrive at the multi-time Schrödinger equation (one power of l^{-1} providing the physical dimension of $\hat{H}_{(k)}$ and $\hat{\mathcal{I}}$) considered by Dirac and

Journal of Physics: Conference Series **701** (2016) 012017 doi:10.1088/1742-6596/701/1/012017

Tomonaga [24, 25]. However, when l is fixed and finite, modifications in the form of powers of $l\mathrm{d}/\mathrm{d}t_k$ arise on its left-hand side, similarly as in single CA case before.

In the present nonrelativistic context, it may be appropriate to identify $t_k \equiv t,\ k = 1, ..., m$, in which case the operator on the left-hand side of Eq. (21), for $l \to 0$, can be simply replaced by $\mathrm{d}/\mathrm{d}t$. This results in the usual (single-time) *many-body Schrödinger equation.*

Finally, the study of the *conservation laws* of the multipartite CA equations of motion can be performed along the lines of Section 2.2. and analogous results have been obtained [17].

2.4.2. Remarks on the Superposition Principle in composite CA The equivalent discrete or continuous many-time equations (20) and (21) are both linear in the CA wave function Ψ. Therefore, superpositions of solutions of these equations also present solutions and the *Superposition Principle* does indeed hold for multipartite Hamiltonian CA.

As in the case of single CA, this entails the fact that these discrete systems – with all variables, parameters, *etc.* presented by Gaussian integers – can produce *interference* effects as in quantum mechanics. Even more interesting, their composites can also show *entanglement*, which is deemed an essential feature of QM. This follows from the form of the equations of motion, which allow for superpositions of factorized states.

For example, in the bipartite case ($k = 1, 2$), assuming that the individual CA are characterized by two degrees of freedom ($\alpha_k = 0, 1$), a time dependent analogue of the totally antisymmetric *Bell state* is given by:

$$\Psi \propto \psi_{n_1}^{\alpha_1=0}\psi_{n_2}^{\alpha_2=1} - \psi_{n_1}^{\alpha_1=1}\psi_{n_2}^{\alpha_2=0} \quad , \tag{22}$$

which may be a solution of appropriate discrete equations of motion.

We conclude with a word of warning concerning the freely used expressions here that we borrowed from QM, such as "wave functions" and "states", in particular. They usually invoke the notion of vectors in a *Hilbert space*, which turns into a complex projective space upon normalization of the vectors. However, as has become obvious in Section 2.2., see Eqs. (8)–(9), and which can be seen similarly in the multipartite CA case [17], as long as the CA are truly discrete ($l \neq 0$), the normalization (squared) of vectors is not among the conserved quantities, hence not applicable, but is replaced by a conserved two-time correlation function instead.

Furthermore, contrary to first impression, perhaps, the envisaged space of states is *not* a Hilbert space, since it fails in two respects: the vector-space and completeness properties are missing. – First of all, the relevant *Gaussian integers* are not complete. Hence the completeness property of the space of states is lacking, which is built with these integers as underlying scalars. Secondly, the integers *only* featuring in all aspects of the CA do not form a field but a commutative ring (for the multiplication of vectors by such scalars there is no multiplicative inverse, such as exists, *e.g.*, for rational, real, or complex numbers). Therefore, we do not have a vector space over a field, as usual in QM. It has to be replaced by a more general structure. This is known as a module over a ring, presently a module over the commutative ring of Gaussian integers. This allows the construction of a linear space endowed with an integer-valued scalar product, *i.e.* a *unitary space.* Taking its incompleteness into account, then, the space of states in the presented CA theory can be classified as a *pre-Hilbert module over the commutative ring of Gaussian integers.*

In any case, we conclude that superpositions of states, interference effects, and entanglement, as in QM, all can be found already on the "primitive" level of the presently considered natural Hamiltonian CA, discrete single or multipartite systems which are characterized by (complex) integer-valued variables and couplings.

3. Conclusion

This presents a brief review of earlier work which has demonstrated surprising quantum features arising in integer-valued, hence "natural", *Hamiltonian cellular automata* [2, 3, 16, 17, 20].

The study of this particular class of CA is motivated by 't Hooft's *Cellular Automaton Interpretation of QM* [1] and various recent attempts to construct models which may eventually lead to demonstrating that the essential features of QM can all be understood to emerge from pre-quantum deterministic dynamics.

The single CA we have considered allow practically for the first time to reconstruct quantum mechanical models with nontrivial Hamiltonians in terms of such systems with a finite discreteness scale. – Furthermore, we have recently extended this study by describing *multipartite systems*, analogous to many-body QM. Not only is this useful for the construction of more complex models *per se* (especially with a richer structure of energy spectra), but it is also necessary, in order to extend the *Superposition Principle* of QM to a description at the CA level. We find that it can be introduced already there to the fullest extent, compatible with a tensor product structure of multipartite states, which entails the possibilities of their *interference* and *entanglement*.

Surprisingly, we have been forced – in our approach employing Sampling Theory for the map between CA and an equivalent continuum picture – to introduce a many-time formulation, which only appeared in relativistic quantum mechanics before, as introduced by Dirac, Tomonaga, and Schwinger [24, 25, 26]. This points towards a crucial further step in these developments, which is still missing, namely a CA model of *interacting quantum fields*. Without the possibility of interacting multipartite CA with quantumlike features, as described here, it is hard to envisage a CA picture of dynamical fields spread out in spacetime. Last not least, a detailed model is still lacking in which the Born rule can be related quantitatively to the "ordinary" statistical mechanics of deterministic dynamical degrees of freedom.

Acknowledgments

It is a pleasure to thank G. Grössing and J. Walleczek for the kind invitation to the conference "Emergent Quantum Mechanics" (Vienna, October 2015) and H. Batelaan, A. Caticha and G. 't Hooft for discussions. Support by the Fetzer-Franklin Fund is gratefully acknowledged.

References

[1] 't Hooft G 2014 The Cellular Automaton Interpretation of Quantum Mechanics. A View on the Quantum Nature of our Universe, Compulsory or Impossible? *Preprint* arXiv:1405.1548
[2] Elze H-T 2014 Action principle for cellular automata and the linearity of quantum mechanics *Phys. Rev.* A **89** 012111
[3] Elze H-T 2014 The linearity of quantum mechanics from the perspective of Hamiltonian cellular automata *J. Phys.: Conf. Ser.* **504** 012004
[4] 't Hooft G 1990 Quantization of discrete deterministic theories by Hilbert space extension *Nucl. Phys.* B **342** 471
[5] 't Hooft G, Isler K and Kalitzin S 1992 Quantum field theoretic behavior of a deterministic cellular automaton *Nucl. Phys.* B **386** 495
[6] 't Hooft G 1997 Quantummechanical behaviour in a deterministic model *Found. Phys. Lett.* **10** 105
[7] Haba Z and Kleinert H 2002 Towards a simulation of quantum computers by classical systems *Phys. Lett.* A **294** 139
[8] Elze H-T and Schipper O 2002 Time without time: a stochastic clock model *Phys. Rev.* D **66** 044020
[9] Grössing G 2004 From classical Hamiltonian flow to quantum theory: derivation of the Schrödinger equation *Found. Phys. Lett.* **17** 343
[10] Khrennikov A 2005 Generalizations of quantum mechanics induced by classical statistical field theory *Found. Phys. Lett.* **18** 637
[11] Margolus N 2008 Mechanical systems that are both classical and quantum *Preprint* arXiv:0805.3357; *do.* 2015 *Lect. Notes Comput. Sc.* **9099** 169

[12] Blasone M, Jizba P, Scardigli F and Vitiello G 2009 Dissipation and quantization for composite systems *Phys. Lett.* A **373** 4106
[13] Sakellariadou M, Stabile A and Vitiello G 2011 Noncommutative spectral geometry, algebra doubling and the seeds of quantization *Phys. Rev.* D **84** 045026
[14] Acosta D, Fernandez de Cordoba P, Isidro J M and Santander J L G 2012 An entropic picture of emergent quantum mechanics *Int. J. Geom. Meth. Mod. Phys.* **9** 1250048
[15] Bisio A, D'Ariano G and Tosini A 2015 Quantum field as a quantum cellular automaton: the Dirac free evolution in one dimension *Ann. Phys.* **354** 244
[16] Elze H-T 2015 Are nonlinear discrete cellular automata compatible with quantum mechanics? *J. Phys.: Conf. Ser.* **631** 012069
[17] Elze H-T 2016 Multipartite cellular automata and the superposition principle *Int. J. Qu. Info. (IJQI)*, to appear
[18] Heslot A 1985 Quantum mechanics as a classical theory *Phys. Rev.* D **31** 1341
[19] Skinner T E 2013 An exact mapping between the states of arbitrary N-level quantum systems and the positions of classical coupled oscillators *Phys. Rev.* A **88** 012110
[20] Elze H-T 2014 Quantumness of discrete Hamiltonian cellular automata *EPJ Web of Conferences* **78** 02005
[21] Shannon C E 1949 Communications in the presence of noise *Proc. IRE* **37** 10
[22] Jerri A J 1977 The Shannon Sampling Theorem — Its Various Extensions and Applications: A Tutorial Review *Proc. IEEE* **65** 1565
[23] Kempf A 2010 Spacetime could be simultaneously continuous and discrete in the same way that information can *New J. Phys.* **12** 115001
[24] Dirac P A M 1932 Relativistic Quantum Mechanics *Proc. Roy. Soc. London* A **136** 453
[25] Tomonaga S 1946 On a Relativistically Invariant Formulation of the Quantum Theory of Wave Fields *Prog. Theor. Phys.* **1** (2) 27
[26] Schwinger J 1948 Quantum Electrodynamics. I. A covariant Formulation *Phys. Rev.* **74** 1439

Atoms of Spacetime and the Nature of Gravity

Thanu Padmanabhan
IUCAA, Pune University Campus, Ganeshkhind, Pune 411007, India.
E-mail: paddy@iucaa.in

Abstract. I clarify the differences between various approaches in the literature which attempt to link gravity and thermodynamics. I then describe a new perspective[1] based on the following features: (1) As in the case of any other matter field, the gravitational field equations should also remain unchanged if a constant is added to the Lagrangian; in other words, the field equations of gravity should remain invariant under the transformation $T^a_b \to T^a_b + \delta^a_b(\text{constant})$. (2) Each event of spacetime has a certain number (f) of microscopic degrees of freedom ('atoms of spacetime'). This quantity f is proportional to the area measure of an equi-geodesic surface, centered at that event, when the geodesic distance tends to zero. The spacetime should have a zero-point length in order for f to remain finite. (3) The dynamics is determined by extremizing the heat density at all events of the spacetime. The heat density is the sum of a part contributed by matter and a part contributed by the atoms of spacetime, with the latter being $L_P^{-4} f$. The implications of this approach are discussed.

1. Linking Gravity with Thermodynamics: Comparison of different approaches

The idea that gravitational field equations could be interpreted using (or derived from) thermodynamic arguments has been explored by many people from widely different perspectives. (See e.g., [1–18]). There is a tendency in the literature to club together these – very different – attempts as essentially the same or, at least, as being very similar. Such a point of view is technically incorrect and, given this tendency, it is useful to clarify the differences between the various approaches, as regards their assumptions, physical motivation and the generality of the results. I will begin with a series of comments aimed at this task:[2]

(1) To begin with, one must sharply distinguish between (i) the attempts concerned with the *derivation* of the field equations by thermodynamic arguments (like e.g., [4, 7–12, 14, 15]) and (ii) the attempts related to the *interpretation* of the field equations in thermodynamic language (like e.g., [1–3, 5, 6, 13, 16–21]). The latter is as important as the former because the existence of a purely thermodynamic interpretation for the field equations is vital for the overall consistency of the programme. It is rather self-defeating to derive the field equations $G^a_b = 8\pi T^a_b$ from thermodynamic arguments and then interpret them in the usual

[1] I thank the organizers for the excellent hospitality during the EmQM15 meeting, and the participants for several useful discussions. In this written version, I elaborate on several conceptual and technical aspects behind the results I presented in my keynote address, which I could only briefly touch upon in my talk. I also address various questions raised by the participants during the discussions.

[2] There are also numerous other attempts which derive/interpret the *linearized* field equations – rather than the *exact* equations – from thermodynamic considerations. In what follows, I am only concerned with attempts which derive/interpret the *exact* field equations.

geometrical language! If gravity is thermodynamic in nature, then the gravitational field equations must be expressible in a *thermodynamic* language. This crucial feature has not been given due recognition in the literature. Unless the final result has an interpretation in thermodynamic language, such a derivation of the field equations is conceptually rather incongruous.

As an example of what I mean by such an interpretation, let me recall the following result. It can be shown that [12] the evolution of geometry can be interpreted in thermodynamic terms, as heating and cooling of null surfaces, through the equation:

$$\int_{\mathcal{V}} \frac{d^3x}{8\pi L_P^2} \sqrt{h} u_a g^{ij} \pounds_\xi p^a_{ij} = \epsilon \frac{1}{2} k_B T_{\text{avg}} (N_{\text{sur}} - N_{\text{bulk}}) \tag{1}$$

where

$$N_{\text{sur}} \equiv \int_{\partial\mathcal{V}} \frac{\sqrt{\sigma}\, d^2x}{L_P^2}; \quad N_{\text{bulk}} \equiv \frac{|E|}{(1/2) k_B T_{\text{avg}}} \tag{2}$$

are the degrees of freedom in the surface and bulk of a 3-dimensional region $\mathcal{V}$ and T_{avg} is the average Davies-Unruh temperature [22, 23] of the boundary. The h_{ab} is the induced metric on the $t =$ constant surface, $p^a_{bc} \equiv -\Gamma^a_{bc} + \frac{1}{2}(\Gamma^d_{bd}\delta^a_c + \Gamma^d_{cd}\delta^a_b)$, and $\xi^a = N u^a$ is the proper-time evolution vector corresponding to observers moving with four-velocity $u_a = -N\nabla_a t$. The factor $\epsilon = \pm 1$ ensures the correct result for either sign of the Komar energy E. The time evolution of the metric in a region (described by the left hand side), which can be interpreted [16] as heating/cooling of the spacetime, arises because $N_{\text{sur}} \neq N_{\text{bulk}}$. In any *static* spacetime [20], on the other hand, $\pounds_\xi(...) = 0$, leading to "holographic equipartition": $N_{\text{sur}} = N_{\text{bulk}}$. This result translates gravitational dynamics into thermal evolution of the spacetime. The validity of Eq. (1) for all observers (i.e., foliations) ensures the validity of Einstein's equations.

In fact, *no* thermodynamic derivation of field equations in the literature actually obtains the *tensorial* form of field equation $G^a_b = 8\pi T^a_b$. What is *always done* is to obtain an equation of the form $G^a_b v_a v^b = 8\pi T^a_b v_a v^b$ (where v^a is either a timelike or null vector) and postulate its validity for all v^a. So it is important to understand the physical meaning of such an equation, especially the left hand side, for a given class of v^a. This will be a recurrent theme which I will elaborate on later sections.

(2) Many thermodynamic derivations of field equations available in the literature, work with the assumption that the entropy of a horizon is proportional to its area (e.g., [4, 10, 11, 14, 15]) and attempt to introduce thermodynamic arguments centered around it. *It is likely that such derivations miss some essential physics.* The connection between gravity and thermodynamics, motivated historically from the laws of black hole mechanics and the membrane paradigm [1–3], transcends Einstein's theory. In a more general class of theories, the (Wald) entropy of the horizon is *not* proportional to its area. One should therefore distinguish approaches in this subject which are specially tuned to Einstein gravity (and uses the entropy-area proportionality) from a broader class of approaches (like e.g. [7, 8, 17, 21]) because the latter ones, being more general, probably captures the underlying physics better. The above criticism is also valid for approaches based on entanglement entropy when it is assumed to be proportional to the horizon area.

(3) Another feature which distinguishes different approaches in the literature is whether the field equations are derived from a variational principle or from some other procedure. I am personally in favour of approaches which use a variational principle because they could offer a better window into microscopic physics. What is more, the approaches which does *not* use a variational principle are very limited in their scope. For example, it is virtually impossible to generalize such models beyond Einstein's theory. (In contrast, the very first approach

Journal of Physics: Conference Series **701** (2016) 012018 doi:10.1088/1742-6596/701/1/012018

which used a thermodynamic variational principle to derive the field equations [7], obtained the field equations for all Lanczos-Lovelock models at one go.) Some of these approaches like, for example, those which use the Raychaudhuri equation also have non-trivial technical issues [24, 25].

Even amongst the approaches which use variational principles, we need to distinguish between (i) those which vary the geometry (viz., the metric in some form, sometimes in a rather disguised manner) and (ii) those which vary some auxiliary vector field, keeping the metric fixed. Many approaches involving holographic concepts and entanglement entropy do vary the geometry in some form and I prefer approaches which vary an auxiliary vector to these approaches. (After all, if you are going to vary the metric/geometry in an extremum principle, why not just use the Einstein-Hilbert action and be done with it ?!) An example [26] of an extremum principle which does *not* vary the metric, is given by the functional

$$Q_{\rm tot} = \int \sqrt{\gamma}\, d^2x\, d\lambda \left(T^a_b \ell_a \ell^b + \left[2\eta\sigma_{ab}\sigma^{ab} + \zeta\theta^2\right]\right) \tag{3}$$

Here, σ_{ab} and θ are the shear and expansion of a null congruence $\ell^a(x)$, $\eta = 1/16\pi L_P^2$, $\zeta = -1/16\pi L_P^2$ are the shear and bulk viscous coefficients of a null fluid [1–3] and the integrand can be interpreted as the rate of generation of heat ('dissipation without dissipation'; see [27, 28]) due to matter and gravity. Varying $Q_{\rm tot}$ with respect to ℓ^a and demanding that the extremum should hold for all ℓ^a (i.e., for all null surfaces) will lead to Einstein's equations. (We will say more about this in Sec. 5.2.) Such a variational principle – and others of similar genre which we will discuss later – treats the geometry as fixed and does not vary the metric.

(4) At a more fundamental level, the horizon entropy cannot be finite unless some kind of discreteness exists in the spacetime near Planck scales. This is clear in the case of entanglement entropy, which is a manifestly divergent quantity (see e.g., [29,30]) and needs to be regularized by some ad-hoc cut-off; but it is implicit in all approaches. So, unless we have a model which captures *at least some of* the quantum gravitational effects on the spacetime, any derivation of field equations using a finite value for entropy is, at best, incomplete.

(5) Finally, let me emphasize that *gravity cannot be an entropic force.* This was ably demonstrated by Matt Visser [31] by an argument which uses (essentially) elementary vector analysis. It is trivial to prove, in the Newtonian limit, that a conservative force $\mathbf{f} = -\nabla\phi$ cannot, in general, be expressed in the entropic form $\mathbf{f} = T\nabla S$ if T is the Davies–Unruh temperature that depends on the magnitude of the acceleration $|\nabla\phi|$. The relation $-\nabla\phi = T\nabla S$ implies that the level surfaces of ϕ coincide with those of S allowing us to introduce a function $S = S(\phi)$. This, in turn, implies $T(dS/d\phi) = -1$ and hence the level surface of ϕ coincide with the level surfaces of T. But since T depends only on $|\nabla\phi|$, this requires the level surfaces of $|\nabla\phi|$ to coincide with those of ϕ. This condition is, in general, impossible to satisfy and can happen only in situations of high symmetry (for example, spherical, cylindrical, planar etc.). It would be preferable if the phrase "entropic gravity" is *not used as a rather generic term* to describe the different approaches in this subject, for the simple reason that gravity cannot be an entropic force.

To summarize, there exist many different attempts in the literature to link gravity and thermodynamics. All of these are *not* equivalent – either conceptually or technically – and it is also likely that at least some of them are fundamentally flawed or incomplete.

The approach I have been pursuing – which I will describe here – is marked by the following features: (1) Much of it works for a wide class of theories, more general than Einstein's gravity. In particular, the results hold for theories in which entropy is *not* proportional to horizon area.

(2) The field equations are derived from a thermodynamic extremum principle in which the geometry is not varied but some other auxiliary vector field is varied. (3) The resulting field equations are interpreted in a thermodynamic language and not in a geometric language. (4) The introduction of a zero-point length to the spacetime by quantum gravitational effects allows us to provide a microscopic basis for the variational principle which is used.

Here, I will concentrate on developing this perspective *from first principles* in a streamlined manner. Obviously this will require us to make some educated guesses but I shall argue that these guesses are well-motivated and the results are quite rewarding. In particular, I will describe the following two aspects:

- I will demonstrate [32] a deep connection between two aspects of gravity which are usually considered in the literature to be quite distinct. The first is the fact that gravity seems to be immune to the shift in the zero level of the energy, i.e, to the shift in the value of cosmological constant. Second is the feature I mentioned above, viz., gravitational dynamics can be reinterpreted in a purely thermodynamic language. I will show how the first feature *leads to* the second and, in fact, provides a simple and natural motivation to consider the heat density of the null surfaces as a key physical entity.
- Much of the previous work treated the spacetime as analogous to a fluid and investigated its properties in the *thermodynamic limit*. The next, deeper, level of description of a fluid will be the *kinetic theory* which recognizes the discreteness and quantifies it in terms of a distribution function for its molecules. I will describe an attempt [32] to do the same for the spacetime by introducing a distribution function for the atoms of spacetime (which will count the microscopic degrees of freedom) and relating it to the extremum principle which, in turn, will lead to the field equations.

2. Is spacetime metric a dynamical variable?

The principle of equivalence, along with principle of general covariance, strongly suggest that gravity is the manifestation of a curved spacetime[3]. described by a non-trivial metric $g_{ab}(x)$. The *kinematics* of gravity, viz. how a given gravitational field affects matter, can then be determined by postulating the validity of special relativistic dynamics in all freely falling frames. This will lead to the condition $\nabla_a T^a_b = 0$ for the energy momentum tensor of matter, which encodes the influence of gravity on matter.

Unfortunately, we do not have any equally elegant guiding principle to determine the *dynamics* of gravity, viz. how matter determines the evolution of the spacetime metric. The dynamics is contained in the gravitational field equation which – in Einstein's theory – is assumed to be given by $G^a_b = 8\pi T^a_b$. (In a more general class of theories, like e.g, Lanczos-Lovelock models, the left hand side will be replaced by a more complicated second rank, symmetric, divergence-free tensor.) One can obtain this equation, as Einstein did, by (i) assuming that the right hand side *must* be T^a_b and (ii) by constructing a second rank, symmetric, divergence-free tensor from the metric containing upto second derivatives. Alternatively, as Hilbert did, one can write down a suitable scalar Lagrangian and vary it with respect to the metric and obtain the field equations.

In either procedure, one tacitly assumes that the spacetime metric is a dynamical variable with a status similar to, say, that of the gauge potential A_j in electromagnetism. This belief is based on the fact that Einstein's equation is a second order differential equation for the metric just as Maxwell's equation is a second order differential equation for A_j. It is in the same spirit that we justify varying the metric in the Hilbert action (as analogous to varying A_j in the electromagnetic action) to obtain the Einstein's equation. Further, once we have a classical

[3] I use the signature $(-+++)$ and will set $\hbar = 1, c = 1$ so that $G = L_P^2$. Occasionally, I will also set $G = 1$ when no confusion is likely to arise.

action $A_H[g_{ab}]$ in which g_{ab} is varied to get the field equations, it is tempting to think of a quantum theory, defined through a path integral over $\exp iA_H[g_{ab}]$ (or in some other equivalent manner) with the metric playing the role of a quantum variable.

But, given the fact that spacetime geometry is conceptually very different from an external field propagating in it, this assumption – viz., that metric is a dynamical variable similar to other fields – is indeed nontrivial. Further, if varying the metric in the Hilbert action is *not* the appropriate way to obtain the classical theory, then one is forced to think afresh about all the quantum gravity programmes. Interestingly enough, this textbook procedure – of treating metric as a dynamical variable, accepted without a second thought – is by no means a unique way to obtain Einstein's equation. In fact, it is probably *not* the most natural or efficient procedure. One can come up with alternative approaches *and physically motivated extremum principles,* leading to Einstein's equation, in which the metric is *not* a dynamical variable. Let me describe one such approach.

The field equations we seek should be a relativistic generalization of Newton's law of gravity $\nabla^2\phi \propto \rho$. A natural way of generalizing this law is to begin by noticing that: (i) The energy density in the right hand side $\rho = T_{ab}u^a u^b$ is foliation/observer dependent where u^i is the four velocity of an observer. There is no way we can keep u^i out of it. (ii) We know from the principle of equivalence that g_{ab} plays the role of ϕ/c^2. So a covariant, scalar generalization of the left hand side, $\nabla^2\phi$, could come from the curvature tensor – which contains the second derivatives of the metric. Any such generalization *must depend on the four-velocity u^i of the observer* since the right hand side does. (iii) It is perfectly acceptable for the left hand side *not* to have second *time* derivatives of the metric, in the rest frame of the observer, since they do not occur in $\nabla^2\phi$.

To obtain a scalar analogous to $\nabla^2\phi$, having *spatial* second derivatives, we first project the indices of R_{abcd} to the space orthogonal to u^i, using the projection tensor $P^i_j = \delta^i_j + u^i u_j$, thereby obtaining the tensor $\mathcal{R}_{ijkl} \equiv P^a_i P^b_j P^c_k P^d_l R_{abcd}$. The only scalar we can get from $\mathcal{R}_{ijkl}$ is $\mathcal{R}^{-2} \equiv \mathcal{R}^{ij}_{ij}$ where $\mathcal{R}$ can be thought of as the radius of curvature of the space.[4] The natural generalization of Newton's law $\nabla^2\phi \propto \rho$ is then given by $\mathcal{R}^{-2} \propto \rho = T_{ab}u^a u^b$. Working out the left hand side (see e.g., p. 259 of [33]) and fixing the proportionality constant from the Newtonian limit, one finds that

$$G_{ab}u^a u^b = 8\pi T_{ab}u^a u^b. \tag{4}$$

If this scalar equation should hold for all observers (general covariance) then we need $G_{ab} = 8\pi T_{ab}$ which is the standard result. Demanding that $\mathcal{R}^{-2} = 8\pi\rho$ holds for each observer, captures the geometric statement – viz. that energy density curves space as viewed by any observer – in a nice manner and is indeed the most natural generalization of Newton's law: $\nabla^2\phi \propto \rho$.

So, in this approach, the fundamental equation determining the geometry is Eq. (4) – which should hold for all normalized, timelike vectors u^i at each event of spacetime – rather than the standard equation $G_{ab} = 8\pi T_{ab}$. While the two formulations are algebraically equivalent, they are conceptually rather different. In the conventional approach to derive $G_{ab} = 8\pi T_{ab}$, we do not invoke any special class of observers. Instead, we *assume* that the right hand side of the field equation *must* be T^a_b and look for a generally covariant, divergence-free, second-rank tensor built from geometry to put on the left hand side. (Alternatively, we look for a scalar Lagrangian made from geometrical variables). But the source in Newtonian gravity is actually $T_{ab}u^a u^b$ which *does* involve an extra four-velocity for its definition. If we introduce observers with four-velocity u^i – and in the end demand that the equation should hold for all u^i – we obtain the same gravitational field equations by a different route. This approach to *dynamics* brings it closer to the way we handled the *kinematics* by introducing the freely falling observers.

[4] This $\mathcal{R}_{ijkl}$ and $\mathcal{R}$ should *not* to be confused with the curvature tensor ${}^3R_{ijkl}$ and the curvature scalar ${}^3R$ of the 3-space orthogonal to u^i.

The real importance of this approach stems from the fact that it allows us to construct different kind of extremum principles which will lead to the gravitational field equations, *without treating the metric as a dynamical variable!* Since this approach introduces an extra vector field u^j into the fray, one can consider an extremum principle in which we vary u^i – which makes physical sense in terms of changing the observer – instead of the metric. It is now possible, for example, to obtain the field equations by varying u^i in a variational principle with the Lagrangian $L_1 \propto (G^a_b - 8\pi T^a_b)u_a u^b$ and demanding that the extremum must hold for all u^i. In fact, we can also use the Lagrangian $L_2 \propto (R^a_b - 8\pi T^a_b)u_a u^b$. Varying u^j in the resulting action, after imposing the constraint $u^2 = -1$ and demanding that the extremum should hold for all u^i, will lead to the equation $R^a_b - 8\pi T^a_b = \lambda(x)\delta^a_b$ where $\lambda(x)$ is the Lagrange multiplier. Using Bianchi identity and $\nabla_a T^a_b = 0$, we will recover the field equations except for an undetermined cosmological constant [34]. Removing a total divergence from $R^a_b u_a u^b$ we see that this is equivalent to a variational principle based on the functional[5]

$$A[u^i] = \int d^4x \sqrt{-g}\, \left[(\nabla_i u^i)^2 - \nabla_j u^i \nabla_i u^j - 8\pi\rho\right] \tag{5}$$

Varying u_i in $A[u^i]$ and demanding the extremum to hold for all u^i will lead to Eq. (4) except for an undetermined cosmological constant.[6] So, one can indeed obtain the classical field equations for gravity *without varying the metric* in any action principle.

The existence of such alternative variational principles takes away the motivation to treat the metric as a dynamical variable either classically or quantum mechanically. If this alternative procedure – or a variant of it – is the correct interpretation of classical gravity, then Hilbert action has no meaning classically (and hence in a quantum mechanical path integral). In the classical theory, what ultimately matters is the field equation and the rest is just window dressing. But the distinction between these two approaches is vital when we want to bring together the principles of quantum theory and gravity. If metric is not a dynamical variable in the classical theory – and the correct classical variational principle involves varying some other auxiliary variable like u^i rather than the metric – it makes no sense to quantise the metric. Such an attempt will, at best, be similar to quantizing the velocity or density field of a material medium. While gravitons will emerge with the same conceptual status as, say, phonons the attempt will not lead to a complete quantum description of the spacetime. So, the alternative paradigm suggests a completely different picture about the quantum nature of spacetime.

This point of view, that the metric is not a dynamical variable, receives independent support from the tantalizing relationship between gravitational dynamics and thermodynamics of null surfaces. As I mentioned earlier, one can provide a purely thermodynamic interpretation to Eq. (4) quite easily but *not* to $G^a_b = 8\pi T^a_b$. We need the extra vector field u^j for this interpretation, just as we need it to define an energy density. We will see later that the situation is still better when we use a null vector in place of the timelike vector.

3. A guiding principle for dynamics and its consequences

One major problem with the conventional approach to gravitational dynamics is that we lack a good guiding principle to determine the field equations. My first task is to take care of this. I

[5] This expression is very similar to the structure seen in ADM Hamiltonian (missing only a ${}^3R$ term) but, of course, here we are varying the vector field u^i and *not* the metric g_{ij}. In fact, one can add any functional of the metric to the Lagrangian and it would make no difference since the metric is not varied.

[6] Usually, if you vary a quantity q_A in an extremum principle, you get an evolution equation for q_A. Here we vary u_i in Eq. (5) but get the equation constraining g_{ab}! This comes about because, after varying u_i, we demand the equation to hold for all u_i to take care of all observers. (Recall that this is also done in *all* attempts to derive field equations by thermodynamic arguments.) While conceptually different from the usual extremum principles, it is perfectly well-defined and makes physical sense.

will begin by postulating a guiding principle [12, 13], which turns out to be as powerful as the principle of equivalence, in obtaining the gravitational dynamics.

Let us recall that the equations of motion for matter, derived from an action principle, remain invariant if we add a constant to the matter Lagrangian, *i.e.*, under the change $L_m \to L_m +$ constant. This encodes the principle that the zero level of energy density does not affect the dynamics. Motivated by this fact, it seems reasonable to postulate that the gravitational field equations should not break this symmetry, which is already present in the matter sector. Since T_{ab} will occur, in one form or another, as the source for gravity (as can be argued from the principle of equivalence and considerations of the Newtonian limit), we postulate that:

- ▶ The extremum principle that determines the dynamics of spacetime must be invariant under the change $T^a_b \to T^a_b +$ (constant) δ^a_b.

This principle leads to two useful results:

First, this principle rules out the possibility of varying the metric tensor g_{ab} in a covariant, local, action principle to obtain the field equations! It is easy to prove [35] that if (i) the action is obtained from a local, covariant Lagrangian integrated over a region of spacetime with the covariant measure $\sqrt{-g}\, d^4x$ and (ii) the field equations are obtained through unrestricted variation[7] of the metric in the action, then the field equations *cannot* remain invariant under $T^a_b \to T^a_b +$ (constant) δ^a_b. In fact, $L_m \to L_m +$ constant is no longer a symmetry transformation of the action if the metric is treated as the dynamical variable. Therefore, any variational principle we come up with cannot have g_{ab} as the dynamical variable.

As I highlighted in the last section, this need not scare us; it is certainly possible to come up with variational principles leading to the field equations in which the metric is not a dynamical variable. This brings us to the second result: The most natural structure, built from T^a_b, which maintains the required invariance under $T^a_b \to T^a_b +$ (constant) δ^a_b, is given by

$$\mathcal{H}_m[\ell_a] \equiv T_{ab}\ell^a\ell^b \tag{6}$$

where ℓ_a is a null vector.[8] This is very similar to the right hand side of Eq. (4) except that we are now using a *null* vector ℓ_a rather than a *timelike* vector u^i to implement our guiding principle. Demanding that the equation

$$G_{ab}\ell^a\ell^b = R_{ab}\ell^a\ell^b = 8\pi T_{ab}\ell^a\ell^b \tag{7}$$

holds for all null vectors ℓ_a at each event will also lead to $G^a_b = 8\pi T^a_b + \Lambda\delta^a_b$ (where Λ is the undetermined cosmological constant) and our guiding principle points towards such an approach. While $T^a_b u_a u^b$, changes under $T^a_b \to T^a_b +$ (constant) δ^a_b, the $T^a_b \ell_a \ell^b$, remains invariant, which is what we want. But, $\rho \equiv T_{ab}u^a u^b$ has a clear physical meaning as the energy density measured by an observer with velocity u^i, but the physical meaning of $\mathcal{H}_m[\ell_a] \equiv T_{ab}\ell^a\ell^b$ is not obvious. Our next task is to clarify that.

In the case of an ideal fluid, with $T^a_b = (\rho + p)u^a u_b + p\delta^a_b$, the combination $T^a_b \ell_a \ell^b$ is actually the *heat density* $\rho + p = Ts$ where T is the temperature and s is the entropy density of the fluid. (The last equality follows from Gibbs-Duhem relation and we have chosen the null vector with $(\ell.u)^2 = 1$ for simplicity.) The invariance of $T^a_b \ell_a \ell^b$ under $T^a_b \to T^a_b +$ (constant) δ^a_b reflects the

[7] The second condition rules out unimodular theories and their variants, in which we vary the metric keeping $\sqrt{-g}$ fixed; I do not think we have a sound physical motivation for this approach.

[8] We want to introduce a minimum number of extra variables. In d-dimensional spacetime, the null vector with $(d-1)$ degrees of freedom is the minimum we need. In contrast, if we use, say, a combination $T^{ab}V_{ab}$ with a symmetric traceless tensor V_{ab}, in order to maintain the invariance under $T^a_b \to T^a_b +$ (constant) δ^a_b, then we need to introduce $(1/2)d(d+1) - 1$ degrees of freedom; in $d = 4$, this introduces nine degrees of freedom, which is equivalent to introducing three null vectors rather than one.

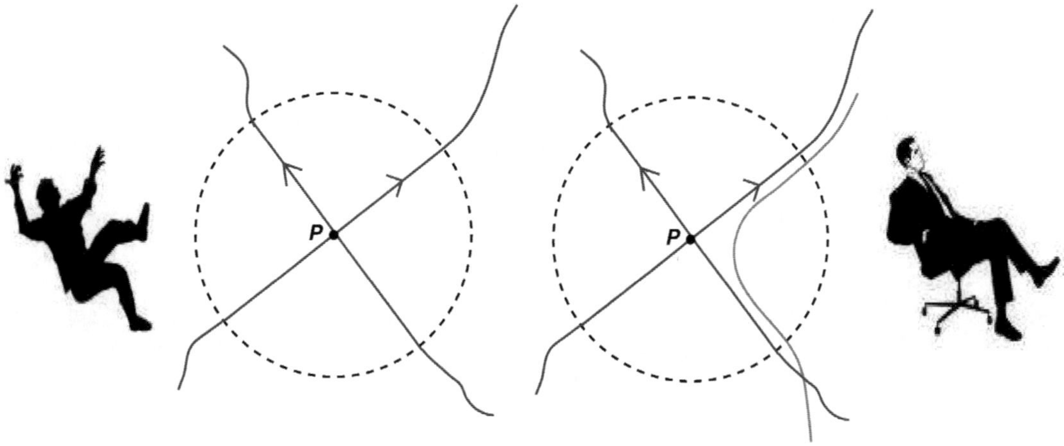

Figure 1. (a) Left: A freely falling observer, with an associated local inertial frame, defined within the region marked by the black circle. The radius of this circle is decided by the curvature of the spacetime at $\mathcal{P}$. Light rays travelling at 45 degrees in the local inertial frame define the light cones at $\mathcal{P}$. (b) Right: A local Rindler observer who is accelerating with respect to the inertial observer. For a sufficiently large acceleration, the trajectory of such an observer will be close to the light cones emanating from $\mathcal{P}$. The local Rindler observer will perceive the light cone as a local Rindler horizon and attribute to it a temperature given by Eq. (8). In other words, the vacuum fluctuations of the local inertial frame will appear as thermal fluctuations in the local Rindler frame.

fact that the cosmological constant, with the equation of state $p + \rho = 0$, has zero heat density. Our guiding principle, as well as Eq. (7), suggests that *it is the heat density rather than the energy density which is the source of gravity.*

But $T^a_b u^b u_a$ is the energy density for *any* kind of T^a_b, not just for that of an ideal fluid. How do we interpret $T^a_b \ell_a \ell^b$ in a general context when T^a_b could describe any kind of source – not necessarily a fluid – for which concepts like temperature and entropy do not exist intrinsically? *Remarkably enough, this can be done.* In any spacetime, around any event, there exists a class of observers (called local Rindler observers) who will interpret $T^a_b \ell_a \ell^b$ as the heat density contributed by the matter to a null surface which they perceive as a horizon. This *motivates* us to introduce [4] the concept of local Rindler frame (LRF) and local Rindler observers which will allow us to provide a thermodynamic interpretation of $T^a_b \ell_a \ell^b$ for any T^a_b. This arises as follows:

In a region around any event $\mathcal{P}$, we first introduce the freely falling frame (FFF) with coordinates $(T, \mathbf{X})$. Next, we boost from the FFF to a local Rindler frame (LRF) with coordinates $(t, \mathbf{x})$ constructed using some acceleration a, through the transformations: $X = \sqrt{2ax}\cosh(at), T = \sqrt{2ax}\sinh(at)$ when $|X| > |T|$ and similarly for other wedges. One of the null surfaces passing though $\mathcal{P}$, will get mapped to the $X = T$ surface in the FFF and will act as a patch of horizon to the $x =$ constant Rindler observers. This construction leads to a beautiful result [22, 23] in quantum field theory. The local vacuum state, defined by the freely-falling observers around an event, will appear as a thermal state to the local Rindler observers with the temperature:

$$k_B T = \left(\frac{\hbar}{c}\right)\left(\frac{a}{2\pi}\right) \tag{8}$$

where a is the acceleration of the local Rindler observer, which can be related to other geometrical variables of the spacetime in different contexts [see Fig. 1]. The existence of the Davies–Unruh temperature tells us that around *any* event, in *any* spacetime, there exists a class of observers who will perceive the spacetime as hot.

Let us now consider the flow of energy associated with the matter that crosses the null surface. Nothing unusual happens when this is viewed in the FFF by the locally inertial observer. But the local Rindler observer attributes a temperature T to the horizon and views it as a hot surface. Such an observer will interpret the energy ΔE, dumped on the horizon, by the matter that crosses the null surface, as energy added to a *hot* surface, thereby contributing a *heat* content $\Delta Q = \Delta E$. (Recall that, as seen by the outside observer, matter actually takes infinite amount of time to cross a *black hole* horizon, thereby allowing for thermalization to take place. Similarly, a local Rindler observer will find that the matter takes a very long time to cross the horizon.) To compute ΔE in terms of T^a_b, note that the LRF provides us with an approximate Killing vector field ξ^a, generating the Lorentz boosts, which coincides with a suitably defined null normal ℓ^a at the null surface. The heat current arises from the energy current $T_{ab}\xi^b$ of matter and hence the total heat energy dumped on the null surface will be:

$$Q_m = \int \left(T_{ab}\xi^b\right) d\Sigma^a = \int T_{ab}\xi^b\ell^a\sqrt{\gamma}d^2x d\lambda = \int T_{ab}\ell^b\ell^a\sqrt{\gamma}d^2x d\lambda \tag{9}$$

where we have used the result that $\xi^a \to \ell^a$ on the null surface. So we find that[9]

$$\mathcal{H}_m[\ell_a] \equiv \frac{dQ_m}{\sqrt{\gamma}d^2x d\lambda} = T_{ab}\ell^a\ell^b \tag{10}$$

can indeed be interpreted as the heat density (energy per unit area per unit affine time) of the null surface, contributed by matter crossing a local Rindler horizon, as interpreted by the local Rindler observer. This interpretation works in the LRF irrespective of the nature of T^a_b. So, the need to work with $\mathcal{H}_m$, forced on us by our guiding principle, *leads to* the introduction of local Rindler observers in order to interpret this quantity as the heat density.

There is an alternative interpretation of $\mathcal{H}_m$ which will prove to be useful. Since the parameter λ (defined through $\ell^a = dx^a/d\lambda$) is similar to a time coordinate, we can also think of $\mathcal{H}_m[\ell_a]$ in Eq. (10) as the heat generated per unit area of null surface per unit time. But since there are null surfaces through any event in the spacetime, we will always have observers who see the matter heating up these surfaces! This is something we probably do *not* want and we will see later on that gravity comes to our rescue. As we will see, contribution to the heating from the microscopic degrees of freedom of the spacetime precisely cancels out $\mathcal{H}_m[\ell_a]$ on all the null surfaces.[10]

4. Heat density of atoms of space

Let us get back to the task of constructing an extremum principle from which we can obtain the field equations. We have argued that matter sector appears in the extremum principle through the combination $\mathcal{H}_m = T_{ab}(x)\ell^a\ell^b$ which has the interpretation of the heat density (or the heating rate) contributed to a null surface by the matter crossing it. We also saw that we cannot vary the metric in the extremum principle. But in any variational principle constructed

[9] Since null vectors have zero norm, there is an overall scaling ambiguity in such expressions. This is resolved by considering the hyperboloids $\sigma^2 \equiv X^2 - T^2 = 2ax$ constant and treating the light cone as the degenerate limit $\sigma \to 0$ of the hyperboloids. We set $\ell_a = \nabla_a\sigma^2 \propto \nabla_a x$ and take the corresponding limit. The motivation for this choice will become clearer later on.

[10] Our use of LRF is *strictly limited* to the purpose of interpreting $\mathcal{H}_m$. I do *not* introduce the notion of entropy for the Rindler horizon (as proportional to its area) or work with its variation.

from $\mathcal{H}_m$, we now have the option of varying ℓ_a and leaving the metric alone. Such a variational principle can take the form:

$$Q_{\rm tot} \equiv \int \sqrt{\gamma}\, d^2x\, d\lambda\, \left(\mathcal{H}_m[x^i,\ell_a] + \mathcal{H}_g[x^i,\ell_a]\right); \qquad \mathcal{H}_m[x^i,\ell_a] \equiv T_{ab}(x)\ell^a\ell^b \tag{11}$$

where we will interpret $\mathcal{H}_g$ as the contribution to the heat density from the microscopic degrees of freedom of geometry ('atoms of space'). This should depend on both x^i and ℓ_a for the variational principle to be well defined. The success of this approach depends on our coming up with a candidate for $\mathcal{H}_g[x^i,\ell_a]$ which is physically well-motivated and also depends on the null vectors at each event. I will now show how this can be achieved.

Since $\mathcal{H}_g$ has the dimension of energy density, it is convenient to write it as $L_P^{-4}f(x^i,\ell_a)$ where we have introduced a length scale L_P to set the dimensions and f denotes the number of atoms of space at an event x^i with an extra attribute characterized by a null vector ℓ_a. So the total heat density now becomes:

$$Q_{\rm tot} \equiv Q_{\rm m} + Q_{\rm g} \equiv \int \sqrt{\gamma}\, d^2x\, d\lambda\, \left(T^a_b(x)\ell_a\ell^b + L_P^{-4}f[x^i,\ell_a]\right) \tag{12}$$

Our next task is to determine f. It appears natural to assume that the number of atoms of space, f, (*i.e.*, the microscopic degrees of freedom, contributing to the heat density) at an event $\mathcal{P}$ should be proportional to either the area or volume (which are the most primitive constructs) we can "associate with" the event $\mathcal{P}$. So we need to give precise meaning to the phrase, "area or volume associated with" the event $\mathcal{P}$. To do this we first introduce the notion of equi-geodesic surface, which can be done either in the Euclidean sector or in the Lorentzian sector; we will work in the Euclidean sector. An equi-geodesic surface $\mathcal{S}$ is the set of all events at the same geodesic distance σ from some event, which we take to be the origin [36–39]. A natural coordinate system to describe such a surface is given by $(\sigma,\theta_1,\theta_2,\theta_3)$ where σ, the geodesic distance from the origin, acts as the "radial" coordinate and θ_α are the angular coordinates on the equi-geodesic surfaces corresponding to $\sigma =$ constant. The metric in this coordinate system becomes:

$$ds_E^2 = d\sigma^2 + h_{\alpha\beta}dx^\alpha dx^\beta \tag{13}$$

where $h_{\alpha\beta}$ is the induced metric[11] on $\mathcal{S}$. The most basic quantities we can now introduce are the volume element $\sqrt{g}\,d^4x$ in the bulk and the area element for $\mathcal{S}$ given by, $\sqrt{h}\,d^3x$. For the metric in Eq. (13), $\sqrt{g} = \sqrt{h}$, and hence, both these measures are identical. Using standard differential geometry, one can show [40] that, in the limit of $\sigma \to 0$, either of these is given by:

$$\sqrt{h} = \sqrt{g} = \sigma^3\left(1 - \frac{1}{6}\mathcal{E}\sigma^2\right)\sqrt{h_\Omega}; \quad \mathcal{E} \equiv R^a_b n_a n^b \tag{14}$$

where $n_a = \nabla_a\sigma$ is the normal to $\mathcal{S}$ and $\sqrt{h_\Omega}$ arises from the standard metric determinant of the angular part of a unit sphere. The second term containing $\mathcal{E}$ gives the curvature correction to the area of (or the volume enclosed by) an equi-geodesic surface. Eq. (14) is a standard result in differential geometry and is often presented as a measure of the curvature at any event.

We can now "associate" an area (or volume) with an event P in a natural way by the following limiting procedure: (i) Construct an equi-geodesic surface $\mathcal{S}$ centered on P at a geodesic distance σ; (ii) compute the volume enclosed by $\mathcal{S}$ and the surface area of $\mathcal{S}$; and (iii) take the limit of $\sigma \to 0$ to define the area (or volume) associated with P. However, as we can readily see from Eq. (14), these measures vanish in the limit of $\sigma \to 0$.

[11] This is the analogue of the synchronous frame in the Lorentzian spacetime, with x^α being the angular coordinates.

This is, however, to be expected. The existence of microscopic degrees of freedom requires some form of discrete structure in the spacetime and they cannot be meaningfully defined if the spacetime is treated as a continuum all the way, just as we cannot have finite number of molecules in a fluid if it is treated as a continuum all the way. Classical differential geometry, which is what we have used so far, knows nothing about any discrete spacetime structure and hence cannot give us a nonzero $\mathcal{H}_g$. To obtain a nonzero $\mathcal{H}_g$ from the above considerations, we need to ask how the geodesic interval and the spacetime metric get modified in a quantum description of spacetime and whether such a modified description will have a $\sqrt{h}$ (or $\sqrt{g}$) which does not vanish in the coincidence limit. I will now turn to this task.

5. Points with finite area!

There is fair amount of evidence (see e.g., [41–46]) to suggest that a primary effect of quantum gravity will be to introduce into the spacetime a zero-point length, by modifying the geodesic interval $\sigma^2(x,x')$ between any two events x and x' (in a Euclidean spacetime) to a form like $\sigma^2 \to \sigma^2 + L_0^2$ where L_0 is a length scale of the order of Planck length.[12]

While we do not know how quantum gravity modifies the classical metric, we have an indirect handle on it if we assume that quantum gravity introduces a zero point length to the spacetime. This arises as follows: Just as the original σ^2 can be obtained from the original metric g_{ab}, we would expect the quantum gravity-corrected geodesic interval $S(\sigma^2)$ to arise from a corresponding quantum gravity-corrected metric [36], which we will call the q-metric q_{ab}. But no such local, non-singular q_{ab} can exist because, for any such q_{ab}, the resulting geodesic interval will vanish in the coincidence limit, by definition of the integral. Therefore, we expect $q_{ab}(x,x')$ to be a bitensor, which should be singular at all events in the coincidence limit $x \to x'$. We can determine [37,38] its form by using two conditions: (i) It should lead to a geodesic interval $S(\sigma^2)$ and (ii) Green's function describing small metric perturbations should have a finite coincidence limit. These conditions determine q_{ab} uniquely [38] in terms of g_{ab} (and its associated geodesic interval σ^2). We find that:

$$q_{ab} = A h_{ab} + B n_a n_b; \qquad q^{ab} = \frac{1}{A} h^{ab} + \frac{1}{B} n^a n^b \tag{15}$$

with

$$B = \frac{\sigma^2}{\sigma^2 + L_0^2}; \qquad A = \left(\frac{\Delta}{\Delta_S}\right)^{2/D_1} \frac{\sigma^2 + L_0^2}{\sigma^2}; \qquad n_a = \nabla_a \sigma \tag{16}$$

where D is the spacetime dimension , $D_k \equiv D - k$. The Δ is the Van-Vleck determinant related to the geodesic interval σ^2 by:

$$\Delta(x,x') = \frac{1}{\sqrt{g(x)g(x')}} \det\left\{\frac{1}{2}\nabla_a^x \nabla_b^{x'} \sigma^2(x,x')\right\} \tag{17}$$

and Δ_S is the corresponding quantity computed by replacing σ^2 by $S(\sigma^2)$ (and g_{ab} by q_{ab} in the covariant derivatives) in the above definition. For our purpose of determining $\mathcal{H}_g$, we will compute the area element $(\sqrt{h}\, d^3x)$ of an equi-geodesic surface and the volume element $(\sqrt{q}\; d^4x)$ for the region enclosed by it, using the renormalized q-metric. (For the q-metric in Eq. (15), corresponding to the g_{ab} in Eq. (13), these two measures will not be equal, because $q_{00} \neq 1$.) If

[12] A more general modification can take the form of $\sigma^2 \to S(\sigma^2)$ where the function $S(\sigma^2)$ satisfies the constraint $S(0) = L_0^2$ with $S'(0)$ finite. Our results are insensitive to the explicit functional form of $S(\sigma^2)$. So, for the sake of illustration, we will use $S(\sigma^2) = \sigma^2 + L_0^2$.

our ideas are correct, we should get a non-zero limit and there must be a valid mathematical reason to prefer one of these measures over the other.

The computation is straightforward and we find (for $S(\sigma^2) = \sigma^2 + L_0^2$ in $D = 4$, though similar results [47,48] hold in the more general case in D dimensions) that:

$$\sqrt{q} = \sigma \left(\sigma^2 + L_0^2\right) \left[1 - \frac{1}{6}\mathcal{E}\left(\sigma^2 + L_0^2\right)\right] \sqrt{h_\Omega} \tag{18}$$

and:[13]

$$\sqrt{h} = \left(\sigma^2 + L_0^2\right)^{3/2} \left[1 - \frac{1}{6}\mathcal{E}\left(\sigma^2 + L_0^2\right)\right] \sqrt{h_\Omega} \tag{19}$$

When $L_0^2 \to 0$, we recover the standard result in Eq. (14), as we should. Our interest, however, is in the limit $\sigma^2 \to 0$ at finite L_0. Something remarkable happens when we take this limit. The volume measure $\sqrt{q}$ vanishes (just as for the original metric) but $\sqrt{h}$ has a non-zero limit:

$$\sqrt{h} = L_0^3 \left[1 - \frac{1}{6}\mathcal{E}L_0^2\right] \sqrt{h_\Omega} \tag{20}$$

The q-metric (which we interpret as representing the renormalized/dressed spacetime) attributes to every point in the spacetime a finite area measure, but a zero volume measure! Since $L_0^3\sqrt{h_\Omega}$ is the volume measure of the $\sigma = L_0$ surface, we define [32] the dimensionless density of the atoms of spacetime, contributing to the gravitational heat density, as:

$$f(x^i, n_a) \equiv \frac{\sqrt{h}}{L_0^3\sqrt{h_\Omega}} = 1 - \frac{1}{6}\mathcal{E}L_0^2 = 1 - \frac{1}{6}L_0^2 R_{ab}n^a n^b \tag{21}$$

So far we have been working in the Euclidean sector with $n_a = \nabla_a\sigma$ being normal to the equi-geodesic surface. The limit $\sigma \to 0$ in the Euclidean sector makes the equi-geodesic surface shrink to the origin. *But, in the Lorentzian sector this leads to the null surface which acts as the local Rindler horizon around this event.* So, in this limit, we can identify n_a with the normal to the null surface ℓ_a and express the gravitational heat density as

$$f(x^i, \ell_a) = 1 - \frac{1}{6}L_0^2 R_{ab}\ell^a \ell^b \tag{22}$$

To see this in some detail, consider the Euclidean version of the local Rindler frame.[14] The local Rindler observers, living on the hyperboloid $R^2 - T^2 = \sigma^2$ perceive local patches of the light

[13] This result is rather subtle. One might think that the result in Eq. (19) (which is $\sqrt{h} = A^{3/2}\sqrt{g}$) arises from the standard result Eq. (14), by the simple replacement of $\sigma^2 \to (\sigma^2 + L_0^2)$. But note that this replacement does *not* work for the result in Eq. (18) (which is $\sqrt{q} = \sqrt{B}A^{3/2}\sqrt{g}$) due to the $\sqrt{B} = \sigma(\sigma^2 + L_0^2)^{-1/2}$ factor which has the limiting form $\sqrt{B} \approx \sigma/L_0$ when $\sigma \to 0$. This is why the each event has zero volume, but finite area associated with it!. A possible insight into this curious feature is provided by the following fact: The leading order dependence of $\sqrt{q}d\sigma \approx \sigma d\sigma$ shows that the volumes scale as σ^2 while the area measure is finite. This, in turn, is related to the fact [47] that *the effective dimension of the renormalized spacetime becomes $D = 2$ close to Planck scales* independent of the original D. This result has been noticed by several people [49–52] in different, but specific, models of quantum gravity. Our approach leads to this result in a *model-independent* manner, which, in turn, is connected with the result that events have zero volume, but non-zero area.

[14] There are two ways of extending the null surface and Rindler observers off the TX plane. One can extend the null lines (45 degree lines in Fig. 1) to the null *plane* $T = X$ in spacetime and similarly for the hyperboloid. Alternatively, one can extend the null lines to the null *cone* by $R^2 - T^2 = 0$ with $R^2 = X^2 + Y^2 + Z^2$ and the hyperboloid $R^2 - T^2 =$ constant will go 'around' the null cone in the spacetime (see the left part of Fig. 2). Observers living on this hyperboloid will use their respective (rotated) X axis to study the Rindler frame physics.

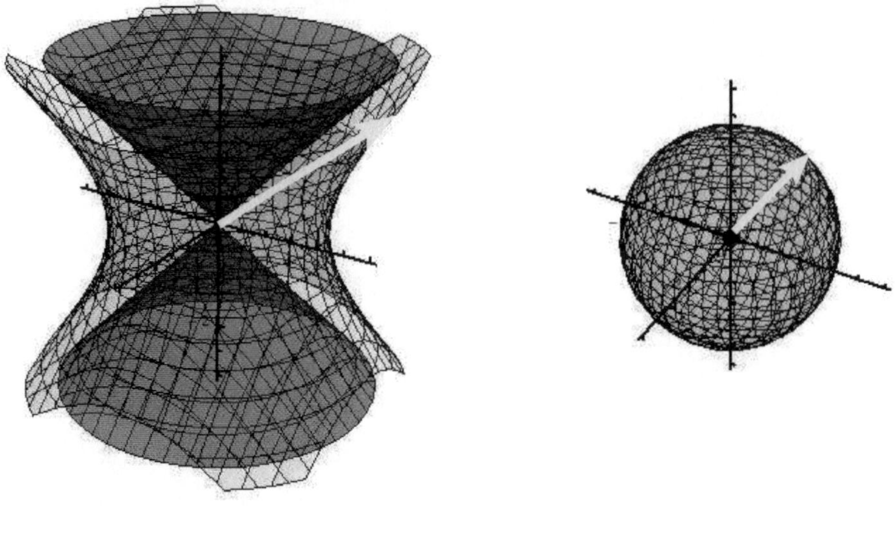

Figure 2. (a) Left: In the local inertial frame, the light cones originating from an event (taken to be the origin) are null surfaces with $R^2 - T^2 = 0$, with a normal ℓ_a. The local Rindler observers live on the hyperboloid $R^2 - T^2 = \sigma^2 =$ constant around these light cones and perceive a patch of the light cone as a local Rindler horizon with a finite temperature. The arrow denotes the normal to the hyperbola. (b) Right: In the Euclidean sector, the hyperboloid $R^2 - T^2 = \sigma^2$ will become a sphere $R^2 + T_E^2 = \sigma_E^2$ and the normal to the hyperboloid becomes the normal to the sphere. The light cone $R^2 - T^2 = 0$ will go over to $R^2 + T_E^2 = 0$ which collapses into the origin. The limit $\sigma_E \to 0$ in the Euclidean sector corresponds to approaching the Rindler horizon in the Lorentzian sector. In this limit the hyperboloid degenerates into the light cones emanating from $\mathcal{P}$. The direction of the normal to the sphere becomes ill-defined in the Euclidean sector when the radius of the sphere shrinks to zero. In the Lorentzian sector we can take it to be the normal to null surface in the limit when the hyperboloid degenerates to the light cone. The dependence of f on n_a in the Euclidean equi-geodesic surface is what translates to its dependence on the null normal ℓ_a in the Lorentzian sector.

cone $R^2 - T^2 = 0$ as their horizon (see the left half of Fig. 2). If we now analytically continue to the Euclidean sector, the hyperboloid $R^2 - T^2 = \sigma^2$ will become a sphere $R^2 + T_E^2 = \sigma_E^2$ (see the right half of Fig. 2). The light cones $R^2 - T^2 = 0$ goes over to $R^2 + T_E^2 = 0$ and collapses into the origin. So, taking the limit $\sigma_E \to 0$ in the Euclidean sector corresponds to approaching the local Rindler horizons in the Lorentzian sector. This is the limit in which the hyperboloid degenerates into the light cones emanating from $\mathcal{P}$. The normal n_a to the Euclidean sphere can then be identified with the normal to the null surface ℓ_a. The dependence of f on n_a in the Euclidean equi-geodesic surface is what translates to its dependence on the null normal ℓ_a in the Lorentzian sector.

So the contribution to the *gravitational* heat density on a null surface in Eq. (12) is obtained by integrating $L_P^{-4} f(x^i, \ell_j)$ over the volume:

$$Q_g = \int \frac{\sqrt{\gamma} d^2x d\lambda}{L_P^4} f(x^i, \ell_j) = \int \frac{\sqrt{\gamma} d^2x d\lambda}{L_P^4} \left[1 - \frac{1}{6} L_0^2 (R_{ab} \ell^a \ell^b)\right] \tag{23}$$

The numerical factor in front of the second term depends on the ratio L_0/L_P which we expect

Journal of Physics: Conference Series **701** (2016) 012018 doi:10.1088/1742-6596/701/1/012018

to be of order unity, if L_P is the standard Planck length. This ratio cannot be determined without better knowledge of quantum gravity. But we get the correct equations with the choice $L_0^2 = (3/4\pi)L_P^2$, which we will make.[15] The full variational principle is then based on the functional:

$$Q_{\rm tot} \equiv \int \sqrt{\gamma}\, d^2x\, d\lambda\, (T^a_b(x)\ell_a\ell^b + L_P^{-4} f[x^i,\ell_a]) = \int \sqrt{\gamma}\, d^2x\, d\lambda \left[\frac{1}{L_P^4} + \left\{T^a_b - \frac{1}{8\pi L_P^2} R^a_b\right\}\ell_a\ell^b\right] \tag{24}$$

Extremising this functional with respect to ℓ_a after introducing a Lagrange multiplier to keep $\ell^2 = 0$ and demanding the extremum holds for all ℓ_a at an event again leads to the result $R^a_b - 8\pi T^a_b = \lambda(x)\delta^a_b$ where $\lambda(x)$ is the Lagrange multiplier. Using Bianchi identity and $\nabla_a T^a_b = 0$, we will recover the field equations except for a cosmological constant. (There is actually a way to determine its value, which I will discuss in Sec. 5.5.)

There are several points which are noteworthy about this result which I will now comment upon:

5.1. *A crucial minus sign*

We defined the heat density of gravity as $L_P^{-4} f(x, n_a)$ with the number of atoms of space f being given by the limit:

$$f(x^i, n_a) \equiv \lim_{\sigma\to 0} \frac{\sqrt{h(x,\sigma)}}{L_0^3\sqrt{h_\Omega}} \tag{25}$$

in a renormalized spacetime with a zero-point length. The result had the combination $R^b_a n_b n^a$, at the relevant order, which is crucial. Further, this term came with a *minus sign* without which the programme would have failed.

These results bring to the center-stage the geodesic interval $\sigma^2(x, x')$ (rather than the metric) as the proper variable to describe spacetime geometry [48]. In a classical spacetime, $\sigma^2(x, x')$ and $g_{ab}(x)$ contain the same amount of information and each is derivable from the other. But $\sigma^2(x, x')$ seems to be better suited to take into account quantum gravitational effects to certain extent.

5.2. *Dissipation without dissipation*

We mentioned earlier that one could have also thought of $\mathcal{H}_m$ as the heating *rate* of the null surface by matter. We see that the corresponding heating rate of the null surface by the atoms of space precisely compensates this on-shell leaving an unobservable constant factor L_P^{-4}. (Since we can add any quantity independent of ℓ_a to the integrand of Q_{tot}, we can even renormalize this away; we leave it because it has an interesting implication for cosmological constant; see Sec. 5.5). So the field equation, which is now in the form of Eq. (7), has a clear physical meaning; both sides represent the heating rate of null surfaces and Einstein's equation arises as a heat balance equation.

This interpretation is reinforced by the fact that the term $R_{ab}\ell^a\ell^b$ is related to the "dissipation without dissipation" [26] of the null surfaces, which can be described as follows: Introduce the second null vector k^a and define the 2-metric on the cross-section of the null surface in the standard manner, $q_{ab} = g_{ab} + k_a k_b + \ell_a \ell_b$. Next define the expansion $\theta \equiv \nabla_a \ell^a$ and shear $\sigma_{ab} \equiv \theta_{ab} - (1/2)q_{ab}\theta$ where $\theta_{ab} = q^i_a q^j_b \nabla_i \ell_j$. (We take the null congruence to be affinely parametrized.) One can then show that (see e.g., eq(A60) of [12]):

$$-\frac{1}{8\pi L_P^2} R_{ab}\ell^a\ell^b = \left[2\eta\sigma_{ab}\sigma^{ab} + \zeta\theta^2\right] + \frac{1}{8\pi L_P^2}\frac{1}{\sqrt{\gamma}}\frac{d}{d\lambda}(\sqrt{\gamma}\theta) \equiv \mathcal{D} + \frac{1}{8\pi L_P^2}\frac{1}{\sqrt{\gamma}}\frac{d}{d\lambda}(\sqrt{\gamma}\theta) \tag{26}$$

[15] If we took the gravitational heat density as $\mu f/L_P^4$ where μ is a factor of order unity, the numerical coefficients will change. We take $\mu = 1$ as a natural choice.

where $\eta = 1/16\pi L_P^2, \zeta = -1/16\pi L_P^2$ are the shear and bulk viscous coefficients [1–3] and $\mathcal{D}$ is the viscous dissipation rate. So ignoring the total divergence term and the constant, the relevant part of Q_{tot} in Eq. (24) can be expressed as

$$Q_{\rm tot} = \int \sqrt{\gamma}\, d^2x\, d\lambda \left(T^a_b(x)\ell_a\ell^b + \left[2\eta\sigma_{ab}\sigma^{ab} + \zeta\theta^2\right]\right) = \int \sqrt{\gamma}\, d^2x\, d\lambda \left(T^a_b(x)\ell_a\ell^b + \mathcal{D}\right) \quad (27)$$

Both terms now have an interpretation of the rate of heating (due to matter or atoms of spacetime).[16] Our extremum principle can be thought of extremising the rate of production of heat on the null surface.

5.3. Connection with the entanglement entropy

The combination $R^a_b n_a n^b$, which is crucial to our results, arises in several geometrical quantities just as it came up in the area measure $\sqrt{h}$ in Eq. (20). In particular, this combination has some relevance to entanglement entropy, in the following sense. In the literature, it is usually claimed that the entanglement entropy S of a field partitioned by, say, a horizon is proportional to the area of the horizon. *To be precise, this statement is meaningless because the entanglement entropy S is a divergent quantity.* For example, for a free, massless, scalar field in D dimensional spacetime, S is given by the expression:

$$S = \frac{A_{D-2}}{12} \int_0^\infty \frac{ds}{s} K_{D-2}(x, x; s) \quad (28)$$

where A_{D-2} is the transverse area and $K(x, y; s)$ is the Schwinger proper time Kernel [53, 54]. In the coincidence limit, the $K_{D-2}(x, x; s) \propto s^{-(D-2)/2}$ and the integral in Eq. (28) diverges as $L_c^{-(D-2)}$ at the lower limit where L_c is a lower cutoff length scale. In $D = 4$ spacetime this gives $S \propto A_\perp / L_c^2$, which diverges quadratically, which is the standard result.

The introduction of a zero-point length into the spacetime corresponds to introducing a regulator $\exp(-L/s)$ into the heat Kernel (with some cutoff scale L) *which will render the entanglement entropy finite* [30] and these calculations meaningful. What is more, the Kernel will also pick up a Van Vleck determinant in a curved spacetime which also contains the factor $R^a_b n_a n^b$ over and above the flat spacetime result. *This suggests that one may be able to relate $f(x^i, n_a)$ in Eq. (21) to the entanglement entropy which could provide an alternative interpretation to our results.* This idea could be tested by computing the entanglement entropy of a field in the renormalized spacetime with zero-point length, which renders it finite in a systematic manner.

5.4. The mechanism that couples matter to spacetime

The coupling between spacetime geometry and matter is now through a vector field ℓ_a and – at the lowest order – they couple to ℓ_a through the terms $R^a_b \ell_a \ell^b$ and $T^a_b \ell_a \ell^b$ respectively. The physical origin of these two couplings is quite distinct. The $T^a_b \ell_a \ell^b$ arises from the behaviour of matter crossing the local Rindler horizon and the ℓ_a in this expression represents the normal to the local Rindler horizon. The $R^a_b \ell_a \ell^b$ term, however, arose from the limit of the area measure $\sqrt{h}$ in a spacetime with zero-point length involving $R^a_b n_a n^b$. The n_a in this case is identified with the normal ℓ_a to the null surface through a limiting process when one takes the limit $\sigma \to 0$ in the Euclidean sector. This, in turn, depends on the fact that the condition $\sigma^2(x, y) = 0$ will lead to $x = y$ in the Euclidean space while it will describe all events connected by a null ray in the Lorentzian space.

[16] Equation (26) is just a restatement of the Raychaudhuri equation. What is relevant in the extremum principle is the *quadratic* terms in shear and expansion while the the term giving the change in the cross-sectional area of the congruence is a total divergence and is irrelevant. This tells us that ignoring the quadratic terms of Raychaudhuri equation can miss a key element of physics; see also [24].

More generally, we can introduce a vector n_a that couples to spacetime curvature through $R^a_b n_a n^b$ and to matter through $T^a_b n_a n^b$ thereby providing an indirect coupling between matter and geometry. In a way this is conceptually rather satisfactory. In the conventional approach to Einstein's theory we actually do not have a *mechanism* which tells us how T^a_b ends up curving the spacetime. The relation $G^a_b = 8\pi T^a_b$ equates apples and oranges; the left hand side is purely geometrical while the right hand side is made of matter with large number of discrete (quantum) degrees of freedom. An equation of the kind, $G^a_b n_a n^b = 8\pi T^a_b n_a n^b$, on the other hand, does better in this regard. We can hope to interpret both sides independently (say, as the heat densities) and think of this equation as a balancing act performed by spacetime. Such a description is reinforced by the extremum principle in Eq. (24) in which both $R^a_b n_a n^b$ and $T^a_b n_a n^b$ can be thought of as distorting the value of f from unity and the gravitational field equations restores the value $f = 1$ on-shell. (I will say a little more about this in the last section.)

Since the classical gravity is obtained by extremising $Q_{\rm tot}$ in Eq. (24) with respect to n_a, we could ask whether it is meaningful to attempt a quantum theory for n_a using a path integral. This would require evaluating the path integral

$$Z = \int \mathcal{D}n_a\, \delta\left[n^2\right]\, \exp iQ_{\rm tot}\left[n_a\right] \tag{29}$$

Incorporating $\delta\left[n^2\right]$ through a functional Fourier transform with respect to a Lagrange multiplier field $\lambda[x]$ we can reduce this to the form:

$$Z \propto \prod_x \int \mathcal{D}\lambda(x) \int \mathcal{D}n_a(x) \exp i\left[(M^a_b + \lambda\delta^a_b)n_a n^b\right]_x \propto \prod_x \int \mathcal{D}\lambda(x)\, \mathrm{Det}^{-1/2}\left[M^a_b + \lambda\delta^a_b\right]_x \tag{30}$$

where $M^a_b \equiv T^a_b - (8\pi L_P^2)^{-1} R^a_b$. This is a rather intriguing theory which will lead to Einstein's equations in the saddle point limit if we demand that the saddle point condition should hold for all n_a. The Euclidean version of this theory suggests working with $\exp(Q_g)$ and it will be interesting to explore this further.[17]

5.5. Relevance to the cosmological constant problem

On shell, when the equations of motion hold, the two terms in the curly brackets in Eq. (24) cancel each other and the net heat density has the Planckian value $1/L_P^4$, which, of course, has no gravitational effect. But it tells us that there *is* a zero-point contribution to the degrees of freedom in spacetime, which, in dimensionless form, is just unity. Therefore, it makes sense to ascribe A/L_P^2 degrees of freedom to an area A, which is consistent with what we know from earlier results in this subject. So a two-sphere $S^{(2)}$of radius L_P has $f_{S^{(2)}} = 4\pi L_P^2/L_P^2 = 4\pi$, which was the crucial input that was used in a previous work to determine the numerical value of the cosmological constant for our universe. (This is similar to assigning $dN/d^3xd^3p = f(x^i, p_a)$ molecules to a phase volume. In kinetic theory, we do not worry about the fact that f is not always an integer. In the same spirit we are not concerned by the fact that 4π is not an integer.). Thus, the microscopic description does allow us to determine [55, 56] the value of the cosmological constant, (which arose as an integration constant), as it should in any complete description.

Let me elaborate a little bit on this aspect, since it can provide a solution to what is usually considered the most challenging problem of theoretical physics today.

[17] If we write the null vector as $\ell_a \equiv q(x)\bar{\ell}_a$ where $\bar{\ell}_a \equiv \nabla_a\sigma$ is an affinely parametrized null normal, then we can upgrade just $q(x)$ to a dynamical field with the Lagrangian $L = (1/2)(\partial q)^2 - q^2[M^a_b\ell_a\ell^b]$. Integrating over $q(x)$, treating it as a rapidly varying fluctuation (with $[M^a_b\ell_a\ell^b]$ approximately constant) will lead to an effective action A_{eff} for $[M^a_b\ell_a\ell^b]$ such that $\delta A_{eff} = 0$ is the same as $\delta Q_{\rm tot} = 0$. I will describe this in detail elsewhere.

Observations suggest that our universe has three distinct phases of expansion: (i) An inflationary phase with an approximately constant density ρ_{inf}, fairly early on. (ii) A phase dominated, first by radiation and then by matter, with $\rho = \rho_{eq}[x^{-4} + x^{-3}]$, where a_{eq} is the epoch at which the matter and radiation densities were equal, $x(t) \equiv a(t)/a_{\rm eq}$ and the ρ_{eq} is a constant equal to the energy density of either matter or radiation at $a = a_{\rm eq}$. (iii) An accelerated phase of expansion at late stages, driven by the energy density ρ_Λ of the cosmological constant. These three constants $[\rho_{inf}, \rho_{eq}, \rho_\Lambda]$ completely specify the dynamics of our universe and act as its signature. Of these ρ_{inf} and ρ_{eq} can, in principle, be determined by standard high energy physics. But we need a *new* principle to fix the value of ρ_Λ, which is related to the integration constant that appears in the field equations in our approach.

It turns out that such a universe, with these three phases, harbors a new *conserved* quantity, which is the number N of length scales (or radial geodesics), that cross the Hubble radius during any of these phases [55, 56]. Any physical principle which can determine the value of N during the radiation-matter dominated phase, say, will fix the value of ρ_Λ in terms of $[\rho_{inf}, \rho_{eq}]$. Taking the modes into the very early phase, we can fix the value of this conserved quantity N at the Planck scale, as the degrees of freedom in a two-sphere of radius L_P. In other words, we take $N = 4\pi L_P^2/L_P^2 = 4\pi$. This, in turn, leads to a remarkable prediction relating the three densities [55, 56]:

$$\rho_\Lambda \approx \frac{4}{27}\frac{\rho_{\rm inf}^{3/2}}{\rho_{\rm eq}^{1/2}} \exp(-36\pi^2) \tag{31}$$

From cosmological observations, we know that $\rho_{eq}^{1/4} = (0.86 \pm 0.09)$eV; if we assume that the range of the inflationary energy scale is $\rho_{\rm inf}^{1/4} = (1.084 - 1.241) \times 10^{15}$ GeV, we get $\rho_\Lambda L_P^4 = (1.204 - 1.500) \times 10^{-123}$, which is consistent with the observations! This approach for solving the cosmological constant problem provides a unified view of cosmic evolution, connecting the three phases through Eq. (31) in contrast with standard cosmology in which the three phases are joined together in an unrelated, *ad hoc* manner.

Moreover, this approach to the cosmological constant problem *makes a falsifiable prediction*, unlike any other approach I am aware of. From the observed values of ρ_Λ and $\rho_{\rm eq}$, we can *predict* the energy scale of inflation within a very narrow band – to within a factor of about five – if we include the ambiguities associated with reheating. If future observations show that inflation occurred at energy scales outside the band of $(1-5) \times 10^{15}$ GeV, our model for explaining the value of cosmological constant is ruled out.

6. Summary of the logical structure

Let me reiterate the logical sequence described in this work which leads to a completely different perspective on gravity and the derivation of its field equations.

- We postulate that the gravitational field equations should arise from an extremum principle which remains invariant under the transformation $T^a_b \to T^a_b + \delta^a_b(\text{constant})$.
- This leads to two conclusions: (a) Metric tensor cannot be a dynamical variable which is varied in the extremum principle. (b) The T^a_b should appear in the extremum principle though the combination $\mathcal{H}_m(x^i, \ell_a) \equiv T^a_b \ell_a \ell^b$ where ℓ_a is a null vector.
- We next look for a physical interpretation of $\mathcal{H}_m(x^i, \ell_a)$ for an arbitrary T^a_b and find that, in any spacetime, the local Rindler observers will interpret $\mathcal{H}_m(x^i, \ell_a)$ as the heat density contributed to a null surface by the matter crossing it. This interpretation works for *any* T^a_b and provides a strong *motivation* for introducing local Rindler observers in the spacetime.
- Since (i) the metric cannot be a dynamical variable and (ii) we now have the auxiliary null vector field ℓ_a arising through $\mathcal{H}_m(x^i, \ell_a)$, we look for an extremum principle in which ℓ_a

is varied. The extremum should hold for all null vectors at any event and constrain the background metric.

- We take the integrand of the extremum principle to be $\mathcal{H}_m(x^i, \ell_a) + \mathcal{H}_g(x^i, \ell_a)$ where $\mathcal{H}_g(x^i, \ell_a) \equiv L_P^{-4} f(x^i, \ell_a)$ is interpreted as the heat density due to microscopic spacetime degrees of freedom. We have introduced a length scale L_P from dimensional considerations and f is the dimensionless count of the microscopic degrees of freedom.
- A discrete count for microscopic degrees of freedom implies a discrete nature for spacetime at Planck scales. We incorporate this fact by an effective, renormalized/dressed metric q_{ab} which ensures that the geodesic distance of the effective metric has a zero-point length and is given by $\sigma^2(x, x') + L_0^2$ where $\sigma^2(x, x')$ is the geodesic distance of the classical spacetime. We take $L_0^2 = (3/4\pi)L_P^2$.
- The spacetime degrees of freedom at an event is taken to be proportional to the area of an equi-geodesic surface centered at that event in the limit of vanishing geodesic distance (see Eq. (25)). With this choice, one obtains an extremum principle based on Eq. (24). Varying this with respect to all null vectors ℓ_a and demanding the equation to hold for all ℓ_a at an event leads to Einstein's equation with an undetermined cosmological constant.
- When equations of motion hold, we can assign A/L_P^2 degrees of freedom with every area element A in spacetime. This, in turn, allows us to fix the value of the undetermined cosmological constant correctly and *provides a solution to the cosmological constant problem.*

7. Conclusions and open questions

The approach outlined here is based on the idea that gravity is the thermodynamic limit of the statistical mechanics of certain microscopic degrees of freedom ('atoms of space'). In the thermodynamic limit, deriving the field equations of classical gravity is algebraically straightforward – one might even say trivial, but let us not shun simplicity! It is obtained from an extremum principle based on the functional:

$$Q_{\text{thermo}} = \int \sqrt{\gamma}\, d^2x\, d\lambda \left\{ T^a_b - \frac{1}{8\pi L_P^2} R^a_b \right\} \ell_a \ell^b \tag{32}$$

Varying ℓ_a in Q_{thermo} with the constraint that $\ell^2 = 0$ and demanding that the result holds for all ℓ_a will lead to Einstein's equation with an arbitrary cosmological constant. The approach works for a wild class of gravitational theories including the Lanczos-Lovelock models.

As far as classical gravity goes, that is the end of the story. But we could enquire about the physical meaning of this extremum principle.[18] The combination $T^a_b \ell_a \ell^b$ is invariant under the shift $T^a_b \to T^a_b + \delta^a_b$ (constant) – which was the original reason to put it in the variational principle. Postulating that the extremum principle must be invariant under the transformation $T^a_b \to T^a_b + \delta^a_b$(constant) naturally *leads to* the introduction of $T^a_b \ell_a \ell^b$ in the variational principle and to the local Rindler observers for its interpretation. As I mentioned earlier, this itself is a valuable insight and shows the connection between two features – viz. the immunity of gravity to shifts in the cosmological constant and the thermodynamic interpretation of gravity – which were considered as quite distinct.

Further $T^a_b \ell_a \ell^b$ *does* have the interpretation as the heat density of matter on any null surface. It is also possible to interpret $R^a_b \ell_a \ell^b$ as the heating rate ("dissipation without dissipation"; see Sec. 5.2) of the null surface [27], thereby providing a purely thermodynamic underpinning for classical gravity. None of this poses any conceptual or technical problem.

[18] We do *not* enquire about the physical meaning of the Einstein-Hilbert action – it has none – in the conventional approach; but then we are now trying to *improve* on the conventional approach!

The real issues arise when we try to go beyond the classical theory and provide a semi-classical or quantum gravitational interpretation of this result. The description in terms of the expression in Eq. (21) is approximate and only valid when $L_P^2 R^a_b n_a n^b \lesssim 1$. The entire description, based on the q-metric acting as a proxy for the effective spacetime metric, cannot be trusted too close to Planck scales. We expect it to capture the quantum gravitational effects to certain extent but at present we have no way of quantifying the accuracy of this approach. (It should also be noted that the identification $n_a \approx \ell_a + ...$ might have further corrections close to Planck scales.)

We also do not know how to deal with matter fields in a quantum spacetime and it is not clear how to introduce T^a_b in a systematic way. In fact, the *only* reason to vary the metric tensor in an extremum principle – a procedure which I have argued against – is to obtain T^a_b classically and $\langle T^a_b \rangle$ semi-classically. None of the thermodynamic derivations, which leads to the exact (rather than linearized) field equations, obtains T^a_b (or $\langle T^a_b \rangle$) from fundamental considerations based on a matter action. It is ironic that the problem arises from the matter sector rather than from the gravity sector!

The ideas outlined in this work suggest a more radical solution. Matter, as we understand it, is quantum mechanical; which essentially means that it is made of *discrete* degrees of freedom. How does such a *discrete* structure end up curving the *continuum* geometry? That is, what is the actual *mechanism* by which T^a_b produces G^a_b? The approach developed here suggests that one needs to introduce certain "hidden variables", viz., the auxiliary vector field n_a, which encodes the discrete nature of geometry, and couple it to T^a_b (which is also fundamentally discrete in nature). The continuum geometry has to be related to n_a, at scales much larger than the Planck scale, by a suitable approximation just as the continuum density or pressure of a fluid arises when we average over the discreteness of the molecules. I have outlined one possible way in which this idea could be implemented but it is by no means unique. Further exploration of this approach could lead to a better understanding of how matter *really* ends up curving the spacetime.

Acknowledgments

My research work is partially supported by J.C.Bose Fellowship of Department of Science and Technology, India. I thank Sumanta Chakraborty, Sunu Engineer, Dawood Kothawala and Kinjalk Lochan for discussions and comments on the first draft.

References

[1] Damour T 1979 Thèse de doctorat d'Etat Université Paris
[2] Damour T 1982 Surface effects in black hole physics *Proceedings of the Second Marcel Grossmann Meeting on General Relativity*
[3] Thorne K S, Price R H and MacDonald D A 1986 *Black Holes: The Membrane Paradigm* (Yale University Press)
[4] Jacobson T 1995 *Phys. Rev. Lett.* **75** 1260
[5] Padmanabhan T 2002 *Class. Quant. Grav.* **19** 5387 (*Preprint* arXiv:gr-qc/0204019)
[6] Padmanabhan T 2004 *Class. Quant. Grav.* **21** 4485 (*Preprint* arXiv:gr-qc/0308070)
[7] Padmanabhan T and Aseem Paranjape 2007 *Phys. Rev.* D **75** 064004 (*Preprint* arXiv:gr-qc/0701003)
[8] Padmanabhan T 2010 *Rep. Prog. Phys.* **73** 046901 (*Preprint* arXiv:0911.5004)
[9] Padmanabhan T 2010 *Mod. Phys. Lett.* A **25** 1129–36 (*Preprint* arXiv:0912.3165)
[10] Verlinde E P 2011 *JHEP* **1104** 029
[11] Smolin L General relativity as the equation of state of spin foam (*Preprint* arXiv:1205.5529)
[12] Padmanabhan T 2014 *Gen. Rel. Grav.* **46** 1673
[13] Padmanabhan T 2015 *Mod. Phys. Lett.* A **30** 1540007
[14] Jacobson T 2015 Entanglement equilibrium and the Einstein equation (*Preprint* arXiv:1505.04753)
[15] Smolin L 2015 The thermodynamics of quantum spacetime histories (*Preprint* arXiv:1510.03858)
[16] Parattu K, Majhi B R and Padmanabhan T 2013 *Phys. Rev.* D **87** 124011 (*Preprint* arXiv:1303.1535)
[17] Chakraborty S and Padmanabhan T 2014 *Phys. Rev.* D **90** 124017 (*Preprint* arXiv:1408.4679)
[18] Chakraborty T and Padmanabhan T 2015 *Phys. Rev.* D **92** 104011 (*Preprint* arXiv:1508.04060)

[19] Chakraborty S, Parattu K and Padmanabhan T 2015 *JHEP* **10**, 097 [arXiv:1505.05297].
[20] T. Padmanabhan, *Phys.Rev.*, **D 81**, 124040 (2010) [arXiv:1003.5665].
[21] Chakraborty S and Padmanabhan T 2014 *Phys. Rev.* **D90** 084021 [arXiv:1408.4791]
[22] Davies P C W 1975 *J. Phys.* **A 8** 609
[23] Unruh W G 1976 *Phys. Rev.* **D 14** 870
[24] Kothawala D 2011 *Phys.Rev.* **D 83** 024026 [arXiv:1010.2207]
[25] Carroll S M. and Remmen G N *What is the entropy in entropic gravity?* [arXiv:1601.07558]
[26] Kolekar S and Padmanabhan T 2012 *Phys.Rev.* **D 85** 024004 [arXiv:1109.5353]
[27] Padmanabhan T 2011 *Phys.Rev.* **D 83** 044048 [arXiv:1012.0119]
[28] Chirco G, Eling C and Liberati S 2011 *Phys.Rev.* **D 83** 024032 [arXiv:1011.1405]
[29] Solodukhin S N 2011 *Living Rev. Relativity* **14** 8 [arXiv:1104.3712]
[30] Padmanabhan T 2010 *Phys.Rev.* **D 82** 124025 [arXiv:1007.5066]
[31] Visser M 2011 *JHEP* **10** 140 [arXiv:1108.5240]
[32] Padmanabhan T 2015 *Entropy* **17**, 7420-52 [arXiv:1508.06286]
[33] Padmanabhan T 2010 *Gravitation: Foundations and Frontiers* (Cambridge University Press)
[34] Padmanabhan T 2008 *Gen.Rel.Grav.* **40** 529-564 [arXiv:0705.2533]
[35] Padmanabhan T 2009 *Adv. Sci. Lett.* **2**, 174–183.
[36] Kothawala D and Padmanabhan T 2014 *Phys. Rev. D* **90** 124060
[37] Kothawala D 2013 *Phys. Rev.* **D 88** 104029
[38] Stargen D J. and Kothawala D 2015 *Phys. Rev.* **D 92** 024046 [arXiv:1503.03793]
[39] Kothawala D and Padmanabhan T 2015 *Phys. Lett. B* **748**, 67–69
[40] Gray A 1974 *Mich. Math. J.* **20** 329–344
[41] DeWitt B S 1964 *Phys. Rev. Lett.* **13**, 114
[42] Padmanabhan T 1985 *Gen. Rel. Grav.* **17**, 215
[43] Padmanabhan T 1985 *Ann. Phys.* **165**, 38
[44] Padmanabhan 1997 *Phys. Rev. Lett.* **78** 1854 [hep-th/9608182]
[45] Garay L 1998 *Phys. Rev. Lett.* **80** 2508 (1998) [gr-qc/9801024]
[46] Garay L 1995 *Int. J. Mod. Phys. A* **10** 145
[47] Padmanabhan T, Chakraborty S and Kothawala D 2015 *Renormalized spacetime is two-dimensional at the Planck scale* [arXiv:1507.05669]
[48] Chakraborty S and Padmanabhan T *work in progress*
[49] Carlip S, Mosna R and Pitelli J 2011 *Phys. Rev. Lett.* **107** 021303
[50] Ambjorn J, Jurkiewicz J and Loll R 2005 *Phys. Rev. Lett.* **95**, 171301
[51] Modesto L 2009 *Class. Quantum Grav.* **26** 242002
[52] Husain V, Seahra S S and Webster E J 2013 *Phys. Rev. D* **88** 024014
[53] See e.g., Toms D J 2007 *The Schwinger Action Principle and Effective Action* (Cambridge University Press)
[54] Sriramkumar L and Padmanabhan T 2002 *Int.J.Mod.Phys.* **D11** 1-34 ([gr-qc/9903054]
[55] Padmanabhan H and Padmanabhan T 2013 *Int. J. Mod. Phys. D* **22** 1342001
[56] Padmanabhan T and Padmanabhan H 2014 *Int. J. Mod. Phys. D* **23** 1430011

Nonlinear Schrödinger equation in foundations: summary of 4 catches

Lajos Diósi
Wigner Research Centre for Physics, Budapest 114, P.O.Box 49, H-1525 Hungary
E-mail: diosi.lajos@wigner.mta.hu

Abstract. Fundamental modifications of the standard Schrödinger equation by additional nonlinear terms have been considered for various purposes over the recent decades. It came as a surprise when, inverting Abner Shimonyi's observation of "peaceful coexistence" between standard quantum mechanics and relativity, N. Gisin proved in 1990 that any (deterministic) nonlinear Schrödinger equation would allow for superluminal communication. This is by now the most spectacular and best known anomaly. We discuss further anomalies, simple but foundational, less spectacular but not less dramatic.

1. Introduction

We discuss major difficulties encountered by nonlinear modifications of the Schrödinger equation. Before we enter the subject, we make it clear that nonlinear Schrödinger equations and nonlinear effective modifications of quantum mechanics are widely used in physics as approximate means. Most typical is the mean-field approximation. It yields the nonlinear Hartree-Fock equation [1] so basic for many-electron systems, and it yields the semi-classical Einstein equation [2] which is indispensable in quantum cosmology. These are approximate (effective) nonlinear quantum mechanics. One can, on the contrary, consider nonlinear modification of quantum mechanics at the fundamental level. As early as in 1952, Jánossy proposed [3] that a simple state-dependent potential ensure localized stationary wave function of a free quantum particle (of mass M):

$$i\hbar\frac{d\Psi(x)}{dt}=-\frac{\hbar}{2M}\Psi''(x)+\frac{1}{2}\alpha^2(x-\langle x\rangle_\Psi)^2\Psi(x), \tag{1}$$

where

$$\langle x\rangle_\Psi=\int x|\Psi(x)|^2dx \tag{2}$$

and α is a certain parameter. The new contractive 'mean-field' potential counters the unlimited dispersion predicted by the standard Schrödinger equation. The stationary states are localized Gaussians (solitons). Jánossy's equation was and remained forgotten until now [4]. A gravity-related version of his nonlinear term was suggested much later independently by the author and by Penrose [5, 6], both being unaware of Jánossy's equation yet seeking the same effect, namely, to ensure localization of quantum objects – massive ones this time. We choose this equation, called the Schrödinger–Newton equation (SNE), to be our testbed to discuss four typical catches otherwise valid for any nonlinear Schrödinger equations.

2. Peaceful coexistence

It has been known since the beginning of quantum theory that it contains a certain latent non-locality. For instance, the collapse of the single particle wave function, occurring in quantum measurement, is to happen instantaneously over whatever large distances covered by the pre-measurement wave function. For two distant particles, Einstein, Podolski and Rosen (EPR) in 1935 [7] constructed the sharpest example of apparent action-at-a-distance and John Bell in 1963 [8] pointed out a specific non-locality. It was also clear that neither the apparent action-at-a-distance nor the quantum non-locality could result in testable dynamical effects like dynamical action-at-a-distance or superluminal communication. The physicist and philosopher Abner Shimony formulated this paradoxical situation as the peaceful coexistence between quantum theory and special relativity [9].

The reason of why formal violations of locality will all cancel from the measurable outcomes of quantum mechanics lie in the linear structure of the mathematical representation of the state space, dynamics, and measurable outcomes themselves. If one adds non-linear terms to the Schrödinger equation the peaceful coexistence ends and the physical violation of special relativity takes over. Interestingly, Jánossy [3] mentions that his non-linear single particle Schrödinger equation suffers of superluminal physical influence. The most spectacular thesis belongs to Gisin [10]. Any nonlinear modification of the single particle Schrödinger equation,

$$i\hbar\frac{d\Psi(x)}{dt} = -\frac{\hbar}{2M}\Psi''(x) + V_\Psi(x)\Psi(x), \tag{3}$$

allows for superluminal communication between the EPR partners for whatever small state-dependent potential $V_\Psi(x)$ (apart from the constant potential, of course).

3. Schrödinger–Newton equation: our testbed

Consider the single-body SNE for the centre-of-mass free motion of a 'large' composite object of mass M:

$$i\hbar\frac{d\Psi(x)}{dt} = -\frac{\hbar}{2M}\Psi''(x) + M\Phi_\Psi\Psi(x), \tag{4}$$

where the state-dependent Newton potential reads

$$\Phi_\Psi(x) = -GM\int\frac{|\Psi(r)|^2}{|x-r|}d^3r. \tag{5}$$

Although this equation is the Newtonian limit of the approximate semiclassical Einstein equation, it was suggested that it might be fundamental [5, 6] and it is currently studied as such, both theoretically and experimentally [11].

Our forthcoming analysis needs the following peculiar features of the SNE. It possesses standing (static) soliton solutions $|\bigcirc\rangle$ of diameter

$$D \sim \frac{\hbar}{GM^3}, \tag{6}$$

c.f., e.g., [5]. We can construct the Schrödinger cat states formed by two separate static solitons:

$$\Psi_\pm = \frac{1}{\sqrt{2}}\left(|\bigcirc_L\rangle \pm |\bigcirc_R\rangle\right). \tag{7}$$

The left and right solitons $\bigcirc_{L/R}$ are separated by a distance $\ell \gg D$ initially. Because of the mean-field $\Phi_\Psi(x)$ in the SNE, the two solitons in $\Psi_\pm$ attract each other and start to move, like this:

$$\bigcirc\!\!\!\!\leftarrow\!-\ell-\!\rightarrow\!\!\!\!\bigcirc \Longrightarrow \bigcirc\bigcirc \Longrightarrow \bigcirc\!\!\!\!\bigcirc \Longrightarrow \bigcirc\bigcirc \Longrightarrow \bigcirc\!\!\!\!\leftarrow\!-\ell-\!\rightarrow\!\!\!\!\bigcirc \Longrightarrow \bigcirc\bigcirc \Longrightarrow \dots$$

They are oscillating as if they were two (interpenetrating) gravitating bodies of mass $M/2$ each, the periodicity is $T = \pi\sqrt{\ell^3/2GM}$ known from the two-body problem.

The feature we should note is the following. Single soliton states $|\bigcirc_L\rangle$ and $|\bigcirc_R\rangle$ are static. The two-soliton Schrödinger cat states $\Psi_\pm$ oscillate. The initial overlap of $\Psi_\pm$ with $\bigcirc_{L/R}$ is $1/\sqrt{2}$, see (7). In standard linear quantum dynamics the overlap would be constant in time. It is not so here, the overlap will oscillate with amplitude $1/\sqrt{2}$. In fact, it is zero for most of the time. The SNE makes $\Psi_\pm(t)$ orthogonal to $|\bigcirc_{L/R}\rangle$ after a few times

$$\delta t \sim \frac{\hbar\ell}{GM^2} \ll T, \tag{8}$$

when the distance differs from ℓ by (a few times) more than the soliton size D. To confirm the guess, consider the relative acceleration $GM\ell^{-2}$ of the two solitons toward each other and calculate the time lapse until they get closer by a length $\sim D$.

4. Schrödinger–Newton equation: four catches

Based on the above features of Schrödinger cat states under SNE, we are going to present four interrelated foundational issues all originating exclusively from the nonlinearity. Our thought experiments adapt Gisin's two-qubit superluminal telegraph [10]. Suppose Alice and Bob are far away from each other. Alice owns a qubit and Bob owns a large mass M, in the following maximally entangled composite state:

$$|\uparrow_z\rangle \otimes |\bigcirc_L\rangle + |\downarrow_z\rangle \otimes |\bigcirc_R\rangle. \tag{9}$$

Three of the forthcoming testable controversies are based on this composite state, the fourth one uses the single particle dynamics in itself.

4.1. Action-at-a-distance from nothing

Alice measures either $\hat\sigma_z$ (case i) or $\hat\sigma_x$ (case ii), each with random outcomes ±1. In case i, Bob's state collapses into the static single soliton states $|\bigcirc_L\rangle$ or $|\bigcirc_R\rangle$, according to the outcomes ±1. Alternatively, in case ii, Bob's state collapses into the Schrödinger cat states $\Psi_\pm$, respectively. So far the story coincides with the standard EPR scenario of standard quantum mechanics. The salient effect of nonlinearity enters from now on. In case i, Bob's states $|\bigcirc_L\rangle$ or $|\bigcirc_R\rangle$ remain static. In case ii, his states $\Psi_\pm$ heavily oscillate. Using no physical interaction with it, Alice could cause a testable effect to the remote object of Bob.

4.2. Superluminal telegraph

This thought experiment is the continuation of the previous one. Bob tests whether his initial state is preserved (case i) or it is changing (case ii). Bob waits until a few times δt (8) and then he measures the projector

$$|\bigcirc_L\rangle\langle\bigcirc_L| + |\bigcirc_R\rangle\langle\bigcirc_R|. \tag{10}$$

The outcome is 1 in case i and 0 in case ii. This confirms that the action-at-a-distance (Sec. 4.1) is testable. Is it superluminal? The answer is immediate if we substitute eq. (8) into the condition $c\delta t < \ell$ of superluminality. Surprisingly, the distance ℓ cancels and we are left with $M > \sqrt{\hbar c/G}$ Accordingly, the action-at-a-distance (in other words: the communication from Alice to Bob) is always superluminal if the mass M is at least a few times larger than the Planck mass.

Journal of Physics: Conference Series **701** (2016) 012019 doi:10.1088/1742-6596/701/1/012019

4.3. Unsuitability for mixed state

The initial setup is the same as before, Alice and Bob are far away from each other but they are not assumed to cooperate this time. They may even not know about each other yet they are supposed to inherit the entangled states (9). We assume that Alice does not measure her qubit at all or, if she measures anything, Bob cannot learn anything about the measurement and the outcome. Bob's local (reduced) state is anyway a mixed state

$$\rho = \frac{1}{2}\Big(|\bigcirc_L\rangle\langle\bigcirc_L| + |\bigcirc_R\rangle\langle\bigcirc_R|\Big). \tag{11}$$

The SNE does, as a matter of fact, not apply to mixed states but pure ones described by a wave function. It is clear now that Bob cannot calculate the dynamical evolution of his local system. This incapacity is general since under natural conditions local quantum systems are never in pure states since they are always entangled with the rest of the world or with their environment at least.

4.4. Breakdown of statistical interpretation

To understand this catch, we need no EPR situation but the single-body SNE in itself. Consider that the SNE evolves a given initial pure state density matrix ρ^i into a given final pure state density matrix ρ^f and note that the map

$$\rho^f = \mathcal{M}[\rho^i] \tag{12}$$

is nonlinear.

We are going to show that any nonlinear map makes the statistical interpretation of quantum theory impossible. The proof is elementary and quick [12]. Consider the weighted statistical mixing of two states ρ_1, ρ_2 with weights $\lambda_1 + \lambda_2 = 1$, the resulting state reads

$$\rho = \lambda_1\rho_1 + \lambda_2\rho_2. \tag{13}$$

In von Neumann standard theory, the order of mixing and dynamical evolution are interchangeable:

$$\mathcal{M}[\lambda_1\rho_1 + \lambda_2\rho_2] = \lambda_1\mathcal{M}[\rho_1] + \lambda_2\mathcal{M}[\rho_2]. \tag{14}$$

Now we recognize that this is the mathematical condition of $\mathcal{M}$'s linearity! From the structure of the proof we see that the interchangeability of mixing and dynamical map excludes any nonlinear Schrödinger equation not just the SNE. Without such interchangeability the statistical interpretation of quantum states totally collapses. Moreover, the principle of interchangeability of mixing and dynamics is not necessarily quantum, it is carved in marble in classical statistical physics.

5. Summary: catches and loopholes

We listed four foundational issues encountered by nonlinear modifications of the Schrödinger equation. Our example was the so-called Schrödinger–Newton equation but all four anomalies are equally valid for any deterministic nonlinear modification of the dynamics of any simple or composite quantum system. The best known and most spectacular anomaly is superluminality, a clear violation of special relativity. The derivation of superluminality exploits nonlinearity together with standard statistical interpretation of the wave function. Less known is that already the statistical interpretation is in ultimate conflict with any (deterministic) non-linearity of quantum dynamics, cf., e.g., [13, 14]. This anomaly is admittedly less spectacular than superluminality. It is, nonetheless, the deepest anomaly of nonlinear quantum mechanics.

Non-linear quantum mechanics, the SNE first of all, attracts growing attention in foundations. Awareness of the also foundational catches has recently motivated a stochastically re-linearized model [15] and theory [16]. The surviving and even growing interest in SNE as it is can be explained and perhaps justified despite the catches if we point out certain loopholes. In summary, nonlinear Schrödinger equations

- allow for
  - fake action-at-a-distance but it may be extremely weak to be detected
  - superluminal communication but it may be too hard to be realized
- do not allow for
  - local dynamics unless a local pure state is prepared
  - standard statistical interpretation hence a new interpretation is needed

The last catch is the major one. *Any* nonlinear quantum theory needs a radical new interpretation of the wave function because *any* (deterministic) nonlinear dynamics makes the Born interpretation contradict the principles of statistics.

Acknowledgments

This work was partially supported by the Hungarian Scientific Research Fund under Grant No. 103917 and EU COST Action MP1209 'Thermodynamics in the quantum regime'. The author is indebted to Fetzer Franklin Fund for its continued generous support of foundational research.

References

[1] McLachlan A D and Ball M A 1964 *Rev. Mod. Phys.* **36** 844
[2] Møller C 1963 *Les theories relativistes de la gravitation* (Paris: CNRS); Rosenfeld L 1963 *Nucl. Phys.* **40** 353
[3] Jánossy L 1952 *Acta Phys. Hung.* **1** 423
[4] Gao S 2015 (lectures)
[5] Diósi L 1984 *Phys. Lett.* **105 A** 199
[6] Penrose L 1996 *Gen. Relativ. Gravit.* **28** 581
[7] Einstein A, Podolsky B, and Rosen N 1935 *Phys. Rev.* **47** 777
[8] Bell J S 1964 *Physics* **1** 195
[9] Shimony A 1978 *Phil. Quat.* **18** 3
[10] Gisin N 1990 *Phys. Lett.* **A 143** 1
[11] Soni V 2002 *Pramana* **59** 375; Geszti T 2004 *Phys. Rev.* **A 69** 032110; van Wezel J and van den Brink J 2008 *Phil. Mag.* **88** 1659; Romero-Isart O 2011 *Phys. Rev.* **A 84** 052121; Yang H, Miao H, Lee D S, Helou B, and Chen Y 2013 *Phys. Rev. Lett.* **110** 170401; Anastopoulos C and Hu B L 2014 *New J. Phys.* **16**, 085007; Colin S, Durt T, and Willox R 2014 *Class. Quant. Grav.* **31** 245003; Bahrami M, Großardt A, Donadi S, Bassi A 2014 *New J. Phys.* **16** 115007; Giulini D and Großardt A 2014 *New J. Phys.* **16** 075005; Bekenstein R, Schley R, Mutzafi M, Rotschild C, and Segev M 2015 *Nat. Phys.* **11** 872; Großardt A, Bateman J, Ulbricht H, and Bassi A 2015 arXiv:1510.01696, arXiv:1510.01262
[12] Diósi L 2007 *A short course in quantum information theory* (Berlin: Springer-Verlag)
[13] Mielnik B 1974 *Commun. Math. Phys.* **37** 221
[14] Diósi L 1986 *Phys. Lett.* **114 A** 451
[15] Nimmrichter S and Hornberger K 2015 *Phys. Rev.* **D 91** 024016
[16] Tilloy A and Diósi L 2016 *Phys. Rev.* **D 93** 024026

All is Ψ

Lev Vaidman

Raymond and Beverly Sackler School of Physics and Astronomy
Tel-Aviv University, Tel-Aviv 69978, Israel

E-mail: vaidman@post.tau.ac.il

Abstract. It is argued that standard quantum theory without collapse provides a satisfactory explanation of everything we experience in this and in numerous parallel worlds. The only fundamental ontology is the universal wave function evolving in a deterministic way without action at a distance.

1. Introduction

This is a contribution to the 3rd International Symposium about Quantum Mechanics based on a "Deeper Level Theory", but the message I want to present is that there is no need for a deeper theory. In my view bare quantum mechanics provides a good explanation of all that we see around us. In fact, it seems that this series of the Symposia leads in this direction. The quote from the objectives of the EmerQuM (2011) reads:

> The theme of "emergent quantum mechanics" is, we believe, an appropriate present-day topic, which can both serve as ... a means to point towards promising future directions in physics. We intend to bring together many of those physicists who are interested in or work on attempts to understanding quantum mechanics as emerging from a suitable classical (or, more generally: deeper level) physics...

The objectives of EmQM13 (2013) were not very different:

> The symposium invites the open exploration of an emergent quantum mechanics, a possible "deeper level theory" that interconnects three fields of knowledge: emergence, the quantum, and information. Could there appear a revised image of physical reality from recognizing new links between emergence, the quantum, and information? Could a novel synthesis pave the way towards a 21st century, "super-classical" physics? ...

However, the first objective of EmQM15 (2015) is changed:

> The symposium invites the open exploration of the quantum state as a reality. The resurgence of interest in ontological quantum theory, including both deterministic and indeterministic approaches, challenges long held assumptions...

I believe the quantum state is a reality. It is the reality. The only fundamental physical ontology is the quantum wave function.

Please allow me a sociology of science speculation for why this opinion is not in the consensus today. At the birth of quantum mechanics there was at least a decade of quantum mechanics without the wave function and the Schrödinger equation of its evolution. During this time

Bohr with his charismatic behaviour convinced the physics community that the reality behind quantum mechanics cannot be grasped by human beings. A half century later, physicists working on the foundations of quantum mechanics expressed growing dissatisfaction with Bohr's view, they followed Bell [1] in search of "local beables". The ingenious model of Spekkens [2] gave a hope that our familiar everyday reality will emerge from a "deeper level theory".

However, the research triggered by these developments cooled down the expectations. It suggested that whatever the "deeper theory" is, it cannot be simpler than the standard quantum theory [3] and it is strongly suggested that the quantum wave function is "ontic" [4, 5], i.e., it is the ontology of quantum mechanics.

Moreover, it was claimed that just a "free will" assumption is enough for a simple proof of an ontic nature of the quantum wave function [6]. In my view, however, this argument is a metaphysical error. In science we *assume* "free will" of an external agent who tests theories we construct, so nothing can be derived from the "free will" assumption.

It is considered "politically incorrect" to claim that physicists, at large, understand Nature. This was the view a little more than a century ago, the view which was proved to be very wrong. Two revolutions in physics took place in the last century: relativity theory and quantum theory. Classical physics provided very plausible explanations of many observable phenomena, but it was shown to be wrong in more precise experiments.

Today there are no paradoxes similar to those encountered by classical physics. However, the ontology of the quantum wave function seems very far from what we observe with our senses, and this is what led the majority to believe that it cannot be the true description of Nature. I, however, think that the classical understanding of Nature at the beginning of the last century is roughly correct also for quantum theory. An atom is not made of point particles moving on definite trajectories, but is a highly entangled quantum wave of electrons, protons, etc. This quantum wave is, however, not relevant for the interaction with other systems, it exerts forces on other systems very much like a point-like body. The wave structure provides an explanation of the spectrum, which classical mechanics does not, but for explaining pressure or other macroscopic features, classical approximations work fine.

In the following I will present the view that all that exists is the wave function of the Universe which evolves according to the Schrödinger equation and without collapses during quantum measurements. This is the Many-worlds interpretation (MWI) of quantum mechanics [7].

2. Many worlds in a single Universe

In my view [8] the MWI is the only interpretation which can be viewed as a possible final theory of physics. It is a deterministic theory without action at a distance as all physics we know today. It is (still) not widely accepted since it requires a radical change in our view of Nature: there are many worlds which exist in parallel at the same space and time as our own.

As a physical theory, the MWI is a theory about the Wave Function of the physical Universe. The equations are those of the standard quantum mechanical formalism: the Schrödinger equation or its relativistic generalizations. The theory postulates that the only ontology is this Wave Function which evolves in a unitary way, i.e., there is no collapse of the Wave Function. Since the collapse process is the only random element in physics and also the only source for action at a distance, randomness and action at a distance are eliminated from physics in the MWI.

Clearly, the most general and complete description of Nature is not the Wave Function of all particles in the Universe. Such a picture does not describe the creation and annihilation of particles and field theory is required. In such a framework a wave functional describing the amplitude for the local fields at each point in space replaces the Wave Function. However, I do not foresee that this entails significant conceptual changes or difficulties. Surprisingly, little effort has been made to look for a rigorous definition of such a functional. Appreciation of the success of the MWI in resolving the paradoxes of standard quantum mechanics may lead to further research in this direction.

In spite of the name, I view the MWI as a theory of a single Wave Function of a single physical Universe. This is the only ontology. Observables, which are frequently considered as basic building blocks of the standard formalism (and which are the basis of algebraic approaches to quantum theory) are not part of the ontology. Thus, the uncertainty relations of quantum observables do not lead to randomness, the theory remains deterministic since the Wave Function evolves deterministically [9].

There are two parts to the MWI: physics and interpretation. Physics, the theory of the evolution of the Wave Function, is rigorously defined and tested up to the maximally possible precision. It is a "good" theory: it has no paradoxes, it is complete, it contradicts neither the spirit nor the letter of the special theory of relativity. As a physical theory it is clearly simpler than the spontaneous collapse theories [10, 11], Bohmian mechanics [12, 13], the consistent histories approach [14], or underlying probabilistic theories [2]. Moreover, recent analyses of possibilities for alternative interpretations show that they cannot be more parsimonious than the theory of the quantum Wave Function, see a recent experimental study [15].

The difficult part of the MWI is the interpretation, the explanation of how the Wave Function of the Universe describes the world we observe in 3-space while the Wave Function "lives" in the configuration space of $3N$ dimensions (N is the number of particles in the world). The role of interpretation is much larger in the MWI than in other interpretations of quantum mechanics. In most other interpretations the connection between ontology, e.g. Bohmian trajectories, or values of observables, is simple, transparent and immediate, so frequently it is not discussed at all, while in the MWI, it is the main part to be analyzed. Interpretation belongs to the realm of human science with different standards and methodology. It has a much wider range of acceptable approaches. FAPP (for all practical purposes) [16] definitions are good enough for discussing our experiences.

The main difficulty of connecting our experience with the ontology in the MWI is that it corresponds not just to our current experience, but also to a multitude of other experiences in parallel worlds. Clearly, there are very many parallel worlds now, although the number of worlds is not rigorously defined. However, even in the gedanken case in the framework of the MWI in which the Wave Function now describes just one world in the Universe, the correspondence to our experience in this world is not obvious. I will start with the analysis of such a single world.

3. A World

The concept of a world in the MWI does not belong to the physics part. It is not like a particle: the form of the Wave Function includes information about the number and the type of particles in the physical Universe. Since the Wave Function is all there is, it must also have information about worlds, but the definition of what a world means is not written in the Wave Function. It belongs to human science. A world is what a layman understands by the word "world". A quantum physicist might have a confusing concept of a world, he can think of a multiverse [17], about particles in a superposition, etc., so the concept might not be clear. A world in the MWI is defined as a world of classical physics:

A world is a collection of objects in the Universe in definite states.

In classical physics we can imagine that all particles have definite positions and velocities. However, the positions of every molecule of Earth's atmosphere are not really relevant for describing a world in the MWI. It is enough to specify definite states of all macroscopic objects. Moving to the quantum domain, we do not even have a description in terms of positions and velocities of particles. In the standard approach, the notion of observables plays an important role. Our experience is described through a set of values of the observables. An observer is aware only of measured observables and their measurements collapse the Wave Function to definite eigenstates which provide the correspondence.

David Bohm, in support of his interpretation, noted that in every experiment we read the values of observables by observing positions in space of the pointers of the measuring devices. Definite temperature corresponds to a particular position of a thermometer, etc. So, the world is specified by well localized positions of all macroscopic objects. The world is a concept in 3-space. It is a crucial point for explaining the connection between the quantum wave and our experience. In a theory with collapses at each quantum measurement there is a single world in which the Wave Function is a product of wave functions of all macroscopic objects in 3-space. The classical picture of interaction according to which an object creates fields in 3-space which cause accelerations of other objects present in the location of the field is valid here. (Recently I found a local explanation of this form for an apparent counter example to this claim, the Aharonov–Bohm effect [18].)

Neuroscience does not have a clear explanation of human experiences. Considering experience as information flow allows to speculate that it might be defined in terms of spin wave function [19]. This can lead to some modification of the picture, but apparently will not change it dramatically and currently there are no signs that our brain operates with spins. The picture I present here is that our experiences supervene upon the space distribution of matter, so the Wave Function which allows us to build a three-dimensional picture of particles in 3-space specifies human experiences. The density of particles is gauge independent and also properly transforms between different Lorentz observers. Thus, the explanation of our experience is unaffected by the "narratability failure" problem [20], the Wave Function description might be different for different Lorentz observers, but the local description remains the same.

In fact, in the standard approach with collapses at every quantum measurement ensuring a single world, the problem of correspondence between the ontology and our experience is not considered problematic. (The collapse process, however, is a very serious problem for the physics part of the theory.) This connection between the collapsed Wave Function and our experience will be the basis of the connection with the experience in the framework of the MWI.

The exact Wave Function of a world is not rigorously defined. It is based on collapse, but there is no precise definition when exactly it happens. The collapse avoids superpositions of macroscopic objects in macroscopically different states, but the word "macroscopic" is a FAPP concept. Von Neumann proved that there is a very large flexibility where we put the cut between quantum and classical, i.e. when exactly the collapse takes place. In the collapsed wave we can see the classical world. The three-dimensional picture of places where the wave density is large is the picture of the world we see around us. Instead of an *ad hoc* von Neumann collapse, we can consider the spontaneously collapsed Wave Function of the GRW-Pearle type theory. The quantum wave resides essentially in 3-dimensions. It is connected to our experience in a transparent way. The macroscopic parts of objects collapse to pure unentangled states due to strong interaction with the environment. States of electrons in atoms, and atom states in molecules, etc. remain entangled, but micro states of particles are not relevant to the description of a world.

The state of an electron in an atom is not relevant for the description of a world, but when we describe a quantum interference experiment with single particles, their state is important. One way to avoid description of micro objects is to note that the macroscopic description of macroscopic preparation and detection devices in the experiment includes the information about these particles. I think that it is more convenient to add a description of such micro particles to the picture of the world since it provides a clear characterization of the weak coupling of the particles with the environment. However, such micro particles require a special treatment, in particular, adding a backward evolving wave function specified by the final measurement [21, 22].

Let me summarize the main points. In a world, by definition, all macroscopic objects have definite macroscopic states. In a world, there cannot be a Schrödinger's cat in a closed box in a superposition of being alive and dead. I add to the mathematical formalism of the Universal Wave Function a postulate regarding the connection to our experience. The connection between experience and the world wave function which is an element of the Universal Wave Function is the same as in the collapse theories in which there is only one world wave function and single experience of every observer. The Wave Function of a world $|\Psi_{\rm world}\rangle$ is a product state of well localized (in 3-space) quantum states of all "macroscopic" objects multiplied by possibly entangled states of microscopic systems irrelevant for the macroscopic description of the world.

4. The Wave Function of the Universe

The correspondence between the Wave Function of a single world and our experience may not be as transparent as the correspondence between our experience and a classical world of particles moving on trajectories, or a world of Bohmian trajectories, but I still find it satisfactory. I do not believe that this is the picture of the Universe because there is no satisfactory physics which leads to such Wave Function. The collapse process which is required here is foreign to all physics that I know. It has some kind of action at a distance and randomness and nobody has found elegant and simple equations for the collapse process. Physics forces me to reject the idea of collapse, and therefore, after every quantum measurement a superposition of wave functions of the single world wave function type is created:

$$|\Psi_{\rm Universe}\rangle = \sum \alpha_i |\Psi_{{\rm world}\ i}\rangle. \tag{1}$$

The Wave Funciton of the Universe is a wave function in a configuration space $|\Psi_{\rm Universe}(x_1, x_2, ...x_N)\rangle$. Each one of the wave functions of the worlds is essentially a product of well localized wave functions of macroscopic objects in 3-space $|\Psi_{{\rm world}\ i}\rangle = \prod_j |\Phi_i(X_j)\rangle\ |\psi_i\rangle$, where $|\psi_i\rangle$ signifies the wave function of other degrees of freedom: relative coordinates of electrons in atoms, weakly coupled particles like neutrinos, etc. Locality and strength of interactions (with a buzz word "decoherence") ensure approximate uniqueness of the decomposition.

Different worlds must have different classical descriptions and therefore they correspond to orthogonal wave functions. Energies are bounded, so even well localized wave functions of macroscopic objects must have nonvanishing tails at macroscopic distances. This makes the above statement of orthogonality not so obvious. Most probably, the world wave functions of different worlds can be constructed to be orthogonal, but even if not, since the concept of a world is a human FAPP concept, a tiny overlap can and should be neglected.

The Wave Function of the Universe $|\Psi_{\rm Universe}\rangle$ can be decomposed into a superposition of the world wave functions $|\Psi_{{\rm world}\ i}\rangle$. The reason we do not experience superpositions is not because they do not exist, but because *we* are not capable of experiencing several different states simultaneously. The phrase "different states" means different places and the locality of physical interactions prevents conscious nonlocal creatures. A nonlocal creature which behaves differently depending on the relative phase ϕ between parts at different locations, $\frac{1}{\sqrt{2}}(|A\rangle + e^{i\phi}|B\rangle)$, cannot be useful because local interactions wash out the relative phase almost immediately. This is the argument for the "preferred basis" of the decomposition of the Wave Function, it has to be the local basis.

Another argument for why the Universe cannot support worlds with nonlocal creatures being in a superposition of two locations is that such a creature will need much more resources. If such a nonlocal creature would like to eat, it will need two dinners instead of one. Both parts of the creature $|A\rangle$ and $|B\rangle$ have to eat, but a superposition of a dinner in two locations $\frac{1}{\sqrt{2}}(|A_{\rm food}\rangle + |B_{\rm food}\rangle)$ cannot provide a dinner for the creature through local interactions. Such interactions cannot transform a product state of the creature and the food into the required state $\frac{1}{\sqrt{2}}(|A\rangle|A_{\rm food}\rangle + e^{i\phi}|B\rangle|B_{\rm food}\rangle)$.

The *stability* of world wave functions $|\Psi_{\text{world } i}\rangle$ is the key issue. Of course, at every quantum measurement the world wave function splits to several different (well localized) world wave functions. But this does not happen too fast and too often. If no observer has the time to be aware of a particular world, we have no reason to define such a world: remember, 'world' is a human concept which is supposed to help explaining our experience. I do want to consider worlds at a particular time, but it is not enough that the picture drawn from a wave function looks like a world (all macroscopic objects are well localized) at this moment. If, immediately after, it does not look like a world, such wave function should not be considered as corresponding to a world.

We can ask what was the past and what will be the future of a world. Every quantum measurement splits the world, so a world in a particular time corresponds to a multitude of worlds in the future. Going backward in time we have a unique past. For many worlds at present there is the same past. So, if we consider worlds for periods of time we obtain "overlapping" worlds: they overlap in the past. (If we want to add some micro particles for describing the world, then splitting for them happens at the measurement *after* the present time of the world we consider [21].)

Another important concept which has to be clarified is "I". In which world am I? If I perform a quantum measurement what happens to me? If somebody else performs a quantum measurement will I be changed?

By definition, I have to be in a well defined state. All parts of my body, neurons in my brain, etc. have to be well localized. When I perform a quantum measurement with a few possible results and I observe an outcome, a few worlds are created with a different "I" in each world. All these "I"s are descendants of the "I" before the measurement, so they have identical memories of what happened before the quantum experiment, but they have different knowledge about the result of the experiment. "I"s which have identical memories, but different locations of bodies, are also considered different. Such situations can be created when I am moved to different places, while asleep, according to the results of a quantum experiment [23]).

If somebody else makes a quantum measurement in a faraway location such that the information about the result does not arrive at my location, then I do not split. The world splits, but I remain part of all newly created worlds. There is no meaning for the question what is the result of this measurement in *my* world. (This is one of the examples in which our language, developed during the time of a firm belief in existence of a single world, has difficulty in describing the situation.) So, it cannot be that different "I"s are present in one world, but it is possible that the same "I" is present in several worlds.

Let me summarize. I build the MWI on the basis of standard quantum mechanics. I ask physicists who accept Wave Function collapse to provide (a rough) description of the Wave Function of the physical Universe. This specifies what might be a world wave function. Next, I decompose the Wave Function of the Universe as a superposition of such world wave functions. I postulate the correspondence with the experience as the one in standard quantum mechanics. Thus, the presence of a term corresponding to a world i in the decomposition, $\alpha_i \neq 0$, ensures the presence of such an experience. It is a question for quantum physicists what should be the world wave function to ensure stability and "classicality" of this world.

The wave function of a world $|\Psi_{\text{world } i}\rangle$ specifies what we "feel" in world i. What is the role of the coefficient α_i? The absolute value of the coefficient specifies the illusion of probability as will be discussed in the next section. The phase is not relevant to us. We assume that worlds are different enough that we cannot make interference experiments between them. The worlds differ by macroscopic differences of states of macroscopic objects. The FAPP meaning of the word "macroscopic" can be defined exactly by this property: the states are macroscopically different if we cannot observe interference between them in a realistic experiment. The phase of α_i can be relevant for a creature having unlimited technological power to perform interference experiments with macroscopic objects, i.e., interference between different worlds.

5. Measure of existence of the world and the illusion of probability

The measure of existence of the world, $\mu_i = |\alpha_i|^2$, provides the illusion of probability in our world. Before discussing the qualitative issue, why $|\alpha_i|^2$ and not something else, I will discuss the meaning of probability in the MWI.

I write "illusion" of probability, because I consider a genuine concept of probability to require that there are several options and only one takes place. In case of quantum measurements all possible outcomes take place.

There is no randomness in the basic laws of physics. But determinism by itself does not prevent the concept of probability. We might be ignorant of some details which specify the outcome. In quantum mechanics there are no such details (which usually are named "hidden variables"). We might have complete information about the system and the measuring device and it still will not help us to know what will be the outcome. More precisely, we know that all outcomes will take place and we say that each outcome will correspond to one of the newly created worlds.

If there is a collapse of the Wave Function to one of the world wave functions $|\Psi_{\text{world } i}\rangle$, then we do have a legitimate concept of probability. An experience of an observer supervenes upon the world wave function. In the MWI, every term in the decomposition (1) corresponds to such experience. By construction, our experience in each particular world is the same as our experience in a physical universe with the collapse law in which only one world exists. So, we have a complete illusion of probability: there is no difference between our experience and the experience of an agent with genuine probability.

Since the MWI is a deterministic theory, there cannot be a genuine chance there, but we can arrange a special situation in which there will be a genuine ignorance probability of an agent about the outcome of a quantum experiment. Before, I pointed out that in quantum mechanics we can know everything about the past without knowing what will be an outcome in the future. Now I will show that in the MWI we might know everything about the present, but still be ignorant about the outcome of already performed experiment.

Such a situation is not easy to understand because we are not used to considering a plurality of worlds. What might help is to imagine first a gedanken situation which is not related to quantum physics. Suppose that creatures with super technology land on Earth. They can do what we would consider to be miracles. The creatures can create copies of Earth with everything on it and add it to the Solar system. They show their ability to people on Earth and they say that tomorrow morning there will be three identical Earth planets with all inhabitants. In the morning I wake up and I do not know if I am the original, or I am one of the two copies. Three Earth worlds exist and there are three "I"s. Each one does not know which "I" he is. The symmetry tells us that the probability to be each particular "I" is one third. Each "I" will bet one third on the fact that he is the original and not a copy.

In the case of a quantum measurement I can arrange a similar situation without super technology [23]. Before performing a quantum experiment which has three possible outcomes I take a sleeping pill and ask my friends to move me while I am asleep to one of three rooms according to the result of the experiment. My friends, instead of performing the experiment themselves, can get instructions using the iPhone application "Universe Splitter" or the Tel Aviv World Splitter [24]. When "I" in a particular room am awake, but still have closed eyes, I do not know in which room am I. It is an unusual situation: I, more precisely, every copy of me, might know everything about the Universe, but still be ignorant about the outcome. I am ignorant about self-location in a particular world. This is my privileged property. Any external observer does not have this probability concept. The question: "In which world am I present?" has no meaning for him. There are different "I"s present in corresponding different worlds, but the external observer, even if he is aware that the splitting has occurred, belongs to all these worlds. Only when he contacts me, will he split his "I" according to the outcome of the quantum experiment.

A widespread approach to probability is the readiness to put an intelligent bet on a particular outcome. Since the probability is relevant for all my descendants as they have a legitimate concept of ignorance probability, I suggest to associate it also with me at the time before the experiment. It is rational for me to place bets since all my descendants would like me to do so. They will get the rewards (and the losses) of the bet. Thus, considering the probability as an amount that an intelligent agent is ready to bet on a particular outcome [25], we have a concept of probability in the framework of the MWI. The rule is that an experimentalist performing a quantum experiment should bet in proportion to the *measure of existence* of the world with this outcome [26].

In classical cases either a world exists or it does not. In case of several planet Earths, there is no way that (at a particular time) one planet exists more than the other. We must associate the same measure of existence with all planets: just 1. The definition of the measure of existence of a quantum world, $\mu_i = |\alpha_i|^2$, allows different values. I postulate, that we should bet according to this measure. This analog of the Born Rule in the framework of the MWI is sometimes named the Born–Vaidman rule [27].

Why accept this postulate? In my view, a good answer is that the records of the outcomes of quantum experiments in our world show that there is no general rule which leads to better results for an experimentalist betting on the outcomes of quantum experiments. One can add that the postulate is natural if we accept the idea of measure of existence. Indeed, arranging (artificially) that all worlds have equal measures of existence leads to a natural property of equal probability for all worlds as in the case of multiple "classical" worlds (compare with my three Earths example). The measure of existence of a world also quantifies the power of a world to interfere with other worlds if a quantum super technology will make interference experiments with macroscopic objects which are interference experiments between different worlds [23].

Another argument for accepting the Born–Vaidman Rule is that all the alternatives I try to imagine lead to contradiction with special relativity. If we have a wave function of a particle distributed in space, the sum of the probabilities to find it in some location adds to 1. Thus, if we can change the probability in one place by some local operation, this will change the probability in another place. Changed probability is a signal, since a large ensemble of identical systems with identical actions allows transmission of a message. Local operations at a local part O can change almost all local aspects of the wave function except for changing its weight, $\int_O \Psi^*\Psi dv$. So, it seems to be that the probability cannot depend on any other function of the wave function but its local weight.

6. Conclusions

The MWI, first, tells us what the ontology is: it is the evolving Wave Function. Then it finds us in this ontology, as stable waves in human shapes, and explains (through some natural postulates) our experience.

The theory of the evolution of the Wave Function is an exact mathematical theory of a single physical Universe. Avoiding discussion of values of variables makes it free of paradoxes, action at a distance, and randomness, the features which are foreign to all known areas of physics.

The second part of the MWI, the interpretation, is good only as a FAPP theory. It allows various modifications, e.g., the original Everett relative state formulations provides a somewhat different decomposition to world wave functions.

The MWI is far from being the most clear explanation of our experience and although it can probably be improved, some other interpretations certainly explain it much better. However, these interpretations pay a very high price in spoiling the physical part of the theory which becomes so unnatural that most physicists are not ready to accept them.

My impression is that the majority of physicists are not ready to accept the MWI because in this framework *we* are not in the center of the theory. The sun does not encircle us. Physicists are still hoping that there will be some new theory (interpretation) which will be better than the MWI. I doubt it, but I think that it should be at least accepted that meanwhile the MWI is the best alternative we have today.

This work has been supported in part by the Israel Science Foundation Grant No. 1311/14.

References

[1] Bell J 1985 The theory of local beables *Dialectica* **39** 86
[2] Spekkens R 2007 Evidence for the epistemic view of quantum states: a toy theory *Phys. Rev. A* **75** 032110
[3] Leifer M 2014 Is the quantum state real? An extended review of ψ-ontology theorems *Quanta* **3** 67
[4] Pusey M F, Barrett J and Rudolph T 2012 On the reality of the quantum state *Nature Phys.* **8** 476
[5] Hardy L 2013 Are quantum states real? *Int. J. Mod. Phys. B* **27** 1345012
[6] Colbeck R and Renner R 2011 No extension of quantum theory can have improved predictive power *Nature Commun.* **2** 411
[7] Everett H 1957 Relative state formulation of quantum mechanics *Rev. Mod. Phys.* **29** 454
[8] Vaidman L 2002 Many-worlds interpretation of quantum mechanics, ed Zalta E N *Stan. Enc. Phil.* `http://plato.stanford.edu/entries/qm-manyworlds/`
[9] Vaidman L 2014 Quantum theory and determinism *Quantum Stud.: Math. Found.* **1** 5
[10] Pearle P 1976 Reduction of the state vector by a nonlinear Schrödinger equation *Phys. Rev. D* **13** 857
[11] Ghirardi G C, Rimini A and Weber T 1986 Unified dynamics for microscopic and macroscopic systems *Phys. Rev. D* **34** 470
[12] de Broglie L 1956 *Une tentative d'interpretation causale et non linéaire de la mécanique ondulatoire (la théorie de la double solution)* (Paris: Gauthier-Villars) English translation: de Broglie L 1960 *Non-linear Wave Mechanics: A Causal Interpretation*, (Elsevier, Amsterdam)
[13] Bohm D 1952 A suggested interpretation of the quantum theory in terms of "hidden" variables, I and II *Phys. Rev.* **85** 166
[14] Griffiths R B 1984 Consistent histories and the interpretation of quantum mechanics *J. Stat. Phys.* **36** 219
[15] Ringbauer M, Duffus B, Branciard C, Cavalcanti E G, White A G and Fedrizzi A 2015 Measurements on the reality of the wavefunction *Nature Phys.* **11**(3) 249–254 `DOI: 10.1038/nphys3233`
[16] Bell J S 1990 Against "measurement" (*NATO ASI Series* no 226), ed Miller A I *Sixty-Two Years of Uncertainty* (Springer) pp 17–31
[17] Deutsch D 2002 The structure of the multiverse *Proc. Roy. Soc. Lon. A* **458** 2911
[18] Vaidman L 2012 Role of potentials in the Aharonov–Bohm effect *Phys. Rev. A* **86** 040101(R)
[19] Albert D 1992 *Quantum mechanics and experience* (Cambridge, MA: Harvard University Press)
[20] Albert D 2013 Physics and narrative, ed Tollaksen J *Quantum Theory: a Two-Time Success Story* (Milan: Springer) pp 171–182
[21] Vaidman L 2010 Time symmetry and the many-worlds interpretation, ed Saunders S, Barrett J, Kent A and Wallace D *Many Worlds? Everett, realism and quantum mechanics* (Oxford University Press) pp 582–586
[22] Vaidman L 2013 Past of a quantum particle *Phys. Rev. A* **87** 052104
[23] Vaidman L 1998 On schizophrenic experiences of the neutron or why we should believe in the many-worlds interpretation of quantum theory *Int. Stud. Phil. Sci.* **12** 245
[24] Tel aviv quantum world splitter `http://qol.tau.ac.il/`
[25] de Finetti B 1974 *Theory of probability* vol 1 (New York: John Wiley and Sons)
[26] Groisman B, Hallakoun N and Vaidman L 2013 The measure of existence of a quantum world and the sleeping beauty problem *Analysis* **73** 695
[27] Tappenden P 2011 Evidence and uncertainty in Everett's multiverse *Brit. J. Phil. Sci.* **62** 99

EPR paradox, quantum nonlocality and physical reality

M Kupczynski

Département de l'Informatique, UQO, Case postale 1250, succursale Hull, Gatineau. QC, Canada J8X 3X 7

E-mail: marian.kupczynski@uqo.ca

Abstract. Eighty years ago Einstein, Podolsky and Rosen demonstrated that instantaneous reduction of wave function, believed to describe completely a pair of entangled physical systems, led to EPR paradox. The paradox disappears in statistical interpretation of quantum mechanics (QM) according to which a wave function describes only an ensemble of identically prepared physical systems. QM predicts strong correlations between outcomes of measurements performed on different members of EPR pairs in far-away locations. Searching for an intuitive explanation of these correlations John Bell analysed so called local realistic hidden variable models and proved that correlations consistent with these models satisfy Bell inequalities which are violated by some predictions of QM and by experimental data. Several different local models were constructed and inequalities proven. Some eminent physicists concluded that Nature is definitely nonlocal and that it is acting according to a law of nonlocal randomness. According to these law perfectly random, but strongly correlated events, can be produced at the same time in far away locations and a local and causal explanation of their occurrence cannot be given. We strongly disagree with this conclusion and we prove the contrary by analysing in detail some influential finite sample proofs of Bell and CHSH inequalities and so called Quantum Randi Challenges. We also show how one can win so called Bell's game without violating locality of Nature. Nonlocal randomness is inconsistent with local quantum field theory, with standard model in elementary particle physics and with causal laws and adaptive dynamics prevailing in the surrounding us world. The experimental violation of Bell-type inequalities does not prove the nonlocality of Nature but it only confirms a contextual character of quantum observables and gives a strong argument against counterfactual definiteness and against a point of view according to which experimental outcomes are produced in irreducible random way.

1. Introduction

Eighty years ago Einstein, Podolsky and Rosen (EPR) [1] demonstrated that an instantaneous reduction of a wave function, describing a couple of entangled physical systems, led to so called EPR paradox.

The paradox disappears in a statistical interpretation of QM according to which a wave function describes only an ensemble of identically prepared physical systems [2–5]. When values of physical observables are measured probabilities of different outcomes depend strongly on experimental context. Quantum probabilities are not degrees of belief of some intelligent agents but are objective properties of physical phenomena and experiments. Whether these probabilities can be deduced from some more detailed description of quantum phenomena is an open question.

In order to reproduce quantum statistics *"sub-quantal"* descriptions have to introduce supplementary parameters which are called hidden variables. Fifty years ago John Bell [6, 7], trying to explain strong correlations between spin polarization projections of two physical systems prepared in a spin singlet state, analyzed a large class of so called local realistic hidden variable models (LRHV). He found that correlations predicted by these models obeyed Bell inequalities (BI) which were violated by some correlations predicted by QM. Five years later John Clauser et al. [8] derived CHSH inequalities which were particularly suited for experimental testing. Using LRHV or stochastic hidden variables models (SHV) several other inequalities were proven and shown to be violated by QM and by data of several ingenious experiments [9–15].

In this paper we concentrate on spin polarization correlation experiments (SPCE) with photons but some of our conclusions apply also to recent experiment with electrons by Hensen *et al.* [15].

In SPCE a source is sending two correlated signals to distant polarization beam splitters (PBS) and detectors. The outcomes on far away detectors, registered in some carefully chosen time windows, are 1=polarization up, -1= polarization down and 0=no count. Outcomes seem to appear randomly but distant time series are correlated more strongly than it is permitted by LRHV and SHV. These correlations are consistent with predictions of QM. One may only conclude that a sub-quantal description of quantum phenomena cannot be based on LRHV or SHV.

However large majority of quantum information community inspired by papers of John Bell, concluded that a local and a causal explanation of correlations observed in SPCE is impossible and that *Nature is nonlocal.*

The violation of Bell type inequalities is believed to imply a violation of local realism (LR) as defined par example by Richard Gill [16]: *"local realism = locality + realism, is closely related to causality"... " events have causes (realism); cause and effect are constrained by time and space (locality)"*.

This belief is incorrect because the assumption of *local realism* or *Bell locality* used in different proofs it is not LR defined above.

Nevertheless *quantum nonlocality* is considered to be a mysterious property of Nature. Apparently a mystery has a seductive power and not only excellent fiction writers but also distinguished scientists allow themselves to over-exaggerate quantum paradoxes. Entangled "photon pairs" are compared to pairs of fair dices which in each trial give perfectly matching outcomes. This is of course impossible. Rolling of fair dices produces random outcomes which cannot be perfectly correlated.

In his recent book Nicolas Gisin [17] explains in a pedagogical way *quantum nonlocality* and reviews results of many experimental and theoretical papers on the subject. He claims that quantum correlations observed in SPCE cannot be explained by causes belonging to the common past: *"Nature does not satisfy the continuity principle ... Nature is nonlocal"*. He advocates a new law of Nature: *"We must accept ... nonlocal randomness, an irreducible randomness that manifests itself in several widely separated places without propagating from one point of space to next"*. At the same time randomly produced events are strongly correlated creating what means inexplicable mystery. In [17] one finds also statements such as: *"A particle passes through two neighboring slits at the same time. Therefore, an electron is indeed both here and a meter to the right of here"*. This is not what QM says!

When promoting his new law of *nonlocal randomness* Gisin doesn't care that quantum field theory (QFT) and a standard model in particle physics are local theories. He seems to forget that all biological phenomena point towards adaptive dynamics and local causality.

For example in their fall migration tiny birds Bar-tailed Godwit (Limosa lapponica baueri) fly from Alaska to New Zealand 11 000 km in about eight days over the open Pacific Ocean, without stopping to rest or refuel. How could it be possible if *nonlocal randomness* was a law of Nature?

The *nonlocal randomness* is not needed in order to understand long range correlations in SPCE. The assumption which is used in LRHV is not LR, as defined above. LRHV use the assumption of *counterfactual definiteness* (CFD) according to which values of quantum observables, including incompatible ones, are predetermined before a measurement and recorded passively by measuring instruments [18].

In SHV it is assumed that results of measurements in distant laboratories are obtained in irreducibly random way what destroys all non-trivial correlations created by a source.

If these assumptions are not used various Bell type inequalities cannot be proven [19–50] and their violations give neither information about the locality of Nature nor about the completeness of QM.

Various probabilistic models used to prove inequalities are not consistent with experimental protocols used in SPCE [31, 37, 44]. If contextual character of quantum observables is properly taken into account correlations may be explained in intuitive way [28-33, 40-47]. Moreover many experiments in quantum optics and in neutron interferometry can be simulated event by event in a local and causal way [51–54].

Unfortunately it seems that these results are ignored or not well understood. This is why in this paper we give a detailed critical analysis of some influential finite sample proofs of Bell type inequalities and we hope that it will finally close the issue.

This paper is organized as follows.

In section 2 we recall EPR paradox and Bohm's spin version of it (EPR-B) and the explanation given by a statistical and contextual interpretation of QM.

In section 3 we explain long range correlations in SPCE and their dependence on how pairing of distant outcomes is done.

In section 4 we analyze finite sample proofs of CHSH and Bell given by Richard Gill [16, 55] and by Sacha Vongher [56] and we discuss impossible quantum Randi challenges proposed by them

In section 5 we show how one may win so called Bell's game discussed in [17] without invoking *nonlocal randomness.*

In section 6 we show how long range correlations in SPCE can be explained in a local and causal way if contextual character of quantum observables is correctly taken into account.

In section 7 we present few results from our paper written with Hans De Raedt [57] showing that *sample homogeneity loophole* was not closed in experiments testing various Bell type inequalities.

In section 8 we present our point of view on the Physical Reality and on its abstract description provided by QT.

The last section contains conclusions.

2. EPR paradox and statistical interpretation

Let us start with few axioms of QM which were believed to be true before the publication of EPR paper:

- A1: Any pure state of a physical system is described by a specific unique wave function Ψ.
- A2: Wave function reduction: any measurement causes a physical system to jump instantaneously into one of eigenstates of the dynamical variable that is being measured. This eigenstate becomes a new wave function describing a state of the system.
- A3: A wave function Ψ provides a complete description of a pure state of an individual physical system.

EPR considered two particular individual systems I+II in a pure quantum state, which interacted in the past, separated and evolved freely afterwards [1]. Using A2 they concluded that:

- A single measurement performed on one of the systems, for example on the system I, gives instantaneous knowledge of the wave function of the system II moving freely far away.

- By choosing two different incompatible observables to be measured on the system I it is possible to assign two different wave functions to the system II (same physical reality: the second system after the interaction with the first).

Since a measurement performed in a distant location on the system I does not disturb in any way the system II thus according to A1 and A3 the system II should be described by a unique wave function not by two different wave functions. Moreover these wave functions are eigenstates of two non-commuting operators representing incompatible physical observables what allows to deduce indirectly the values of these incompatible physical observables for the system II without disturbing it in any way what contradicts Heisenberg uncertainty relations.

Bohr [58] promptly reacted to EPR paper and pointed out that it was not possible to assign two different wave functions to the same reality (the second system after the interaction with the first) since the different wave functions could be assigned to the system II only in two different incompatible experiments in which both systems were exposed to different influences before the measurement on the system I was performed.

Bohr's arguments show that different eigenfunction expansions of the same wave function Ψ provide probabilistic predictions for the behavior of identically prepared physical systems in different mutually excluding (complementary) experimental contexts. Therefore Ψ is not an attribute of a single couple of these systems but only a mathematical tool used to deduct statistical regularities in various quantum experiments performed on identically prepared couples of these systems.

In 1936 Einstein [2] noticed that EPR paradox disappeared if purely statistical interpretation of QM was used: "*Ψ function does not, in any sense, describe the state of one single physical system. Reduced wave functions describe different sub-ensembles of the systems*" [2]. The statistical interpretation has been promoted with success by Leslie Ballentine [3, 4]: "*the habit of considering an individual particle to have its own wave function is hard to break ... though it has been demonstrated strictly incorrect*".

According to the statistical contextual interpretation of QM (SCI) [4, 5, 31, 40, 41]:

(i) A state vector Ψ is not an attribute of a single electron, photon, trapped ion, quantum dot etc. A state vector Ψ or a density matrix ρ describe only an ensemble of identical state preparations of some physical systems.

(ii) A mysterious wave function reduction is neither instantaneous nor non-local. In EPR experiment a state vector describing the system II obtained by the reduction of the entangled state of two physical system I+II describes only an sub-ensemble of systems II being partners of those systems I for which a measurement of some observable gave the same specific outcome. Different sub-ensembles are described by different reduced state vectors.

(iii) A value of a physical observable, such as a spin projection, is not a predetermined attribute of a system but is created in the interaction of a system with a measuring instrument [5, 36, 44].

Let us recall below EPR-B paradox [4, 59] in which a source produces pairs of particles in a spin singlet state.

According to the orthodox interpretation of QM each pair of photons is described by a state vector

$$\Psi = (|+\rangle_P|-\rangle_P - |-\rangle_P|+\rangle_P)/\sqrt{2} \tag{1}$$

where $|+\rangle_P$ and $|-\rangle_P$ are state vectors corresponding to photon states in which their spin is "up" or "down" in the direction $\mathbf{P}$ respectively. If we measure a spin projection of a photon I on the direction $\mathbf{P}$ we have an equal probability to obtain a result "1" or "-1". If we obtain "1" a reduced state vector of the photon II is $|-\rangle_P$, if we obtain "-1" a reduced state vector of the photon II is $|+\rangle_P$. By choosing a direction $\mathbf{P}$, for the measurement to be performed on

Journal of Physics: Conference Series **701** (2016) 012021 doi:10.1088/1742-6596/701/1/012021

the photon I, when " photons are in flight and far apart" we can assign different incompatible reduced state vectors to the same photon II. In other words: we can predict with certainty, and without in any way disturbing the second photon, that the **P**-component of the spin of the photon II must have the opposite value to the value of the measured **P**-component of the spin of the photon I. Therefore for any direction **P** the **P**-component of the spin of the photon II has unknown but predetermined value what contradicts QM.

The solution of EPR-B paradox given by SCI is simple the wave function reduction is not instantaneous and a reduced one particle state $|+\rangle_P$ describes only an ensemble of partners of the particles I which were found to have "spin down" by a spin polarization analyzer pointing in the direction **P**. For various directions **P** it is a different sub ensemble of particles II. Strong correlations between distant outcomes in EPR experiments are due to various conservation laws. More detailed discussion of EPR and EPR-B paradoxes may be found for example in [40].

3. Long distance correlations in SPCE and their interpretation

For singlet state (1) QM predicts strong correlations between spin projections (denoted A and B) measured in different directions characterized by angles θ_A and θ_B namely:

$$E(AB|\Psi) = -\cos(\theta_A - \theta_B) \tag{2}$$

Equation (2) seems to predict strict anti-correlation of all **P**-components of the spin when the coincidence measurements are performed in the same direction **P** on both photons in far-away locations.

Since a choice of a direction **P** can be made when "photons are in flight" it seems impossible to keep strict anti-correlations of spin projections on all possible directions unless QM is incomplete and spin projections on all possible directions are predetermined by a source. In this case we have a mixed statistical ensemble of photon pairs characterized by correlated and predetermined spin projections on all possible directions which are recognized by polarization analysers and registered by the detectors. This CDF assumption motivates a use of joint probability distributions in LRHV [22, 23, 37, 44].

Correlations which one may obtain using LRHV or SHV always satisfy Bell type inequalities [6, 7]. These inequalities are for some directions violated by quantum correlations [2] and by experimental data [9–15].

This is why some scientists started to consider supra-luminal influences between members of the photon pairs or between distant experimental settings and when these speculations failed *quantum nonlocality* became a mystery of Nature.

Let us show below that, in spite of equation (2), QM predicts only imperfect correlations for SPCE and a causal explanation of observed correlations is possible.

(i) In SPCE a pulse from a laser hitting the nonlinear crystal produces two correlated signals propagating in opposite directions. Clicks on the distant detectors are correlated.

(ii) Bohr strongly insisted on a wholeness of quantum phenomena. We do not see pairs of photons when they are created and when they travel across the experimental set-up. We only register clicks on the detectors.

(iii) Quantum field theory tells us only that a photon is a one photon state of quantized electromagnetic field and how it can be used in quantum calculations without giving us any intuitive picture of photons or virtual photons.

(iv) QM mechanics does not predict strict anti-correlations [36, 39–41] because directions of spin polarization analyzers are not sharp and can only be defined by some small intervals I_A and I_B containing angles close to θ_A and θ_B, respectively. Therefore even if detection efficiencies were perfect and if we dealt with a perfect singlet spin state a correct prediction of QM for expectation values in SPCE is:

$$E(AB|\Psi) = -\int_{I_A}\int_{I_B} \cos(\theta_1-\theta_2)\, d\rho_A(\theta_1)\, d\rho_B(\theta_2) \tag{3}$$

(v) In order to estimate correlations one has to define specific time windows in distant locations and define a pairing of clicks observed. There are many cases when a click is observed only in one location or no clicks are observed at all.

(vi) To explain outcomes of SPCE, instead of a singlet state, more complicated mixed quantum states have to be used and strict anti-correlations are not predicted even for sharp directions. For more details see a paper by Köfler *et al.* [60].

(vii) LRHV models fail because they neglect a contextual character of QT: "*The measuring instruments must always be included as part of the physical situation from which our experience is obtained*" [61].

In SPCE we have two far away laboratories performing experiments x and y on two physical signals S_1 and S_2 produced by some source S. Outcomes of experiments (x, y) in each particular time window are (a, b) where $a = \pm 1$, 0 and $b = \pm 1, 0$. The outcomes form two ordered samples of data: $S_A = \{a_1, a_2, \dots a_n \dots\}$ and $S_B = \{b_1, b_2, ..bn \dots\}$. These outcomes are observations of two time series of random variables $\{A_1, A_2 \dots A_n \dots\}$ and $\{B_1, B_2 \dots B_n \dots\}$ called sampling distributions. If all A_i are independent and identically distributed (i.i.d) as some random variable A and all B_i are i.i.d as some random variable B then outcomes of experiments x and y can be completely described by conditional generalized joint probability distributions (GJPD): $P(A = a, B = b|x, y, S_1, S_2)$.

GJPD, describing outcomes of distant experiments, have different properties than standard joint probability distributions of a multivariate random variable and they strongly depend on how pairing of distant outcomes is made [44].

For example let us define two pairings:

- Systematic pairing: $S_{AB}(1k) = \{(a_1, b_k), (a_2, b_{k+1}), (a_3, b_{k+2}) \dots\}$
- Random pairing: $S_{AB}(R) = \{(a_s, b_t)|s \leq t$ and s and t are chosen at random$\}$.

If $S_A = \{-11-11-11-1\dots\}$ and $S_B = \{1-11-11-11\dots\}$ then for k-odd paring we have perfect anti-correlations, for k-even pairing we have perfect correlations and for random pairing there are no correlations at all...

In SPCE we have two synchronized clocks in both laboratories, time windows are chosen in function of them and an appropriate systematic pairing is used. There are several difficulties and uncertainties related to this procedure, called the coincidence-time loophole, which have to be closed [62].

It is important to underline that in statistics:

(i) Correlations do not prove any causal relation between x and y.

(ii) No communication or direct influence between x and y is needed for their existence.

Now let us describe a plausible contextual model, consistent with local causality, able to describe imperfect long range correlations observed in SPCE.

(i) Signals S_1 and S_2, correlated at the source, when arriving to respective experimental settings x and y, are described by supplementary parameters $\lambda_1 \in \Lambda_1$, $\lambda_2 \in \Lambda_2$ and $P(\lambda_1, \lambda_2)$.

(ii) A choice of a setting x has no influence on measurements performed using a setting y in a distant location (no signaling, parameter independence).

(iii) Experimental settings in each location can be chosen randomly or in a systematic way and observed correlations do not depend on how a choice is made.

(iv) Measuring devices, as perceived by incoming signals, are described by supplementary parameters $\lambda_x \in \Lambda_x$, $\lambda_y \in \Lambda_y$, $P_x(\lambda_x)$ and $P_y(\lambda_y)$.

(v) To preserve a partial memory of correlations created by a source outcomes a and b have to be produced in a local and deterministic way in function of local supplementary parameters $\lambda = (\lambda_1, \lambda_2, \lambda_x, \lambda_y)$ describing signals and measuring devices in successive time windows. Namely $a = A_x(\lambda_1, \lambda_x)$ and $b = B_y(\lambda_2, \lambda_y)$ where A_x and B_y are functions equal ± 1 or 0.

(vi) Correlations in this local and causal model are described by:

$$E(AB|x,y) = \sum_{\lambda \in \Lambda_{xy}} P(\lambda)\, A_x(\lambda_1, \lambda_x) B_y(\lambda_2, \lambda_y) \tag{4}$$

where $P(\lambda) = P(\lambda_1, \lambda_2) P_x(\lambda_x) P_y(\lambda_y)$ and $\Lambda_{xy} = \Lambda_1 \times \Lambda_2 \times \Lambda_x \times \Lambda_y$ depend on (x, y).

Let us draw a causal graph representing how successive results $(a,\ b)$ are produced:

$$x \rightarrow \Lambda_x \rightarrow \lambda_x \rightarrow a \leftarrow \lambda_1 \leftarrow \Lambda_1 \leftarrow S_1 \leftarrow S \rightarrow S_2 \rightarrow \Lambda_2 \rightarrow \lambda_2 \rightarrow b \leftarrow \lambda_y \leftarrow \Lambda_y \leftarrow y. \tag{5}$$

It is important to underline that assuming a free will a choice of experimental settings (x, y) does not depend on supplementary parameters $\lambda = (\lambda_1, \lambda_{2,} \lambda_{x,} \lambda_y)$ but of course parameters (λ_x, λ_y) strongly depend on the choice of the settings. The reasoning based on the symmetry which tries to prove that the assumption of a free will implies that λ do not depend on the choice of settings is simply incorrect.
If model (4) is used different Bell type inequalities may not be proven because there is no common probability space for all settings (x, y) [44].
One does not need to evoke *nonlocal randomness* and a quantum magic to give an intuitive causal interpretation of Hansen *et al.* [15] experiment. We base our discussion below on a pedagogical description of this experiment given by Howard Wiseman [63]. Alice and Bob in successive time windows create entangled states of their electrons with photons. Photons are sent to Juanita's laboratory. Alice and Bob randomly choose settings for measurements of their respective electrons. They obtain their measurement outcomes and Juanita performs a joint measurement of photons sent by Alice and Bob.
If Juanita registers undistinguishable photons it means that the electrons in distant locations were prepared in particular physical states in which outcomes of measurements of their spin polarisation projections are correlated. In this experiment the cause of correlations is not a partial memory of a common past or instantaneous communications coming from Juanita but similar physical conditions created independently in far-away locations when "successful" measurements were performed. Outcomes of Juanita's experiments give only information whether such conditions were created or not in a given time-window. The decisions which measurement outcomes are post-selected to estimate correlations depend on rare "successful" Juanita's results.
A detailed discussion of an intimate relation of experimental protocols and probabilistic models and why (4) cannot be derived by partial integration from some larger common probability space was given in [44]. Let us mention here an important paper by Andrei Khrennikov [33] who constructed a rigorous Kolmogorov model for SPCE in which CHSH could not be proven.
In the next section we examine some finite sample proofs of Bell and CHSH inequalities and show that they contain flaws and use assumptions not valid for SPCE.

4. Finite sample proofs of Bell type inequalities and quantum Randi challenges

Finite samples due to statistical fluctuations may violate Bell and CHSH inequalities even if a corresponding probabilistic model never violates them. Assuming particular experimental protocols Richard Gill [16, 55] found probabilistic bounds on possible violations of CHSH in function of a sample size N.

In his first influential paper [16] he studies an experimental set-up of five computers: O, x, y, R_A and R_B.

- A computer O, called a source, sends two correlated messages: strings of the length N containing approximately 50% of 0s and 1s.
- Computers x and y are "measurement stations" producing outputs ± 1 in function of the messages received. Produced strings contain approximately 50% of -1 and 1.
- Computers R_A and R_B, called randomizers, send randomly (as they were results of independent fair coin tosses) setting labels "1" or"2" for each pair of outcomes produced by x and y. This is the only source of randomness.
- Regrouping pairs of outputs corresponding to four possible choices of pairs of labels: (1,1), (1,2), (2,1) and (2,2) we obtain 4 samples of sizes $\approx N/4$ formed as by a random pairing $S_{AB}(R)$ of distant outcomes ± 1 produced by x and y. As we saw such pairing destroys pre-existing correlations between messages as in SHV [44] and remaining correlations satisfy CHSH inequality.

Using this idealized model of SPCE he proves that a probability, of finding a violation of CHSH, so large as predicted by QM and reported by Weihs *et al.* [11], is smaller than 10^{-32}.

The protocol described above can be modified in a way that a sample from some joint probability distribution of the outcomes $(\pm 1, \pm 1; \pm 1, \pm 1)$ corresponding to 4 possible experimental settings is prepared. From this sample, using the labels sent by randomizers, marginal samples for particular settings are extracted.

This protocol is described in detail in [55]. An idealized SPCE with N subsequent "*photon pairs*" is analyzed. There are no losses and 4 possible experimental settings are chosen in a random way.

For clarity of an argument we replace N by 4N. Possible outcomes ± 1 are values of 4 random variables A, A', B, B'. Assuming that "*measurements which were not done also have outcomes; actual and potential measurement outcomes which are independent of the measurement settings actually used by all the parties*" Gill describes $4N$ subsequent pairs using $4N \times 4$ spreadsheet of numbers ± 1. The rows are labelled by an index $j = 1, 2, \ldots, 4N$ and columns by A, A', B, and B'.

This spreadsheet defines a random sample of size $4N$ drawn from some joint probability distribution of 4 random variables (A, A', B, B') and marginal expectation values $E(AB)$ values can be estimated by $\langle AB \rangle = \frac{1}{4N} \sum_{j=1}^{4N} A_j B_j$. Similarly one finds estimates $\langle A'B \rangle$, $\langle AB' \rangle$ and $\langle A'B' \rangle$. Since for any row we have $A_j B_j + A_j B'_j + A'_j B_j - A'_j B'_j = \pm 2$ thus for each value of N we obtain CHSH inequality:

$$\langle AB \rangle + \langle AB' \rangle + \langle A'B \rangle - \langle A'B' \rangle \leq 2. \tag{6}$$

In SPCE settings are chosen at random thus Gill constructs finite samples of expected size N for each experimental setting in the following way: "Suppose *that for each row of the spreadsheet, two fair coins are tossed independently of one another, independently over all the rows. Suppose that depending on the outcomes of the two coins, we either get to see the value of A or A', and either the value of B or B'. We can therefore determine the value of just one of the four products AB, AB', A'B, and A'B', each with equal probability 1/4, for each row of the table. Denote by* $< AB >_{obs}$ *the average of the observed products of A and B ("undefined" if the sample size is zero). Define* $< AB' >_{obs}$, $< A'B >_{obs}$ *and* $< A'B' >_{obs}$ *similarly. When N is large one*

Journal of Physics: Conference Series **701** (2016) 012021 doi:10.1088/1742-6596/701/1/012021

would expect $< AB >_{obs}$ to be close to $< AB >$ and the same for the other three averages of observed products. Hence the equation (6) should remain approximately true when we replace the averages of the four products over all 4N rows with the averages of the four products in each of four disjoint subsamples of expected size N each".
Following this construction Gill finds some probabilistic bound on the violation of (6) and makes a conjecture:

$$\Pr\left(\langle AB\rangle_{obs} + \langle AB'\rangle_{obs} + \langle A'B\rangle_{obs} - \langle A'B'\rangle_{obs} \geq 2\right) \leq \frac{1}{2} \tag{7}$$

He proposes also a Quantum Randi Challenge (QRC): "*Construct 4Nx4 spreadsheet deduce from them 4 marginal samples of the expected size N, find the expectation values: $\langle AB\rangle_{obs}$, $\langle A'B\rangle_{obs}$, $\langle AB'\rangle_{obs}$, and $\langle A'B'\rangle_{obs}$ and check the CHSH inequality. If the program reproducibly, repeatedly (significantly more than half the time) violates CHSH, then the creator has created a classical physical system which systematically violates the CHSH inequalities, thereby disproving Bell's theorem.*
Of course this QRC is impossible since finite samples of expected size N, extracted as above, from a counterfactual 4Nx4 spreadsheet may not, as it was proven, violate CHSH significantly and more than half the time. It does not mean that CHSH may not be consistently violated by finite experimental samples or samples generated using a specific local contextual model (4).
CFD is also used by Sacha Vongher [56]. He considers 800 tennis balls with instructions written on them. Each ball results in a measurement 0 or 1 according to the angle it encounters and instructions (HV) it carries. Angles at the measuring stations and HV may change randomly for each pair. Vongher uses angles $\alpha = a\pi/8$ ($a = 0$ or 3) and $\beta = b\pi/8$ ($b = 0$ or 2). The instructions on each ball tell what to output for any choice of the angles it encounters: (A_0, A_3) and (B_0, B_2) for Alice's and Bob's balls respectively.
For each pair of balls we can fill one row of 800×4 counterfactual spreadsheet but now instead of -1 we put 0. Vongher adds an additional constraint the strict anti-correlation for $a = b = 0$ what reduces degrees of freedom to 3 and he chooses for remaining independent variables: A_3, B_0 and B_2. Different settings are labelled by $d = |b-a|$ and $N_d \approx 200$ for $d = 0, 1, 2, 3$. Outcomes are counted by 8 counters $N_d(X)$ where $X = E$(equal) or $X = U$(unequal). In his model $N_0(E) = 0$ and $N_0(U) \approx 200$ because of imposed strict anti-correlations. Using this notation and the 800×4 spreadsheet Vongher proves a finite version of Bell inequality:

$$N_1(U) \leq N_2(E) + N_3(U) \tag{8}$$

He simulates his experiment with tennis balls running a computer program 1000 times. Results are revealing statistical fluctuations present in any finite sample and are very instructive.

- No violation of Bell and CHSH inequalities was observed for 800 pairs.
- For second model, in which strict anti-correlations for $d = 0$ were kept but some pairs were not prepared, Bell inequality (8) and CHSH were violated 50% of the time in 1000 runs.
- For third model, in which anti-correlations occurred only 87% of time, Bell inequality was violated 87% of time and CHSH still only 50% of time.
- For simulations based on the idealized quantum model (2) Bell inequality was violated 91% and CHSH was violated 99% of time in 1000 runs.

QRC proposed by Vongher is the following. Write a computer program preserving the idea of CDF and strict anti-correlations for $d = 0$, simulate 1000 finite samples and show that Bell and CHSH inequalities are violated so consistently as by samples generated using the quantum model (2). Since finite samples constructed Gill and Vongher are similar to those drawn from the populations described by probabilistic models for which Bell, CHSH and Eberhard inequalities are satisfied therefore their QRC are impossible challenges.

In SPCE 4Nx4 spreadsheets do not exist. Outcomes are not predetermined by a source and in some time-windows there are no counts or only single counts are detected. Thus random variables A, A', B and B' take 3 values ± 1 and 0 and a post-selection is needed in order to extract interesting finite samples containing the outcomes obtained when particular experimental settings are used. Correlations estimated in different incompatible experimental settings do not need to satisfy Bell, CHSH or CH inequalities [20–50].
As we explained in the discussion following the equation (3) strict anti-correlations are neither predicted by QM nor observed in SPCE. "Photons" are neither tennis balls with the instruction written on them (LRHV) nor perfect dices (SHV).
A strong argument against *nonlocal randomness* was given by Hans de Raedt and Kristel Michielsen et al. [51–54] who simulated event by event in a local and causal way several experiments in quantum optics and in neutron interferometry.
One could try to reproduce outcomes of various SPCE using a particular contextual model (4) but it was not done.

5. How to win Bell's game without nonlocal randomness

In order to explain in a pedagogical way *quantum nonlocality* Nicolas Gisin discusses a particular Bell's game [17].
Alice and Bob have two identical boxes each equipped with a joystick and a screen. If the joystick is pushed to the left or to the right from a vertical (neutral position) a result appears on a screen. The results are binary: 0 or 1. Alice and Bob synchronize their watches and move some distance apart. Starting at 9am every minute they push the joysticks on their boxes and record joystick positions and the results on the screen. Alice does not know the choice of the position of the joystick made by Bob and vice versa.
Let $x = 0$ or 1 denote positions of the joystick on the Alice's box and $a = 0$ or 1 displayed results. Let $y = 0$ or 1 denote positions of the joystick on the Bob's box and b=0 or 1 displayed results.
The rules of Bell's game are the following:

- Settings (positions) x and y are chosen randomly.
- Outcomes a and b are determined locally in function of x and y.
- If $(x, y) = (1, 1)$ and $a{\neq}b{\rightarrow}1$ point gained.
- If $(x, y){\neq}(1, 1)$ and $a = b{\rightarrow}1$ point gained.
- Otherwise no point is gained.
- The game is won if the average score is greater than 3.

These rules can be summarized by a simple equation: $[a + b]_2 = xy$ (a sum modulo 2 of a and b is equal to a normal product of x and y). A point is gained if the equation is obeyed. A claim is made that S is always smaller than 3 thus the winning of Bell's game is impossible.

Let us analyze a proof given in [17]. Since the results a and b are determined locally in function of x and y thus each box may use only 4 available local programs which are denoted i and j.

- $i = 1$: $a = 0$ for all x $\quad$ $j = 1$: $b = 0$ for all y
- $i = 2$: $a = 1$ for all x $\quad$ $j = 2$: $b = 1$ for all y
- $i = 3$: $a = x$ $\quad$ $j = 3$: $b = y$
- $i = 4$: $a = 1 - x$ $\quad$ $j = 4$: $b = 1 - y$

We have 16 combinations of programs (i, j). Programs can change at each minute. Programs (i, j) determine outcomes (a, b) for settings (x, y). Gisin displays all possible outcomes for 16

(i, j)	(a,b) for (0,0)	(a,b) for (0,1)	(a,b) for (1,0)	(a, b) for (1,1)	S
(1,1)	(0,0)	(0,0)	(0,0)	(0,0)	3

Table 1. Counterfactual calculation of scores of Bell's game in [17]

Time	(i,j)	(x,y)	(a,b)	S
1	(1,1)	(0,0)	(0,0)	1
2	(2,2)	(0,1)	(1,1)	1
3	(4,3)	(1,1)	(0,1)	1
4	(3,4)	(1,0)	(1,1)	1

Table 2. Calculation of scores in 4 consecutive minutes of Bell's game

possible pairs of programs (i, j) in a 16 row table. In Table 1 we reproduce the first row of Gisin's table (on page 17 of his book) using slightly different notation.

For 3 first settings $a = b$ thus 1 point is gained, no point is gained for the setting $(1, 1)$ for a total score $S = 3$. This is the maximal score possible. For example if $(i, j) = (1, 2)$ then $a = 0$ and $b = 1$ for all settings and a total score is $S = 1$, Thus the average $< S >$ of scores obtained, in this way, for any set of lines is necessarily smaller than 3.

The reasoning given above is counterfactual and incorrect since programs chosen on each side may change every minute. In Table 1 it is assumed that the same couple $(1, 1)$ of programs is used for all possible 4 settings chosen one after another. This is not what is happening, minute after minute, during Bell's game thus one cannot use scores of each line in order to estimate $< S >$ for the game.

We display below in Table 2 possible scores in 4 consecutive minutes of Bell's game

Even if there are no common causes responsible for choices of settings one can, for a particular finite sample, obtain $< S > \approx 4$ consistent with no-signalling. If (i, j) and (x, y) were chosen at random we would expect to obtain $< S > \approx 2$ in a long run.. Therefore it is clear that in order to win Bell's game consistently outcomes are neither predetermined as in Table 1 nor produced randomly.

The experiment described above is called Bell's game because it resembles an idealised SPCE for which QM, using (2), predicts an average score $< S > \approx 3.41$. Let us notice that if outcomes for each setting are predetermined the experimental protocol of Bell's game is similar to the protocol used by Vongher [56] in which (i, j) are instructions written on pairs of tennis balls.

If we reject *nonlocal randomness* we have to use a principle of a common cause, local causality and contextuality in order to explain how a protocols $(i,\ j)$ may depend on settings chosen randomly in distant locations.

We follow the same logic which led us to equations (4) and (5). Every minute settings are chosen randomly and their microstates are described by two local parameters (λ_x, λ_y) drawn from $\Lambda_x \times \Lambda_y$. Two correlated signals, described by parameters (λ_1, λ_2) drawn from $\Lambda_1 \times \Lambda_2$, arrive to the Alice's and Bob's boxes. The outcomes $(a, b) = (i(x), j(y))$ where $i = f(\lambda_1, \lambda_x)$ and $j = g(\lambda_2, \lambda_y)$. Thus protocols used by distant boxes are correlated and correlations depend on settings and on correlated signals arriving to boxes.

Calculations in the equation $[a + b]_2 = xy$ cannot be done locally what means that neither Alice nor Bob can find both values a and b having only their local information. It does not mean that Nature does not satisfy the continuity principle and is nonlocal. The outcomes (a, b) are not results of magical nonlocal calculations but they are produced in locally causal interactions of correlated signals with experimental devices in different experimental contexts.

6. Sample homogeneity loophole and violation of Bell type inequalities

Bell-CHCH-CH inequalities cannot be proven if a contextual character of physical observables is correctly taken into account [22–47]. Therefore it is not strange for us that they are violated in SPCE, in experiments from different domains of science [31, 32] and even in classical mechanics [42].

After several ingenious experiments and after closing several experimental loopholes [9–15] there seems to be strong experimental evidence that Bell type inequalities are violated in Nature.

However there are two problems troubling us. Both of them are related to finite statistics.

(i) As Vongher demonstrated when a particular local realistic model was used to generate 1000 samples of size 800 it turned out that Bell inequality was violated 87% of time and CHSH 50% of time. To compare in the experiment of Weihs *et al.* [11] two long runs were analysed and in only one of them a significant violation of CHSH was observed. In Giustina *et al.* [13] and in Christenson *et al.* [14] experiments only one long run was analysed and a significant violation found. In Hensen *et al.* [15] we have only one sample containing 245 data items.

(ii) Statistical analysis made in these experiments assumed that studied samples were homogeneous. Recently we reported with Hans de Raedt [57] a dramatic breakdown of statistical inference due to sample inhomogeneity. Therefore if sample homogeneity is not checked results of significance tests may not be trusted and we say that the *sample homogeneity loophole* (SHL) was not closed [64].

In order to show how detrimental SHL can be we quote here one example from our paper [57]. We simulated a random experiment in which a measuring device, operating according to some specific internal protocol, was outputting one of 6 possible discrete values. We generated 100 runs (each run containing 10^5 data items). Using these large samples we made a significance test of a null hypothesis $H_0 : 1 - B \geq 0$. When three runs 25, 50 and 75 were used the inequality was violated for each run by more than 2000 SEM (standard error of the mean) and one could with a great confidence reject the null hypothesis. When we performed the average over 100 runs (10^7 data items) $1 - B = +0.95$ SEM and of course the null hypothesis could not be rejected. The reason was that samples produced by our device were not homogeneous.

Let us now comment on the experiment of Giustina *et al.* [13]. In this experiment a source is sending polarization-entangled photons which after passing by one of four possible polarization measuring settings (α_i, β_j) are sent to two detectors one operated by Alice and another by Bob. Clicks on the detectors are registered and the coincidence counts determined.

Each setting defines a different random experiment and in order to check Eberhard's inequality one estimates a value of a random variable J being a particular combination of values of 6 random variables (4 coincidence counts and 2 single counts) deduced from data gathered in all four experimental settings. If the local realism (CFD) was true J should be always positive thus the null hypothesis tested is $H_0 : J \geq 0$.

Data gathered during 300 seconds of recording per setting were divided into 30 bins and a sample S of 30 different J-values was obtained. From this sample a value of the mean $< J >$ together with its standard mean error SEM= $s/n^{0.5}$ ($n = 30$, s=sample standard deviation) were estimated and 67σ (67 SEM) violation of Eberhard's inequality was reported.

Usually we use this terminology if we are convinced that the central limit theorem can be used for a finite sample and $< J >$ is normally distributed. Since the normality of this distribution cannot be proven Khrennikov *et al.* [65] used Chebyshev inequality and concluded that the null hypothesis can be still rejected at the confidence level of 99.95%.

However this null hypothesis test is based on only one (1) sample containing 30 data items. Since a whole set of data contains outcomes from 4 different random experiments it would be surprising if it was homogeneous. A sample of observed values of J could only be homogeneous if data sets obtained for each fixed setting were homogeneous and it was not tested carefully enough.

Giustina *et al.* [13] closed the " fair-sampling loophole". Köfler *et al.* [61] showed that the experiment was immune to the "production-rate loophole" and that the results were consistent with quantum theory. Larsson *et al.* [62] proved that the experiment was not vulnerable to the "coincidence-time loophole". It is clear that it is an excellent experiment performed with a great scrutiny. Nevertheless unless the additional sample homogeneity tests are performed and SHL is not closed its impact on conclusions of significance tests is unknown.

Similar opinion has Andrei Khrennikov [66] who postulates that the open access data-base for, e.g., Bell tests should be created and points out that the statistical analysis of data, which was reported as violating Bell's inequality, suffers of a number of problems.

One could wrongly understand that we do not believe in the experimental evidence indicating the violation of Bell type inequalities. As we already told above it is just the opposite.

7. Physical reality

All our science is built on the assumption that there exists an objective external world governed by some laws of Nature which we want to discover and to harness [67–70].

In Physics we construct idealized mathematical models in order to explain in qualitative and quantitative way various phenomena which we observe or we create in our laboratories.

Our perceptions are biased by our senses and by our brain and depend on the location from where they are made. For example motions of the planets are complicated when observed from the Earth and it took many centuries to explain them in a simple way using the heliocentric system and Newton's equations of motion. Stars and planets are perceived by a naked eye as clinking points on a sky. These images are created by our brain when the light hits the retina of our eyes but of course there is something real behind the scenes causing these perceptions, The stars and planets existed before the life existed on the Earth thus statements such as: "A Moon does not exist if we don't look at it", made in order to impress a general public, are incorrect and misleading.

Similarly in quantum phenomena, which we observe and create, there should be something behind the scenes which is causally responsible for statistical regularities we discover. In our opinion quantum probabilities neither correspond to irreducible propensities of individual physical systems nor to beliefs of some human agents but they are objective properties of quantum phenomena as a whole [4, 5, 31, 37, 67, 68, 70]. Necessity of probabilistic description is due to impossibility to see and control what is going behind the scenes.

Even in macroscopic physics we are unable to describe all observed phenomena. For example we neither can explain all ripples on a surface of a lake after passing of a boat nor motions of successive water droplets in a waterfall. In quantum phenomena we see some "ripples" but not "a lake"?

In contrast to classical physics measuring instruments do not register, in general, pre-existing values of physical observables characterizing physical systems. Value of a physical observable is obtained as a result of an incontrollable interaction of a physical system with a measuring device and any attempt to provide more details about what is going on behind the scenes has to include a description of a microscopic state of a measuring device in a moment of measurement.

Taking into account enormous successes of QM and QFT a more detailed unambiguous description of quantum phenomena does not seem to be needed or possible. However telling that there is nothing behind the scenes is in our opinion naïve, unproven and unproductive.

The efforts to prove that QM is an emergent theory and to understand better its foundations may help to dismiss the quantum magic and speculations that Nature is non local.

If locality of interactions and causality were violated in the micro-world how could we have locally causal macro-world and how could we reconcile QM and QFT with the General

Relativity?

Let us finish this section with three remarks.

- QM and QFT describe statistical properties of quantum phenomena in a way consistent with Einsteinian locality.
- According to fathers of QM a question what is a wave function of the Universe or a question by which slit electron passes in two slit interference experiment are meaningless. Therefore a statement that an electron is at the same time here and a meter from here is nonsense.
- *Quantum teleportation* is an implementable experimental protocol in quantum information but statements that Alice has two entangled photons, sends one of them to Bob and after makes a joint measurement of a remaining photon with a photon to be teleported are incorrect. Three photons cannot be manipulated as described above. Such statements have nothing to do with QM and create a confusion among general public.

The quantum magic is born if exotic interpretations of QM are adopted or if naïve or incorrect models of sub-quantum phenomena are constructed and examined.

When we start to evoke magic it is the end of physical explanation and an admission that we will never understand. Fortunately some papers presented at this conference and at previous Emergent Quantum Mechanics conferences show that we can get some intuitive insight on what might be going under the scenes. Let us cite here for example a paper by Grössing *et al.* [71].

8. Conclusions

Paradoxes are created if incorrect interpretation of QT is adopted or if inappropriate sub-quantal description of quantum phenomena is used. In our opinion a statistical contextual interpretation of QM reconciles the ideas of Bohr and Einstein and is free of paradoxes.

QM, QFT and a standard model in elementary particle physics are consistent with Einsteinian locality. The proofs of various Bell type inequalities are based on CFD or on the assumption that the experimental outcomes are produced in irreducibly random way. If these assumptions are not valid Bell-type inequalities cannot be simply proven.

Therefore paraphrasing Howard Wiseman [63] the violation of these inequalities observed in several experiments hammers, in our opinion, the final nail in the coffin of *counterfactual definiteness* and *irreducible randomness.*

In order to explain long range correlations observed in SPCE we do not need to postulate a new law of Nature called *nonlocal randomness.* An intuitive, contextual and causal explanation may be given: signals are correlated by a source, keep partial memory of it when flying to distant laboratories and outcomes are produced in locally deterministic way in function of micro-states of the signals and of measuring devices at the moment of the measurement.

Therefore the violation of Bell type inequalities gives an additional argument in favor of a point of view that quantum probabilities might emerge from some underlying more detailed and causal description of quantum phenomena to be discovered.

We have recently demonstrated with Hans de Raedt [57] that significance tests may dramatically break down if studied samples are not homogeneous. We do not doubt that various Bell type inequalities are violated in SPCE however the results of various significance tests may not be trusted since sample homogeneity was not or could not be tested carefully enough.

Concerning the Physical Reality we strongly believe that there exists an external world whose existence does not depend on whether it is observed or not. This external world is governed by laws of Nature which we try to discover. Quantum phenomena which we create depend on the devices used to probe this external world and on detailed contexts of our experiments. The information we get is contextual and complementary but quantum probabilities are objective properties of quantum phenomena and not beliefs of some human agents.

An additional argument in favor of the idea that QM might be an emergent theory would be given
by a discovery of some fine structures in experimental time series of data which were not predicted by the theory. It would not only prove that QM may not provide the most complete description of the individual physical systems but it would also prove that QM is not predictably complete [35, 40, 72, 73].

Let us finish this article with words of Einstein [2]:"*Is there really any physicist who believes that we shall never get any insight into these important changes in the single systems, in their structure and their causal connections... To believe this is logically possible without contradiction; but, it is so very contrary to my scientific instinct that I cannot forego the search for a more complete description*".

Acknowledgments
I would like to thank Gerhard Grössing and Fetzer Franklin Fund for the invitation and their hospitality extended to me during this interesting conference.

References

[1] Einstein A, Podolsky B and Rosen N 1935 *Phys. Rev.* **47** 777
[2] Einstein A 1936 Physics and Reality *Journal of the Franklin Institutee* **221** 349
[3] Ballentine L E 1970 The statistical interpretation of quantum mechanics *Rev. Mod. Phys.* **42** 358
[4] Ballentine L E 1998 *Quantum Mechanics: A Modern Development* (Singapore: World Scientific)
[5] Allahverdyan A E, Balian R and Nieuwenhuizen T M 2013 Understanding quantum measurement from the solution of dynamical models *Phys. Rep.* **525** 1
[6] Bell J S 1964 On the Einstein Podolsky Rosen paradox *Physics* **1** 195
[7] Bell J S 2004 *Speakable and Unspeakable in Quantum Mechanics* (Cambridge: Cambridge UP)
[8] Clauser J F, Horne M A, Shimony A and Holt R A 1969 *Phys. Rev. Lett.* **23** 880
[9] Aspect A, Grangier P and Roger G 1982 *Phys. Rev. Lett.***49** 91
[10] Aspect A, Dalibard J and Roger G 1982 *Phys. Rev. Lett.* **49** 1804
[11] Weihs G et al 1998 *Phys. Rev. Lett.* **81** 5039
[12] Rowe M A et al 2012 *Nature* **409** 791
[13] Giustina M et al 2013 *Nature* **497** 227
[14] Christensen B G et al. 2013 *Phys. Rev. Lett.* **111** 1304.
[15] Hensen B et al 2015 Loophole-free Bell inequality violation using electron spins separated by 1.3 kilometres *Nature* **526** 682–686
[16] Gill R D 2003 Time, finite statistics, and Bell's fifth position in *Proc. Found. Probability Phys.*-2 179 (Växjo: Univ. Press)
[17] Gisin N 2014 *Quantum Chance* (Dordrecht: Springer)
[18] Żukowski M and Brukner Č 2014 Quantum non-locality—it ain't necessarily so *J. Phys. A: Math. Theor.* **47** 424009
[19] Accardi L 1981 Topics in quantum probability *Physics Reports* **77** 169
[20] Accardi L 2005 Some loopholes to save quantum nonlocality *AIP Conf. Proc.* **750** p1
[21] Accardi L and Uchiyama S 2007 Universality of the EPR-chameleon model *AIP Conf. Proc.* **962** 15
[22] Fine A 1982 Hidden variables, joint probability and the Bell inequalities *Phys. Rev. Lett.* **48** 291
[23] Fine A 1982 Joint distributions, quantum correlations and commuting observables *JMP* **23** 1306
[24] Hess K and Philipp W 2005 Bell's theorem: critique of proofs with and without inequalities *AIP Conf. Proc.* **750** 150
[25] Hess K, Michielsen K. and De Raedt H 2009 Possible Experience: from Boole to Bell *Europhys. Lett.* **87** 60007
[26] Hess K, De Raedt H and Michielsen K 2012 Hidden assumptions in the derivation of the theorem of Bell *Phys. Scr.* **T151**, 014002
[27] Khrennikov A Yu 1999 *Interpretation of Probability* (Utrecht: VSP)
[28] Khrennikov A Yu 2007 Bell's inequality: nonlocality, "death of reality", or incompatibility of random variables *AIP Conf. Proc.* **962** 121
[29] Khrennikov A Yu 2009 Violation of Bell's Inequality and nonKolmogorovness *AIP Conf. Proc.* **1101** 86
[30] Khrennikov A Yu 2019 Bell's inequality: physics meets probability *Inf. Science* **179** 492
[31] Khrennikov A Yu 2009 *Contextual Approach to Quantum Formalism* (Dortrecht: Springer)

[32] Khrennikov A Yu 2010 *Ubiquitous Quantum Structure* (Berlin: Springer)
[33] Khrennikov A Yu 2015 CHSH inequality: Quantum probabilities as classical conditional probabilities *Found. Phys.* **45** 711
[34] Kupczynski M 1984 New test of completeness of quantum mechanics *Preprint* IC/84/242
[35] Kupczynski M 1986 On some new tests of completeness of quantum mechanics *Phys. Lett.* A **116** 417
[36] Kupczynski M 1987 Pitovsky model and complementarity *Phys. Lett.* A **121** 51
[37] Kupczynski M 1987 Bertrand's paradox and Bell's inequalities *Phys. Lett.* A **121** 205
[38] Kupczynski M 2002 On the completeness of quantum mechanics *Preprint* arXiv:quant-ph/0208061
[39] Kupczynski M 2005 Entanglement and Bell inequalities *J. Russ. Laser Res.* **26** 514
[40] Kupczynski M 2006 Seventy years of the EPR paradox *AIP Conf. Proc.* **861**
[41] Kupczynski M 2007 EPR paradox,locality and completeness of quantum *AIP Conf. Proc.* vol **962** 274
[42] Kupczynski M 2012 Entanglement and quantum nonlocality demystified *AIP Conf. Proc.* **1508** 253
[43] Kupczynski M 2014 Causality and local determinism versus quantum nonlocality *J. Phys. Conf. Ser.* **504** 012015
[44] Kupczynski M 2015 Bell Inequalities, Experimental Protocols and Contextuality *Found. Phys.* **45** 735
[45] De Muynck V M, De Baere W and Martens H 2001 Interpretations of quantum mechanics, joint measurement of incompatible observables and counterfactual definiteness *Found. Phys.* **24** 1589
[46] Nieuwenhuizen T M 2009 Where Bell went wrong *AIP Conf. Proc.* **1101** 127
[47] Nieuwenhuizen T M 2011 Is the contextuality loophole fatal for the derivation of Bell inequalities *Found. Phys.* **41** 580
[48] Pitovsky I 1983 Deterministic model of spin statistics *Phys. Rev.* D **27** 2316
[49] Pitovsky I 1994 George Boole 's conditions of possible experience and the quantum puzzle *Brit. J. Phil. Sci.* **45** 95
[50] De Raedt H, Hess K and Michielsen K 2011 Extended Boole-Bell inequalities applicable to Quantum Theory *J. Comp. Theor. Nanosci.* **8** 10119,
[51] Michielsen K, Jin F, and De Raedt H 2011 Event-based corpuscular model for quantum optics experiments *J. Comp. Theor. Nanosci.* **8** 1052
[52] De Raedt H, Jin F and Michielsen K 2012 Discrete-event simulation of neutron interferometry experiments *AIP Conf. Proc.* **1508** 172
[53] De Raedt H, Jin F and Michielsen K 2013 Data analysis of Einstein-Podolsky-Rosen-Bohm laboratory experiments *Proc. SPIE* **8832** 88321N
[54] Michielsen K. and De Raedt H 2014 Event-based simulation of quantum physics experiments *Int. J. Mod. Phys. C* **25** 1430003
[55] Gill R D 2014 Statistics, Causality and Bell's Theorem *Stat. Science* **29** 512
[56] Vongehr S 2013 Exploring inequality violations by classical hidden variables numerically *Annals of Physics* **339** 81
[57] Kupczynski M and De Raedt H 2016 Breakdown of statistical inference from some random experiments *Computer Phys. Commun.* **200** 168
[58] Bohr N 1935 *Phys. Rev.* **48** 696
[59] Bohm D J 1951 *Quantum Theory* (Erglewood Cliffs: Prentice-Hall)
[60] Kofler J, Ramelow S, Giustina M and Zeilinger A On Bell violation using entangled photons without the fair-sampling assumption *Preprint* arXiv:1307.6475 [quant-ph]
[61] Furry W H 1936 Note on the Quantum-Mechanical Theory of Measurement *Phys. Rev.* **49** 395
[62] Larsson J-A et al 2014 Bell-inequality violation with entangled photons, free of the coincidence-time loophole *Phys. Rev. A* **90**
[63] Wiseman H 2015 Death by experiment for local realism Nature **526** 649–650
[64] Kupczynski M 2015 Significance tests and sample homogeneity loophole *Preprint* arXiv:1505.06349 [quant-ph]
[65] Khrennikov A et al 2014 On the equivalence of Clauser-Horne and Eberhard inequality based tests *Phys. Scr.* **T163** 014019
[66] Khrennikov A 2015 Unuploaded experiments have no result *Preprint* arXiv:1505.04293v2 [quant-ph]
[67] Kupczynski M 2015 Greeks were right: critical comments on Qbism *Preprint* arXiv:1505.06348v2 [quant-ph]
[68] De Raedt H., Katsnelson M I and Michielsen 2014 Quantum theory as the most robust description of reproducible experiments *Ann. Phys.* **347** 45
[69] Marchildon L 2004 Why should we interpret quantum mechanics? *Found. Phys.* **34** 1453
[70] Marchildon L 2015 Why I am not a QBist? *Found. Phys.* **45**(7) 754–761
[71] Grössing G, Fussy S, Mesa Pascasio J and Schwabl H 2013 "Systemic nonlocality" from changing constraints on sub-quantum kinematics *J. Phys. Conf. Ser.* **442** 012012

[72] Kupczynski M 2009 Is the quatum theory predictably complete *Phys. Scr.* **T135** 014005
[73] Kupczynski M 2010 Time series, stochastic processes and completeness of quantum theory *AIP Conf. Proc.* **1327** 394

Journal of Physics: Conference Series **701** (2016) 012022 doi:10.1088/1742-6596/701/1/012022

Dirac neutrino mass from a neutrino dark matter model for the galaxy cluster Abell 1689

Theodorus Maria Nieuwenhuizen

[1]Institute for Theoretical Physics, University of Amsterdam, Science Park 904, 1090 GL Amsterdam, The Netherlands

[2]International Institute of Physics, Federal University of Rio Grande do Norte, Natal, Brazil

Abstract. The dark matter in the galaxy cluster Abell 1689 is modelled as an isothermal sphere of neutrinos. New data on the $2d$ mass density allow an accurate description of its core and halo. The model has no "missing baryon problem" and beyond 2.1 Mpc the baryons have the cosmic mass abundance. Combination of cluster data with the cosmic dark matter fraction – here supposed to stem from the neutrinos – leads to a solution of the dark matter riddle by left and right handed neutrinos with mass $(1.861 \pm 0.016)h_{70}^{-2}\mathrm{eV}/c^2$. The thus far observed absence of neutrinoless double beta decay points to (quasi-) Dirac neutrinos: uncharged electrons with different flavour and mass eigenbasis, as for quarks. Though the cosmic microwave background spectrum is matched up to some 10% accuracy only, the case is not ruled out because the plasma phase of the early Universe may be turbulent.

Nothing matters more than dark matter[1]

1. Introduction

The status of dark matter (DM) is paradoxical. While the cosmic microwave background (CMB) [1, 2] and baryon acoustic oscillations [3] bring strong support for a WIMP (weakly interacting massive particle), the axion or perhaps a sterile neutrino, three decades of (in)direct searches have consistently left the researchers empty handed. Recent experiments include LUX [4], an underground Xe experiment, the Large Hadron Collider (LHC) in its 2012-2014 run at 7-8 GeV [5], the Alpha Magnetic Spectrometer at the International Space Station measuring electrons, positrons and (anti)protons [6, 7], and the Fermi X-ray telescope [8, 9, 10]. None of them yielded a decisive clue. Hopes are now set on the 13 TeV run at the LHC, till now again without dark matter hints, the next one of Lux and the first of Xenon1T.

In parallel with this, the 2-12 GeV gamma-ray excess radiation in the Galactic center is likely not due to dark matter but of astrophysical origin [11, 12], while the possibility of a 7 keV warm dark matter particle (sterile neutrino) related to a 3.5 keV X-ray line observed first in Andromeda and the Perseus cluster [13] is fading now that it is not observed in Draco [14, 15].

But are we perhaps barking up the wrong tree? Various tensions with Λ cold dark matter (ΛCDM) cosmology are known, let's mention a few. Planck confirms the existence of anomalies, in particular the cold spot [16], which may be aligned with a supervoid [17]; on a scale of nearly a giga parsec, 19 quasars have been observed with spins likely aligned with their hosts large-scale structures [18]; a ring of nine gamma ray bursts spreads more than 5 billion light

[1] And nothing energises more than dark energy

years across, exceeding the theoretical limit of 1.2 billion light years for the largest possible structures expected from the cosmological principle [19]; for the "cosmic train wreck" galaxy cluster Abell 520, which has a central $3-4\,10^{13}M_\odot$ mass clump with the enormous mass-to-light ratio $800M_\odot/L_{R\odot}$, it is difficult to explain why weakly interacting DM would have separated itself from the galaxies [20, 21, 22]; in the cluster Abell 3827 an offset between baryonic and dark mass occurs [23], which is an order of magnitude larger than ΛCDM can explain [24]. Global issues exist as well: the predicted transition for the most massive galaxies to transform from their initial halo assembly at redshifts $z = 8 - 4$ to the later baryonic evolution seen in star-forming galaxies and quasars is not observed [25]; the galaxy power spectrum deduced from SDSS-III observations fits well to the stretched exponential $\exp[-(k/k_b)^{1/2}]$ predicted by turbulence [26].

These and further issues [27, 28, 29] make us reflect on the whole DM riddle. Of course, the most natural source for DM is neutrinos. In ΛCDM they cannot make up the DM due to free streaming: structure formation is hindered when the total mass in the neutrino families exceeds a bound that is maximally 0.72 eV/c^2 [2]. This corresponds to a cosmic mass fraction $\Omega_\nu < 1.6\%$, far below the cosmic dark matter fraction $\Omega_c = 24.2 \pm 0.2\%$ [2]. However, there is an argument against this *linear* structure formation: *the Jeans instability is absolute* beyond linear perturbations, since any domain of "Jeans-stable" plane waves contains pairs of plane waves which produce at the bilinear level a beat with wavelength inside the Jeans-unstable regime [30]. Hence the plasma may be (weakly) turbulent, produce nonlinear structures like galaxy clusters [31], explain many perplexing observations [27, 28] and leave signatures of turbulence in the galaxy distribution [26]. Indeed, the quark-gluon plasma has a low viscosity, which leads to a very large Reynolds number Re $\sim 10^{19}$ for a patch of acoustic horizon size [32, 33]. Since this horizon is larger than the causal horizon, turbulence can not develop – at *this* scale. But *turbulence can develop on subhorizon scale* – and hence it should if there is enough time. The plasma keeps large Reynolds numbers till ~ 100 at the recombination, so it is expected to remain turbulent until then [32, 33]. Hence, lacking further study, the option that neutrinos make up the dark matter, is not ruled out by the neutrino free streaming in linear approaches. We shall not dwell further into this, but rather test whether the case is indeed worth the effort.

Neutrino oscillations yield small m^2 differences, so if they are heavy, they have nearly the same mass. The upper limit of the electron antineutrino mass is 2.0 eV/c^2 [34], and for the 3 families (ν_e, $\bar\nu_e$; ν_μ, $\bar\nu_\mu$; ν_τ, $\bar\nu_\tau$) the thermal densities add up to $\Omega_\nu = \rho_\nu/\rho_c \leq 0.13$, less than Ω_c. Hence if ν's are the source of DM, also right-handed ones (sterile ν's, not involved in weak processes) must exist. Reactor experiments favor them too [35, 36, 37].

Cosmology offers further evidence. While the CMB has achieved the status of "precision cosmology", gravitational lensing can reach this status too. Indeed, the quite relaxed galaxy cluster Abell 1689 offers a stringent test. There are well constrained data for the 2*d* mass density $\Sigma(r)$, primarily from strong lensing (SL), that is to say, from background galaxies lensed into several (≤ 5) pieces of an arc, up to some 200 kpc off the cluster centre. Further out, weak lensing (WL) occurs, that is, randomly oriented background galaxies get a systematic tangential shear $g_t(r)$ towards a circle around the cluster centre [38]. The effect can be deduced by averaging over the galaxies within a suitably defined field. The cluster galaxies can of course be observed easily. Finally, X-ray observations provide the electron density profile $n_e(r)$ and thus the mass density of the X-ray gas. These inputs provide a unique basis for an accurate reconstruction of the mass profile.

The setup of this paper is as follows. First we shortly discuss A1689 and the data sets which are analyzed. Next we introduce a model for the baryons in the Brightest Cluster Galaxy (the central one), the only relevant galaxy for our purpose, and we fit the gas data to a Sérsic profile. This is followed by a fit to the NFW profile of dark matter. Next, a model for neutrino dark

[2] Unless indicated otherwise, we adopt the cosmology $H_0 = h_{70}70$ km/s Mpc, $h_{70} = 1$, $\Omega_m = 0.30$ and $\Omega_\Lambda = 0.70$. One arc minute corresponds at the redshift $\bar z = 0.183$ of A1689 to $1' = 184.5$ kpc.

matter is introduced and developed and a fit is presented. The results are combined with results from Planck and neutrinoless double beta decay. The paper closes with a summary.

2. The galaxy cluster Abell 1689

Abell 1689 is a rather large, well relaxed galaxy cluster that has long been studied. An initiating WL study of A1689 was performed in [39]. Ref. [40] reports a SL analysis yielding Σ for radii up to 270 kpc. Ref. [41] employs an entropy-regularized maximum likelihood analysis of the lens magnification and the distortion of red background galaxies detected with Subaru. Ref. [42] unveils the $3d$ structure of galaxy clusters such as A1689 and resolves the discrepancy between X-ray and lensing masses. Ref. [43] fits the NFW profile of cold dark matter [44] with the concentration parameter $c_{200} = 8 - 9$ [43]. However, it is known that modelling only the central data $r < 300$ kpc leads to smaller values, $4 - 5$. An NFW fit of the combined data leads to $c_{200} = 4.5$ [45]. Nowadays the focus is on the triaxiality of A1689 [42, 46, 43]. However, the spherical approximation will serve our goals, since most of the matter is dark and more spherical than the gas.

Most authors focus on the NFW profile, so that no information about the mass of the DM particle is obtained. Still, some groups consider isothermal fermions. Cowsik & McClelland [47] model the DM of the Coma cluster as an isothermal sphere of neutrinos, which leads to a mass[3] of $\simeq 2$ eV. Treumann et al. study 2 eV thermal neutrinos next to thermal CDM and X-ray gas for clusters like Coma [48]. We apply an isothermal fermion model for a single type of dark matter, for the galaxies and the X-ray gas; a fit to lensing data of the Abell 1689 cluster works well and yields as best case the neutrino with mass ~ 1.5 eV [49]. This approach has been followed up whenever new data became available. In [50] we apply the model to SL data by [51] and X-ray data by [52], confirming the result. In [45] we adopt the galactic matter profile of [53], to find that neutrinos perform better ($\chi^2/\nu \sim 0.7$) than the NFW profile ($\chi^2/\nu \sim 2.2$).

2.1. Data sets to be employed

Recently, Umetsu et al. perform a WL analysis combined with some magnification properties [43], which provides important information as we shall discuss below. The authors provide 14 data points Σ^i_{wl} at $124 < r_i < 2970$ kpc and the correlation matrix C^{wl}_{ij}. The latter needs no regularisation, because each bin has $S/N > 1$. The methods for combining the WL shear and magnification effects to derive the radial mass density profiles are explained in [46, 54]. Between 124 and 270 kpc the data overlap with those of [40].

We take additional data for $\Sigma(r)$ from SL analysis by Limousin et al [40]. For description of the analysis we also refer to [45]. There result 12 data points (r_i, Σ^i_{sl}) for $3 < r_i < 270$ kpc and their covariance matrix C^{sl}_{ij}, which has eigenvalues between $0.03\Sigma_c^2$ and $1.4\,10^{-8}\Sigma_c^2$. To account for information due to (implicitly assumed) priors and for further scatter, we add a constant σ_{sl}^2 to the diagonal elements, so that the role of the small eigenvalues is suppressed. This leads to the modified covariance matrix $\overline{C}_{sl} = \sigma_{sl}^2 I + C_{sl}$. A model for $\Sigma(r)$ then involves $\chi^2_{sl}(\Sigma) = \sum_{i,j=1}^{12}[\Sigma(r_i) - \Sigma^i_{sl}](\overline{C}_{sl}^{-1})_{ij}[\Sigma(r_j) - \Sigma^j_{sl}]$. We choose σ_{sl} on an empirical basis. In order not to erase the information coded in C_{sl}, one needs $\sigma_{sl}^2 \ll 0.03\Sigma_c^2$. In [45] we take the average of the diagonal elements, now we even allow half of it: $\bar{\sigma}_{sl}^2 = \mathrm{tr}\, C_{sl}/24 = (0.0375\Sigma_c)^2$.

We also employ the 13 data points for g_t between 200 kpc and 3 Mpc from the WL study [43]. While based on better statistics, they overlap with the data of [41].

The data for n_e we take from Morandi et al [52], who analyze Chandra X-ray observations. The density profile is recovered in a non-parametric way by rebinning the surface brightness into circular annuli via spherical deprojection [55], see also the discussion in [45].

[3] In the text we employ the short hand eV for eV$/c^2$.

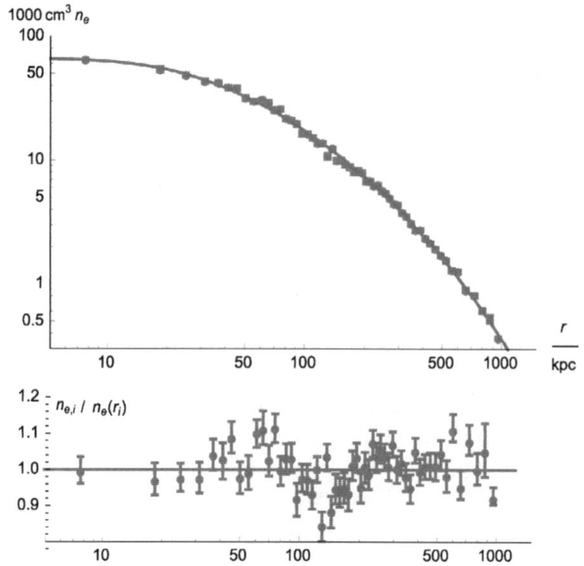

Figure 1. Electron density as function of radius. Data from [52], full line: cored Sérsic profile (2). Lower pane: relative values.

3. Modelling the baryons

In a galaxy cluster there are three components: Galaxies, X-ray gas and dark matter. The galaxy mass density is dominated by the brightest cluster galaxy (BCG, "central galaxy" cg). An adequate profile with mass M_G, core size R_G^i and scale R_G^o is [53]

$$\rho_G(r) = \frac{M_G(R_G^i + R_G^o)}{2\pi^2(r^2 + R_G^{i\,2})(r^2 + R_G^{o\,2})}. \tag{1}$$

If $R_G^i \ll R_G^o$ it is a cored, truncated isothermal sphere.

The mass density of the X-ray gas follows [52] as $\rho_g(r) = 1.167\, m_N n_e(r)$ for a typical $Z = 0.3$ solar metallicity [49]. The 56 data points for n_e with the maximum of the upper and lower errors fit quite better to a cored Sérsic profile than to a β-profile. The Sérsic profile has the form

$$n_e(r) = n_e^0 \exp\left[k_g - k_g\left(1 + \frac{r^2}{R_g^2}\right)^{1/(2n_g)}\right]. \tag{2}$$

The best fit parameters are

$$n_e^0 = 0.0670 \pm 0.0028\,\mathrm{cm}^{-3}, \quad k_g = 1.98 \pm 0.25, \quad R_g = 21.6 \pm 2.7\,\mathrm{kpc}, \quad n_g = 2.97 \pm 0.14. \tag{3}$$

The value $\chi^2(n_e)/\nu = 1.71$ with $\nu = 52$ implies the marginally acceptable q-value 0.0010 for our spherical approximation. A fit to a β model yields $\chi^2(n_e)/\nu = 6.3$, far from acceptable.

3.1. Towards data fits

SL data yield the $2d$ mass density

$$\Sigma(r) = \int_{-\infty}^{\infty} \mathrm{d}z \rho \left(\sqrt{r^2+z^2}\right). \tag{4}$$

Within a disc of radius r it has the average $\overline{\Sigma}(r) = 2r^{-2}\int_0^r \mathrm{d}r'\, r'\Sigma(r')$. In weak lensing one determines the transversal shear, which is related to $\overline{\Sigma}$ and Σ as

$$g_t(r) = \frac{\overline{\Sigma}(r) - \Sigma(r)}{\Sigma_c - \Sigma(r)}. \tag{5}$$

The data of [43] involve $\Sigma_c = 0.682\, h_{70}\, \mathrm{gr\, cm}^{-2}$.

We fit our parameters by minimizing $\chi^2(\Sigma, g_t, n_e) = \chi^2(\Sigma, g_t) + \chi^2(n_e)$, where the lensing errors add up to

$$\chi^2(\Sigma, g_t) = \chi^2_{sl}(\Sigma) + \chi^2_{wl}(\Sigma) + \chi^2_{wl}(g_t). \tag{6}$$

4. NFW fitting

Minimising only $\chi^2_{wl}(\Sigma)+\chi^2_{wl}(g_t)$ leads to $\nu = 27-2$ and $\chi^2/\nu = 1.25$, implying the good q-value 0.18, in support of a recent Bayesian analysis [43]. However, the SL data put stress on this. Ref. [45] reports $[\chi^2_{sl}(\Sigma) + \chi^2_{wl}(g_t)]/\nu = 2.2$ for its regulator $\sigma_{sl} = \sqrt{2}\bar{\sigma}_{sl}$ and $\nu = 20$ so that $q = 1.5\, 10^{-3}$, marginally acceptable; this case now yields $\chi^2(\Sigma, g_t)/\nu = 2.21$ for $\nu = 37$ with a bad q-value $3.3\, 10^{-5}$. Our present $\bar{\sigma}_{sl}$ yields $\chi^2(\Sigma, g_t)/\nu = 2.98$ with $\nu = 37$ and $q = 3.3\, 10^{-9}$, very bad. All by all, NFW fits are less satisfactory for the combined SL and WL data.

5. Neutrino model

For the DM we now consider g identical, nonrelativistic thermal fermion modes, denoted by ν, with mass m_ν and chemical potential $m_\nu\mu$, at temperature $T_\nu = m_\nu\sigma_\nu^2$. With potential energy $m_\nu\varphi(r)$ due to all gravitating matter, its mass density reads

$$\rho_\nu(r) = \int \frac{\mathrm{d}^3p}{(2\pi\hbar)^3} \frac{g m_\nu}{\exp\{[p^2/2m_\nu + m_\nu\varphi(r) - m_\nu\mu]/m_\nu\sigma_\nu^2\}+1}. \tag{7}$$

Because positive energy particles will have escaped from the cluster, the integral is restricted to the domain $p^2/2m_\nu + m_\nu\varphi(r) < m_\nu\varphi(\infty) \equiv 0$. The gravitational potential φ is solved from the Poisson equation $\varphi'' + 2\varphi'/r = 4\pi G\rho$, with the total mass density $\rho = \rho_G + \rho_g + \rho_\nu$.

The data in fig. 1 expose that Σ does not decay fast for $r > 300$ kpc. Hence our previous fits [49, 50, 45] with a large chemical potential μ and fast decay of $\rho_\nu(r)$ do not apply. But this turns out to be a blessing: isothermal fermions can produce a good fit to the data *including the tails*. Then, when plotting the baryon fraction $f_b = M_b/(M_\nu + M_b)$, with $M_b = M_G + M_g$, as function of r, we notice around $r_{ms} = 2250$ kpc a quadratic maximum $f_b(r_{ms}) = 0.16 \pm 0.01$, overlapping with the cosmic value $f_b^c = 0.1580 \pm 0.0014$ [2]. The absence of a "missing baryon problem" around r_{ms} allows to eliminate it theoretically: we take the gas density from the observations and continue its fit (2) beyond 1 Mpc, solve ρ_ν from the Poisson equation up to some *mass separation radius* r_{ms} and impose beyond r_{ms} the cosmic ratio $\rho_\nu(r) = (1/f_b^c - 1)\rho_g(r)$, so that it holds there that $M_\nu(r) = (1/f_b^c - 1)M_b(r)$ and $f_b(r) = f_b^c$ (see Fig. 3). To achieve the matching, we set $f_b^c = \bar{f}_b^c \pm \delta f_b^c$ and add to $\chi^2(\Sigma, g_t, n_e)$ the combination

$$\chi^2_{ms}(r_{ms}) = \frac{1}{\delta f_b^{c2}}\left(\frac{M_b}{M} - \bar{f}_b^c\right)^2 + \frac{1}{\delta f_b^{c2}}\left(\frac{\rho_b}{\rho} - \bar{f}_b^c\right)^2, \tag{8}$$

with $M_b = f_G^b M_G + M_g$, $M = M_\nu + M_b$, all taken at r_{ms}, and likewise for the ρ's. f_G^b is the baryon fraction of the BCG, which we take as 1, though $f_b^G = \bar{f}_b^c$ would hardly change the fit. The best minimum yields $\chi^2(\Sigma, g_t, n_e) + \chi^2_{ms}(r_{ms}) = 131.9$ for $\nu = 97 - 11$ d.o.f., so that its $q = 0.0011$ slightly improves the gas-only value. The $\chi^2_{sl}(\Sigma) = 5.63$, $\chi^2_{wl}(\Sigma) = 8.33$, $\chi^2_{wl}(g_t) = 28.8$ add up to $\chi^2(\Sigma, g_t) = 42.7$. Their $\nu = 39 - 6$ d.o.f. imply $\chi^2(\Sigma, g_t)/\nu = 1.29$ and the good lensing-alone q-value 0.12. The fits for Σ and g_t are represented in figs. 1 and 2, respectively.

Accounting for the variation of r_{ms} causes conservative error bars. The BCG has mass and inner and outer radius

$$M_G = 3.2 \pm 1.0\, 10^{13} M_\odot, \qquad R_G^i = 7.2 \pm 1.4\,\text{kpc}, \qquad R_G^o = 129 \pm 40\,\text{kpc}. \tag{9}$$

The gas fit (3) does not get altered within the error bars when combining its data with the lensing data. The velocity dispersion and chemical potential read

$$\sigma_\nu = 1330 \pm 40\,\frac{\text{km}}{\text{s}}, \quad \mu - \varphi(0) = 5.0 \pm 0.9\ 10^6 \frac{\text{km}^2}{\text{s}^2}. \tag{10}$$

Finally, the from (7) the fermion mass comes out in the combination $gm_\nu/\lambda_\nu^3 \sim gm_\nu^4$ due to the thermal length $\lambda_\nu = \sqrt{2\pi}\hbar/m\sigma_\nu$. The fit yields

$$\left(\frac{g}{12}\right)^{1/4} m_\nu = 1.92^{+0.13}_{-0.16}\,\frac{\text{eV}}{c^2}. \tag{11}$$

This exceeds the 1.45 ± 0.03 of [49], the 1.55 ± 0.04 of [50] and the 1.51 ± 0.04 eV of [45]. However, in those works an additional type of dark matter is assumed, and modeled explicitly in Ref. [45].

The depth of the potential well $\varphi(0) = -19.4 \pm 0.4\ 10^6\ \text{km}^2/\text{s}^2$ has also a comparable systematic error from the extrapolation of the gas data beyond 1 Mpc with the Sérsic profile.

At $r_{200} = 2.37$ Mpc the cluster has overdensity 200 with $M_{200} = (18.1 \pm 0.8)10^{14} M_\odot$, to be compared with the $(1.32 \pm 0.09)/h\, 10^{15} M_\odot = (18.9 \pm 1.3)\ 10^{14} M_\odot$ of [43].

At $r_{ms} = 2.05 \pm 0.39$ Mpc the Newton acceleration

$$\frac{GM(r_{ms})}{r_{ms}^2} = 5.4\,10^{-11}\frac{\text{m}}{\text{s}^2} = 0.92\frac{cH_0(\bar{z})}{4\pi}, \tag{12}$$

with $M(r_{ms}) = (16.4 \pm 0.7)10^{14} M_\odot$, leads to a consistent picture: Beyond r_{ms} the cosmic acceleration dominates the attraction to the cluster, preventing mass separation. Hence, unlike r_{200}, r_{ms} has a clear physical meaning.

5.1. The cosmic budget and neutrinoless double beta decay

The mass (11) depends on the degeneracy factor g: In the cluster a smaller DM particle mass can be compensated by a larger degeneracy, at fixed $g^{1/4}m$. At the cosmic scale gm gets fixed, lifting the degeneracy. Each active (anti)neutrino has a number density of 56.0 cm^{-3}, and we assume that sterile ν's are also thermal. Supposing there are $N_f = 3 + \Delta N_f$ neutrino families with $g = 2N_f$ degrees of freedom, all having basically a mass of 1.92 eV, the cosmic budget

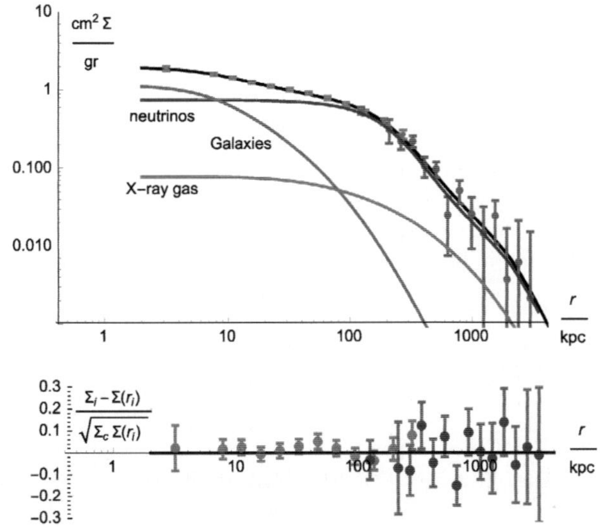

Figure 2. $2d$ mass density Σ and the contributions from its components as function of r. Left data from [40] with error bars $(\overline{C}_{ii}^{sl})^{1/2}$. Right data from [43] with error bars $(C_{ii}^{wl})^{1/2}$. Black line: best fit to our model. The kink in the slope at $r_{ms} = 2.05$ Mpc reflects the mass separation: below r_{ms} the neutrinos deplete towards the interior. Lower pane: Relative deviations.

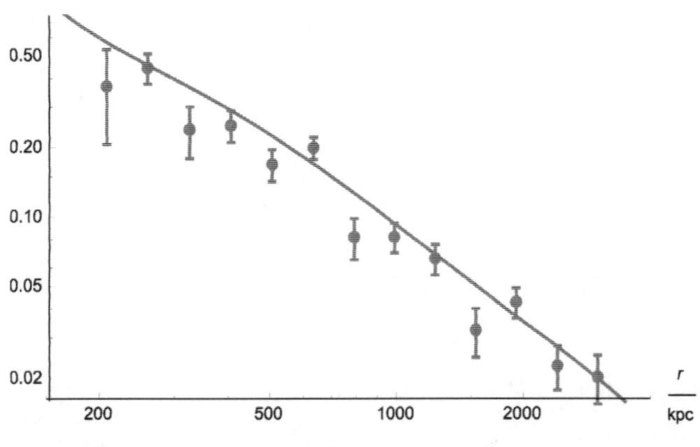

Figure 3. Weak lensing shear g_t as function of r. Data from Umetsu et al. [43]; full line: best fit to the model.

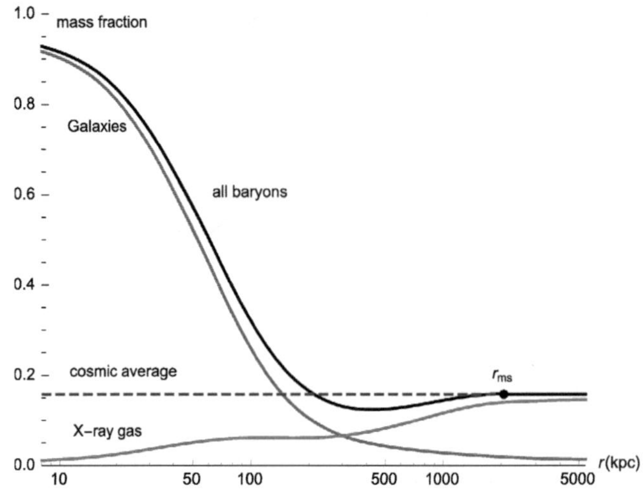

Figure 4. Baryonic mass fractions as function of r. In the centre the brightest cluster galaxy dominates. The depletion beyond 300 kpc is standard. Beyond 700 kpc the neutrinos are relatively depleted, allowing the baryons to reach the cosmic fraction 0.158 at the mass separation radius $r_{ms} = 2.05$ Mpc. From there on baryons and neutrinos occur at cosmic ratios.

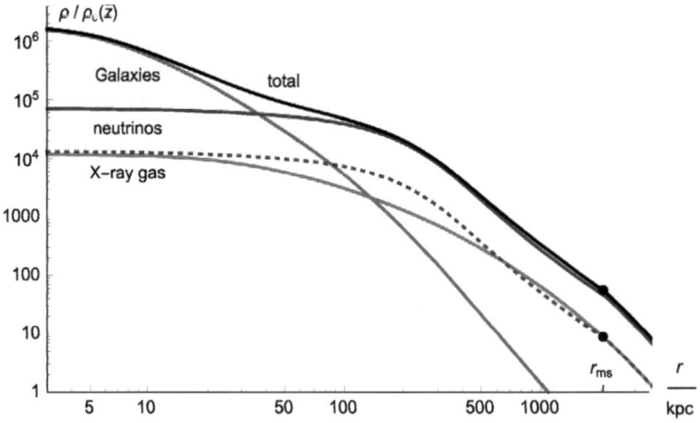

Figure 5. Mass densities of Galaxies, X-ray gas and neutrinos in A1689, normalised to the critical density $\rho_c(\bar{z})$, as function of r. The dotted line presents the gas density if it had the cosmic fraction of the neutrino mass density. Its overlap beyond r_{ms} indicates that our model avoids a missing baryon problem; its near-overlap at the origin a is surprise. The surplus from 20 to 650 kpc and the depletion from 650 to $r_{ms} = 2050$ kpc show that the neutrinos are cored.

comes out as $\Omega_\nu = (N_f/6)^{3/4}(0.251 \pm 0.019)h_{70}^{-3/2}$, where we reinserted Hubble's constant [49]. The physical density parameter thus reads $\Omega_\nu h^2 = (N_f/6)^{3/4}(0.123 \pm 0.009)h_{70}^{1/2}$. The Planck cold dark matter fraction is $\Omega_c h^2 = 0.1188 \pm 0.0010$ [2]. Neutrinos can exactly account for this in case $N_f = 6$, $g = 12$, which amounts to the 3 active and 3 sterile ones (3+3 model), all having nearly the same mass. That case for $\Delta N_f = 2$ needs $m_\nu = 2.14$ eV which is ruled out [34].

In general it is not required that the masses are equal, and neither the chemical potentials nor the velocity dispersions. While neutrino oscillations have shown small mass differences between the active neutrinos, reactor experiments expose hints for one, two or possibly 3 sterile neutrinos with eV mass differences; at least two sterile species are needed to explain

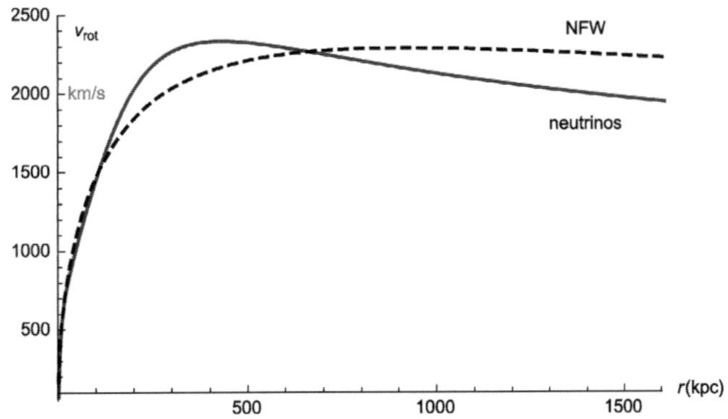

Figure 6. Full line: Circular rotation velocity of galaxies as function of r; the bump around 430 kpc is caused by the neutrino core. Dashed: NFW prediction.

CP violation [35, 36, 37]. Neutrinoless double beta decay ($0\nu\beta\beta$) implies for the known mixing angles [34] a sharpest lower bound when both Majorana phases are equal to π, namely $m_{\beta\beta}^{0\nu} \geq 0.33 m_\nu = 0.64 \pm 0.05$ eV. Upper bounds for $m_{\beta\beta}^{0\nu}$ are 0.45 eV by EXO-200, 0.28 eV by KamLAND-Zen [56], 0.4 eV by GERDA [57] and of $0.27 - 0.76$ eV by CUORE [58], respectively. Hence Majorana neutrinos are disfavoured and we arrive at Dirac neutrinos: uncharged electrons with different flavour and mass eigenbases, like quarks (or: quasi-Dirac neutrino's, nearly having these properties.) Detection of $0\nu\beta\beta$, though, would lead to a smaller mass of the active ν's and a higher mass(es) of the sterile ν's.

We considered only neutrinos bound to the cluster. Accounting, as in our earlier works, also for the unbound ones would overlook that they probably had time enough to escape to infinite. The case yields anyhow a mass and other parameters compatible within the error bounds.

We noticed the overlap of our Ω_ν with the Ω_c from the CMB. Equating them and taking the smaller error bar of the latter [2] offers a sharper prediction for the 3+3 Dirac mass:

$$m_\nu = (1.861 \pm 0.016) \frac{1}{h_{70}^2} \frac{\text{eV}}{c^2}. \tag{13}$$

The Dirac nature implies that the $\Delta m^2 \sim 1\ \text{eV}^2$ effects seen so far at 2σ–3σ evidence in neutrino oscillations must be statistical flukes. Much effort will be needed to test this prediction and establish its Dirac mass differences $\Delta m_{12} = 1.1\,10^{-5} m_\nu$ and $|\Delta m_{13}| = 3.5\,10^{-4} m_\nu = 1.3\,10^{-9} m_e$.

It is generally expected that active neutrinos have sub-eV masses, implying that they hardly contribute to the cosmic mass budget. Within our approach this case leads to 5 or 6 sterile neutrino families (3+5, 3+6 models). These extra modes are massless during big bang nucleosynthesis, so they while they may put stress on observations of ^{4}He and D, they may actually help to soften the ^{7}Li problem [59].

Massive neutrinos are known not to describe the CMB correctly. Our 3+3 Dirac model with $m_\nu = 1.86\,/h_{70}^2$ eV can be fit to the Planck CMB data [2]. With the CLASS-code [60] we find that the parameter set $\Omega_b h^2 = 0.0245$, $h = 0.725$, $\tau_{\text{reio}} = 0.175$, $A_s = 2.19\,10^{-9}$, $n_s = 1.33$ and $Y_{\text{He}} = 0.25$ achieves to have the peaks at the right positions with amplitude at 70% of the first acoustic TT peak and within some relative 10% below or above the other TT, TE and EE peaks. It remains to be seen whether the deviations can be repaired by the effects of turbulence.

6. Summary

The dark matter of the galaxy cluster A1689 is modelled by isothermal neutrinos. The fit works well and is remarkably consistent, without "missing baryons". Beyond the *mass separation radius* $r_{ms} = 2.1$ Mpc the baryon fraction is cosmic, consistent with observations [61]; the Newton force, weaker than the cosmic acceleration $cH_0/4\pi$, is unable to achieve mass separation. When for $r < r_{ms}$ the actual ratio of the mass densities of the ν's and the X-ray gas is compared to the cosmic ratio, the "excess mass" of the ν-core is compensated in the outskirts [61] by an ν under-concentration between 0.7 and 2.1 Mpc, rather than by a baryon over-concentration [62].

The rotation around the cluster centre has a maximum of 2330 km/s at $r = 430$ kpc due to the neutrino core (Fig. 6), a further test for the model. The central X-ray gas density appears to be near the cosmic baryon ratio w.r.t. the local neutrino dark matter density (Fig. 5).

It is desirable to have gas data beyond 1 Mpc, so that the extrapolation employed in our model is better constrained. Likewise, a new strong lensing analysis may update the 2007 Limousin et al. result [40], which is central for sharply fixing the model parameters, and known to lead to NFW fits different from the ones for weak lensing.

Our cluster results stem with the cosmic ones if there exist 3 Dirac neutrino families with mass of approximately 1.86 eV. This is not automatically ruled out by free streaming because the plasma may have been turbulent, as expected from its high Reynolds number and from the stretched exponential shape of the galaxy power spectrum [26]. Perhaps the ^{7}Li problem [59] can be solved by small-scale temperature differences arising from turbulence. Support for the neutrino picture may come from the "cosmic train wreck" galaxy cluster Abell 520, where a central clump with huge mass-to-light ratio $800M_\odot/L_{R\odot}$ exhibits DM separated from the galaxies [20, 21, 22]. This is a puzzle for ΛCDM, but colliding cores of degenerate neutrinos may partly end up in the center due to the exclusion principle. After all, "neutrino stars" (i.e., neutrino cores) may act as nuclei in heavy ion collisions. Likewise, such cores may hinder each other and cause an offset between baryonic and dark mass as in Abell 3827 [23], where it is larger than ΛCDM can explain [24].

Neutrinos of mass 1.86 eV$/c^2$ become nonrelativistic at redshift 7900, so near the recombination they behave like cold dark matter and explain most of the CMB data, up to some 10% accuracy. However, the standard theory of linear fluctuations does not account for turbulence effects, so that a match may still arise. The KATRIN experiment searches the neutrino mass down to 0.2 eV$/c^2$. Most likely they are Dirac particles with mass in the $\bar{\nu}_e$ itself, so that the detection or rejection of our prediction will be achievable.

Acknowledgments

We thank Marceau Limousin, Andrea Morandi and Keiichi Umetsu for supplying data, and Keiichi Umetsu, Andrea Morandi, Armen Allahverdyan, Erik van Heusden and Remo Ruffini for discussion.

References

[1] Hinshaw G, Larson D, Komatsu E, Spergel D, Bennett C, Dunkley J, Nolta M, Halpern M, Hill R, Odegard N *et al.* 2013 *The Astrophysical Journal Supplement Series* **208** 19
[2] Planck-Collaboration 2015 *Preprint arXiv:1502.01589*
[3] Anderson L, Aubourg E, Bailey S, Bizyaev D, Blanton M, Bolton A S, Brinkmann J, Brownstein J R, Burden A, Cuesta A J *et al.* 2012 *Monthly Notices of the Royal Astronomical Society* **427** 3435–3467
[4] Akerib D, Araujo H, Bai X, Bailey A, Balajthy J, Bedikian S, Bernard E, Bernstein A, Bolozdynya A, Bradley A *et al.* 2014 *Physical Review Letters* **112** 091303
[5] Baer H, Barger V, Mickelson D and Padeffke-Kirkland M 2014 *Physical Review D* **89** 115019
[6] Accardo L, Aguilar M, Aisa D, Alpat B, Alvino A, Ambrosi G, Andeen K, Arruda L, Attig N, Azzarello P *et al.* 2014 *Physical Review Letters* **113** 121101
[7] Giesen G, Boudaud M, Génolini Y, Poulin V, Cirelli M, Salati P and Serpico P D 2015 *Journal of Cosmology and Astroparticle Physics* **2015** 023

[8] Ackermann M, Ajello M, Albert A T, Atwood W, Baldini L, Ballet J, Barbiellini G, Bastieri D, Bechtol K, Bellazzini R *et al.* 2011 *Physical Review Letters* **107** 241302
[9] Fermi-LAT Collaboration: Ackermann M *et al.* 2015 *Preprint arXiv:1503.02641*
[10] Wood M, Anderson B, Drlica-Wagner A, Cohen-Tanugi J *et al.* 2015 *Preprint arXiv:1507.03530*
[11] Lee S K, Lisanti M, Safdi B R, Slatyer T R and Xue W 2015 *Preprint arXiv:1506.05124*
[12] Bartels R, Krishnamurthy S and Weniger C 2015 *Preprint arXiv:1506.05104*
[13] Boyarsky A, Ruchayskiy O, Iakubovskyi D and Franse J 2014 *Physical Review Letters* **113** 251301
[14] Jeltema T E and Profumo S 2015 *Preprint arXiv:1512.01239*
[15] Ruchayskiy O, Boyarsky A, Iakubovskyi D, Bulbul E, Eckert D, Franse J, Malyshev D, Markevitch M and Neronov A 2015 *Preprint arXiv:1512.07217*
[16] Rassat A, Starck J L, Paykari P, Sureau F and Bobin J 2014 *Journal of Cosmology and Astroparticle Physics* **2014** 006
[17] Szapudi I, Kovács A, Granett B R, Frei Z, Silk J, Burgett W, Cole S, Draper P W, Farrow D J, Kaiser N *et al.* 2015 *Monthly Notices of the Royal Astronomical Society* **450** 288–294
[18] Hutsemékers D, Braibant L, Pelgrims V and Sluse D 2014 *Astronomy & Astrophysics* **572** A18
[19] Balázs L, Bagoly Z, Hakkila J, Horváth I, Kóbori J, Rácz I and Tóth L 2015 *Monthly Notices of the Royal Astronomical Society* **452** 2236–2246
[20] Jee M, Mahdavi A, Hoekstra H, Babul A, Dalcanton J, Carroll P and Capak P 2012 *The Astrophysical Journal* **747** 96
[21] Clowe D, Markevitch M, Bradač M, Gonzalez A H, Chung S M, Massey R and Zaritsky D 2012 *The Astrophysical Journal* **758** 128
[22] Jee M J, Hoekstra H, Mahdavi A and Babul A 2014 *The Astrophysical Journal* **783** 78
[23] Massey R, Williams L, Smit R, Swinbank M, Kitching T D, Harvey D, Jauzac M, Israel H, Clowe D, Edge A *et al.* 2015 *Monthly Notices of the Royal Astronomical Society* **449** 3393–3406
[24] Schaller M, Robertson A, Massey R, Bower R G and Eke V R 2015 *Monthly Notices of the Royal Astronomical Society: Letters* **453** L65–L69
[25] Steinhardt C L, Capak P, Masters D and Speagle J S 2015 *Preprint arXiv:1506.01377*
[26] Bershadskii A 2015 *Preprint arXiv:1510.01909*
[27] Nieuwenhuizen T M, Gibson C H and Schild R E 2009 *EPL (Europhysics Letters)* **88** 49001
[28] Nieuwenhuizen T M, Schild R E and Gibson C H 2010 *Preprint arXiv:1011.2530*
[29] Nieuwenhuizen T M 2011 *Preprint arXiv:1103.6284*
[30] Nieuwenhuizen T M 2016 *to appear*
[31] Gibson C H 1996 *Applied Mechanics Reviews* **49** 299–315
[32] Nieuwenhuizen T M 2014 Gravitational hydrodynamics as an application of turbulence in the early universe. talk at 'turbulence in the sky as on earth', iip-ufrn, natal, brazil
[33] Nieuwenhuizen T M 2015 Neutrino dark matter and turbulence in the plasma: A perfect couple. talk at fqmt'15, prague, czech republic
[34] Olive K *et al.* (Particle Data Group) 2014 *Chin. Phys.* **C38** 090001
[35] Giunti C and Kim C W 2007 *Fundamentals of neutrino physics and astrophysics* (Oxford university press)
[36] Schwetz T, Tortola M and Valle J 2011 *New Journal of Physics* **13** 063004
[37] Kopp J, Maltoni M and Schwetz T 2011 *Physical Review Letters* **107** 091801
[38] Bartelmann M and Schneider P 2001 *Physics Reports* **340** 291–472
[39] Tyson J A and Fischer P 1995 *The Astrophysical Journal* **446** L55
[40] Limousin M, Richard J, Jullo E, Kneib J P, Fort B, Soucail G, Elíasdóttir Á, Natarajan P, Ellis R S, Smail I *et al.* 2007 *The Astrophysical Journal* **668** 643
[41] Umetsu K and Broadhurst T 2008 *The Astrophysical Journal* **684** 177
[42] Morandi A, Pedersen K and Limousin M 2011 *The Astrophysical Journal* **729** 37
[43] Umetsu K, Sereno M, Medezinski E, Nonino M, Mroczkowski T, Diego J M, Ettori S, Okabe N, Broadhurst T and Lemze D 2015 *The Astrophysical Journal* **806** 207
[44] Navarro J F, Frenk C S and White S D 1997 *The Astrophysical Journal* **490** 493
[45] Nieuwenhuizen T M and Morandi A 2013 *Monthly Notices of the Royal Astronomical Society* stt1216
[46] Sereno M and Umetsu K 2011 *Monthly Notices of the Royal Astronomical Society* **416** 3187–3200
[47] Cowsik R and McClelland J 1973 *The Astrophysical Journal* **180** 7–10
[48] Treumann R, Kull A and Böhringer H 2000 *New Journal of Physics* **2** 11
[49] Nieuwenhuizen T M 2009 *EPL (Europhysics Letters)* **86** 59001
[50] Nieuwenhuizen T M and Morandi A 2011 *Preprint arXiv:1103.6270*
[51] Coe D, Benítez N, Broadhurst T and Moustakas L A 2010 *The Astrophysical Journal* **723** 1678
[52] Morandi A, Pedersen K and Limousin M 2010 *The Astrophysical Journal* **713** 491
[53] Limousin M, Kneib J P and Natarajan P 2005 *Monthly Notices of the Royal Astronomical Society* **356**

309–322
[54] Umetsu K, Broadhurst T, Zitrin A, Medezinski E and Hsu L Y 2011 *The Astrophysical Journal* **729** 127
[55] Morandi A, Ettori S and Moscardini L 2007 *Monthly Notices of the Royal Astronomical Society* **379** 518–534
[56] Asakura K, Gando A, Gando Y, Hachiya T, Hayashida S, Ikeda H, Inoue K, Ishidoshiro K, Ishikawa T, Ishio S *et al.* 2015 Results from kamland-zen *XXVI INTERNATIONAL CONFERENCE ON NEUTRINO PHYSICS AND ASTROPHYSICS: Neutrino 2014* vol 1666 (AIP Publishing) p 170003
[57] Majorovits B 2015 The search for $0\nu\beta\beta$ decay with the gerda experiment: Status and prospects *Low Radioactivity Techniques 2015 (LRT 2015): Proceedings of the 5th International Workshop in Low Radioactivity Techniques* vol 1672 (AIP Publishing) p 110003
[58] CUORE Collaboration: C Alduino e a 2015 *Physical Review Letters* **115** 102502
[59] Cyburt R H, Fields B D, Olive K A and Yeh T H 2015 *Preprint arXiv:1505.01076*
[60] Lesgourgues J 2011 *Preprint arXiv:1104.2932*
[61] Morandi A, Sun M, Forman W and Jones C 2015 *Monthly Notices of the Royal Astronomical Society* **450** 2261–2278
[62] Rasheed B, Bahcall N and Bode P 2011 *Proceedings of the National Academy of Sciences* **108** 3487–3492

Journal of Physics: Conference Series **701** (2016) 012023 doi:10.1088/1742-6596/701/1/012023

Can a sub-quantum medium be provided by General Relativity?

Thomas C Andersen

gravityphysics.com - Meaford, Ontario, Canada

E-mail: tom@palmerandersen.com

Abstract. Emergent Quantum Mechanics (EmQM) seeks to construct quantum mechanical theory and behaviour from classical underpinnings. This paper explores the possibility that the field of classical general relativity (GR) could supply a sub-quantum *medium* for these sub-quantum mechanics. Firstly, I present arguments which show that GR satisfies many of the a priori requirements for a sub-quantum medium. Secondly, some potential obstacles to using GR as the underlying field are noted, for example field strength (isn't gravity a very weak force?) and spin 2. Thirdly, the ability of dynamical exchange processes to create very strong effective fields is demonstrated through the use of a simple model, which solves many of the issues raised in the second section. I conclude that there appears to be enough evidence to pursue this direction of study further, particularly as this line of research also has the possibility to help unify quantum mechanics and general relativity.

1. The Sub-quantum Medium

In emergent QM the sub-quantum medium is the field out of which quantum behaviour emerges. There have been other investigations into the physical make up of a sub-quantum medium, such as stochastic electrodynamics (SED)[1] and Brady's compressible inviscid fluid - an entirely new classical field that is posited to underpin quantum mechanics and electromagnetism.[2]

This paper proposes a sub-quantum medium that is already experimentally confirmed and is somewhat surprisingly stronger and more flexible than usually thought - classical general relativity (GR). Other proposals that are similar in some ways are Wheeler's geons of 1957 - constructed of source free electromagnetic fields and gravity under the laws of standard QM[3] and Hadley's 4-geons[4]. Hadley's proposal is perhaps the most similar to that here, but Hadley assumes the independent reality of an electromagnetic field. This paper instead uses only GR as the fundamental field.

General relativity has some qualities that lend itself to consideration as a sub-quantum medium: its practically frictionless (i.e. inviscid), obviously covariant, couples to everything and its non linear regime allows for a rich variety of behaviour, most of it unexplored.

2. Potential Problems

(i) Gravity is weak. Solution: While the e^-e^- force provided by electromagnetism is 10^{42} stronger than simple gravitational attraction, GR can transfer much more energy than electromagnetism. At optical frequencies gravitational waves can transfer 10^{30} times the power of light at the Schwinger limit. So only the '$0Hz$' component of gravity is weak.

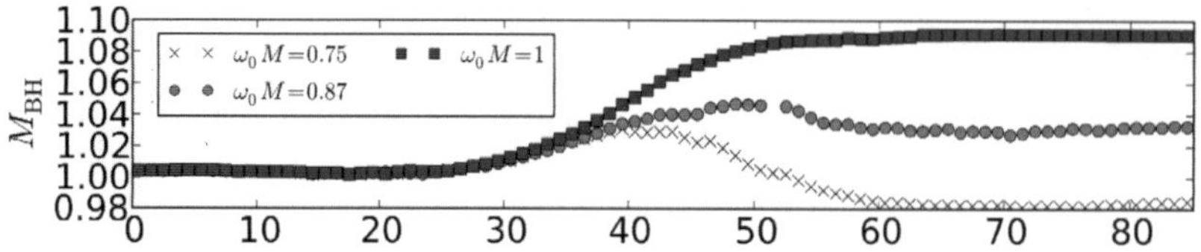

Figure 1. From East[6]: Mass change over time, for incident gravitational waves with three different frequencies. $\omega_0 M = 0.75$ is superradiant, while $\omega_0 M = 1$ shows complete absorption. The entire energy change happens over 10's of wavelengths.

(ii) Gravity has a weak coupling. Solution: In order to model a quantum system (say a hydrogen atom), we require the quantum effects to be much stronger than the electromagnetic forces. Yet the coupling of gravity to the electron is much weaker than even the electromagnetic force. The solution to this problem lies in realizing that gravity can couple not only through '$0Hz$' effects but also through the exchange of gravitational wave energy. The *Possible Mechanisms* section below outlines how this could happen.

(iii) Gravity is quadrupole (spin 2). Solution: Emergence - underlying fields can give rise to apparent net fields of different spin. E.g. Monopole gravitational waves[5].

3. Possible Mechanisms

This paper posits that EmQM's sub-quantum zero point field is the result of a run away superradiance - like effect (limited by non linear mechanics). Is the universe a black hole bomb?

This superradiant (see figure 1) energy exchange of the particle with its surroundings causes the particle to be subjected to huge forces - superradiance for example allows for a substantial fraction of the mass of a rotating black hole to change over time scales a few times the light travel time across the hole. The recent paper by East et al. studies black holes undergoing superradiance using a numerical method.[6]. It seems that the superradiance is on a knife edge with absorption - these effects happen at only slightly different frequencies. These large energy transfers could be the ZPF.

4.Discussion

If the pursuit of an emergent quantum mechanics is to prove fruitful, then the idea that a field like general relativity does not hold on the microscale may have to be re-considered, as with any EmQM theory there is no overarching 'quantum regime' for GR to stop working at. With general relativity still on the stage at $10^{-17}m$ and below, Occam's razor perhaps suggests that we prove that general relativity is not the sub-quantum medium before a new field is invented. The complete poster and paper is available from the author.

References

[1] de la Pena L and Cetto A M 2005 Contribution from stochastic electrodynamics to the understanding of quantum mechanics *pre-print* (*Preprint* arXiv:quant-ph/0501011)
[2] Brady R 2013 The irrotational motion of a compressible inviscid fluid *pre-print* (*Preprint* arXiv:1301.7540)
[3] Misner C W and Wheeler J A 1957 Classical physics as geometry *Ann. Phys.* **2** 525–603 ISSN 00034916
[4] Hadley M J 2007 A gravitational explanation for quantum theory – non-time-orientable manifolds, *AIP Conf. Proc.* vol 905 (AIP) pp 146–152 ISSN 0094243X (*Preprint* arXiv:gr-qc/0703150)
[5] Kutschera M 2003 Monopole gravitational waves from relativistic fireballs driving gamma-ray bursts *Mon. Not. Roy. Astron. Soc.* **345** L1–L5 ISSN 00358711 (*Preprint* arXiv:astro-ph/0309448)
[6] East W E, Ramazanoglu F M and Pretorius F 2014 Black hole superradiance in dynamical spacetime *Phys. Rev. D* **89** 061503 ISSN 1550-7998 (*Preprint* arXiv:1312.4529 [gr-qc])

Journal of Physics: Conference Series **701** (2016) 012024 doi:10.1088/1742-6596/701/1/012024

Local ontology for a dual-rail qubit

Pawel Blasiak
Institute of Nuclear Physics, Polish Academy of Sciences
ul. Radzikowskiego 152, 31-342 Kraków, Poland
E-mail: pawel.blasiak@ifj.edu.pl

Abstract. We show that quantum predictions for the dual-rail realisation of a qubit can be faithfully simulated with classical stochastic gates and particles which interact entirely in a local manner. In the presented model 'non-locality' appears only on the epistemic level of description.

We consider single quantum particle in a circuit which consists of two spatially separated paths ($i = 0, 1$) and a sequence of gates which include phase shifters $P(\omega)$ and detectors D_i attached to each path separately, and beam splitters $B(\xi)$ on which the paths cross; cf. Fig. 1 (on the left). Quantum description of such a dual-rail system boils down to a qubit.

In this note we report construction of a classical local stochastic model which faithfully reconstructs quantum predictions for the dual-rail qubit [1]. It shows that this kind of framework is not enough to establish genuine quantum non-locality, since for a dual-rail qubit alleged 'non-locality' can be explained as an epistemic effect due to restricted means of investigating the system. In the following, we sketch out the main concepts of the model.

Ontology of the model. We postulate existence of two kinds of local particles carrying inner degrees of freedom which propagate along the paths: *real* ones described by a unit vector $\vec{n} \in \boldsymbol{S}^2$ in 3D and *ghosts* carrying a phase $\varphi \in \boldsymbol{S}^1$. The crucial difference between the particles is that detectors register only *real* particles and remain blind to *ghosts*. For the purpose at hand, we restrict our attention to situations in which there is a single *real* particle present in the system with a *ghost* in the other path or the other path being *empty*; see Fig. 1 (on the left). Hence, at each time the system is fully characterised by a triple $(i, \vec{n}, \varphi)$ in the ontic state space:

$$\Lambda \equiv \{0, 1\} \times \boldsymbol{S}^2 \times \boldsymbol{S}^{1^*}, \tag{1}$$

where $\boldsymbol{S}^{1^*} \equiv \boldsymbol{S}^1 \cup \{\varnothing\}$. The first component $i = 0, 1$ specifies position of the *real* particle carrying vector $\vec{n}$ and the other path featuring a *ghost* with phase φ or being *empty* $\varphi = \varnothing$. In general, the system is described by a probability distribution over the ontic states, i.e. a point in $\mathcal{P}(\Lambda)$.

Building blocks of the model. To complete description of the model we need to specify the stochastic counterparts of the interferometric gates by defining action on the ontic states $\Lambda \to \mathcal{P}(\Lambda)$ which does not violate the locality principle. Here are the definitions:

– *Detector* $\mathbb{D}_j$ placed in the j-th path is a deterministic gate:

$$(i, \vec{n}, \varphi) \xrightarrow{\mathbb{D}_j} \begin{cases} \delta_i\, \delta_{\hat{z}}\, \delta_\varphi & \text{for} \quad i = j \quad \text{(detector 'CLICKS')}, \\ \delta_i\, \delta_{\vec{n}}\, \delta_\varnothing & \text{for} \quad i \neq j \quad \text{(detector doesn't 'CLICK')}. \end{cases} \tag{2}$$

Note that in either case the detector affects only the particle in the j-th path, i.e. *real* particle is set to point north $\vec{n} \to \hat{z}$ and the *ghost* is removed from the path $\varphi \to \varnothing$.

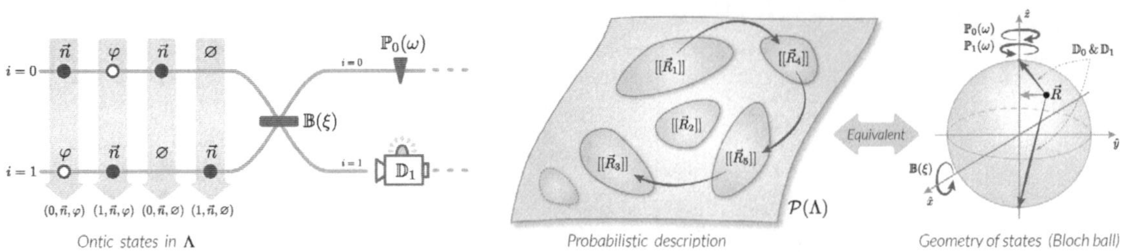

Figure 1. On the left: Possible configurations of *real* (•) and *ghost* (○) particles in the model which traverse the circuit built from phase shifters $\mathbb{P}_j(\omega)$, detectors $\mathbb{D}_j$ and beam splitters $\mathbb{B}(\xi)$ defined in Eqs. (2)-(4). On the right: Agent investigating the system explores only a restricted set of distributions in $\mathcal{P}(\Lambda)$. This set can be coarse-grained into classes $[[\vec{R}]]$ indistinguishable to the agent and their geometry is shown to be equivalent to the Bloch ball representation.

– *Phase shifter* $\mathbb{P}_j(\omega)$ placed in the j-th path is a deterministic gate:

$$(i,\vec{n},\varphi) \xrightarrow{\mathbb{P}_j(\omega)} \begin{cases} \delta_i\,\delta_{\vec{n}'}\,\delta_\varphi & \text{for} \quad i=j\,, \\ \delta_i\,\delta_{\vec{n}}\,\delta_{\varphi'} & \text{for} \quad i\neq j\,, \end{cases} \tag{3}$$

where $\vec{n}' = R_z\big((-)^{\bar{j}}\omega\big)\,\vec{n}$ and $\varphi' = \varphi + (-)^j\omega$ are obtained by appropriate rotation by angle $\pm\omega$ about the $\hat{z}$-axis, and in the case of *empty* path $\varphi = \varnothing$ we put $\varphi' = \varnothing$.

– *Beam splitter* $\mathbb{B}(\xi)$ is a stochastic which acts as follows:

$$(i,\vec{n},\varphi) \xrightarrow{\mathbb{B}(\xi)} \cos^2\left(\tfrac{\theta'}{2}\right)\,\delta_i\,\delta_{\vec{n}'}\,\delta_0 \;+\; \sin^2\left(\tfrac{\theta'}{2}\right)\,\delta_{\bar{i}}\,\delta_{-\vec{n}'}\,\delta_0\,, \tag{4}$$

which means that particles remain in their respective paths ($i \to i$) or get swapped ($i \to \bar{i} \equiv 1-i$) with respective probabilities $\cos^2\left(\frac{\theta'}{2}\right)$ and $\sin^2\left(\frac{\theta'}{2}\right)$. The vector $\vec{n}' = (\theta',\phi') = R_x(\xi)\,R_z(-\varphi)\,\vec{n}$ obtains by appropriate rotation about $\hat{z}$ and $\hat{x}$-axes, and in the case of *empty* path $\varphi = \varnothing$ we put $\vec{n}' = (\theta',\phi') = R_x(\xi)\,\hat{z}$. In either case, notice presence of the *ghost* $\varphi = 0$ after transformation.

Operational description of the model. Imagine an agent without any prior notion of the model trying to make sense of how it works only on the basis of experiments she performs. Clearly, the agent acts under *epistemic constraints* which confine her perception of the system under study – she has a limited choice of gates for building circuits and her detectors 'see' only *real* particles; cf. Eqs. (2)-(4). That being the case the agent is legitimate to be concerned only with situations within her reach. In other words, for all practical purposes it is sufficient to account for a limited set of preparation, transformation and measurement procedures that arise in any experimental circuit that she can build according to the rules of the model.

Following this reasoning, we have shown in [1] that such an agent in all her actions remains confined within a certain non-trivial subset of distributions in $\mathcal{P}(\Lambda)$. This set has an interesting geometry which can be analysed by coarse-graining distributions into classes indistinguishable to the agent. It can be shown that these classes have the geometry of the Bloch ball, i.e. they are in one-to-one correspondence with points in the Bloch ball, transform congruently under the action of stochastic gates defined above, and these transformations are the same as for the quantum interferometric gates, see Fig. 1 (on the right). This means that from the agent's perspective predictions for the quantum case of a dual-rail qubit and classical circuits built from the stochastic counterparts defined in the model are the same, and thus for all practical purposes both systems are equivalent. We emphasise that the model is local on the ontological level, and alleged 'non-locality' is only an epistemic effect caused by restrictions on gaining knowledge.

References

[1] Blasiak P 2015 Local model of a qubit in the interferometric setup *New. J. Phys.* **17** 113043

Journal of Physics: Conference Series **701** (2016) 012025 doi:10.1088/1742-6596/701/1/012025

On Undecidability and the Laws of Physics

Erik F G van Heusden

Institute for Theoretical Physics, University of Amsterdam, Science Park 904, 1090 GL Amsterdam, The Netherlands

E-mail: Erik.vanHeusden@student.uva.nl

Abstract. We argue that insights in fundamental physics could be provided by fields such as mathematical logic, and, as an example, briefly discuss some of the issues that arise from the incompleteness theorems. The advantage of this type of approach would be that it could perhaps give access to information about physics at the most fundamental level, without having to rely on increasingly accurate although always approximate empirical models alone.

An assumption that underlies virtually all theories in physics is that nature is sufficiently structured or ordered to be described by some set of laws, such that phenomena may be considered to be consequences of these laws. Just as one tests theories for consistency with mathematics and observations, so too could this presumed underlying structure be investigated this way. This type of approach might provide insights in physics at the most fundamental level, instead of indefinitely expanding empirical models. As an example, we will discuss a few implications for theoretical physics that come from the incompleteness theorems. Most theories of physics intend only to treat phenomena that may be described by mathematics entirely, while some even argue for theories where the universe is entirely mathematical in nature, or can even be described as a simulation of some sort. Any such approach that would require not only the theory, but also the behaviour of the reality the theory intends to describe, to be at least entirely isomorphic to some mathematical formalism, must require physics itself, and not just our description thereof, to be subject to constraints that generally apply to mathematical formalisms. This could be one of the very few ways to not only obtain information about a phenomenological theory or a theory of physics at some deeper level, but about the nature of physics at, by construction, the most fundamental level. One issue of this type is produced by the so-called incompleteness theorems.[1] In an appropriate English translation, these state[2]:

(i) Any consistent formal system F within which a certain amount of elementary arithmetic can be carried out is incomplete; i.e., there are statements of the language of F which can neither be proved nor disproved in F.

(ii) For any consistent system F within which a certain amount of elementary arithmetic can be carried out, the consistency of F cannot be proved in F itself.

If nature, or any fully independent part of it, such as a causal patch in a local theory, may be represented as some formal system, the options thus can be roughly categorized as follows:

(i) The fundamental laws of physics cannot be represented as a (consistent) formal system.

(ii) The fundamental laws of physics behave as a formalism constructed to not be subject to the incompleteness theorems.

(iii) Any axiomatic representation of the laws of physics requires an infinite number of axioms.

(iv) There exist real phenomena that do not follow from the laws of physics. Moreover, the consistency of the laws of physics is one of these phenomena.

Note that the issue will not only occur in models or theories, but also in the actual laws of physics they intend to describe, if such things exist. They too are subject to these considerations, just as much as a theory would be. This could be a much larger problem than it is in pure mathematics, where the consistency of one theory may simply be proven in another [3], since one may not expect nature to simply switch to different laws of physics whenever convenient.

The most straightforward choice would seem to be option (i). One must, however, note that even a theory that does not seem to rely on any formal construction is not enough, the construction does not necessarily have to be one similar to conventional mathematics: as long as an apparently non-formal system can somehow be precisely represented as some formal system with a finite number of sufficiently precise axioms, nothing is gained. Only a construction that can fundamentally not be represented as a formal system would qualify as a solution of type (i).

Option (ii) would seem to be the most viable approach: sufficiently weak formal systems do not generally need to be subject to the incompleteness theorems, and investigating the possibility of a theory of everything constructed from such a relatively simple system might prove productive indeed. This would, however, require laws of physics that cannot be used to generate phenomena that can exactly represent the properties of a sufficiently strong theory of arithmetic. This could have interesting consequences, since one cannot obtain information about even purely mathematical statements that fundamentally, no physical phenomenon may ever represent: the information produced by such a system and the statement would be independent.

Option (iii) could involve infinitely many boundary conditions, phenomena that cannot be described as operating along some finite set of laws, or even effectively reduce to option (iv).

Option (iv) would seem to be prefered by most theories. Note that it is not necessarily problematic if not all physical truths are implications of the laws of physics: if, for example, the laws of physics fail to state that they will never cause a collision at some precise centre-of-mass energy E, this is not necessarily a problem. The second incompleteness theorem, however, states that the consistency of such laws of physics is one of these statements that cannot be determined from, or considered as consequences of, this kind of laws of physics. The laws of physics could still simply happen not to cause any contradictions, but this is not a consequence of them, and no physical system could ever know with certainty. Issues with mathematics appear here also: should a physical system be capable of proving the consistency of a theory this type, which itself is capable of describing at least this specific system, then the theory could describe the proof of its own consistency, as produced by the relevant physical system. This would be in violation of the second incompleteness theorem: either the proof of consistency is incorrect, the physical system is actually not capable of representing part of the alleged proof, and, therefore, of obtaining useful information from it, or the system used to represent the proof cannot be described by the theory. One could disprove a theory by proving its consistency. Thus, option (iv) would require a theory of everything to be constructed in a manner that leaves its consistency undecidable in any formalism one requires some physical systems to provide information about.

Acknowledgements: We would like to thank Theo Nieuwenhuizen for discussion.

References

[1] Gödel K 1930 *Über formal unentscheidbare Sätze der Principia Mathematica und verwandter Systeme* Monatshefte für Math. u. Physik **38** 173–198

[2] Raatikainen P 2015 *Gödel's Incompleteness Theorems*, ed. Zalta E N *The Stanford Encyclopedia of Philosophy* `http://plato.stanford.edu/archives/spr2015/entries/goedel-incompleteness`

[3] Gentzen G 1935 Die Widerspruchfreiheit der reinen Zahlentheorie *Math. Annalen* **112** 493–565

Contextually in a Peres–Mermin square using arbitrary operators

A Laversanne-Finot[1], A Ketterer[1], M R Barros[3], S P Walborn[3], T Coudreau[1], A Keller[2] and P Milman[1]

[1]Univ. Paris Diderot, Sorbonne Paris Cité, Laboratoire Matériaux et Phénomènes Quantiques, UMR 7162, CNRS, F-75205, Paris, France

[2]Univ. Paris Sud, Institut des Sciences Moléculaires d'Orsay (UMR 8214 CNRS), F-91405 Orsay, France

[3]Instituto de Física, Universidade Federal do Rio de Janeiro, Caixa Postal 68528, Rio de Janeiro, RJ 21941-972, Brazil

E-mail: adrien.laversanne-finot@univ-paris-diderot.fr

Abstract. The contextuality of quantum mechanics can be shown by the violation of inequalities based on measurements of well chosen observables. These inequalities have been designed separately for both discrete and continuous variables. Here we unify both strategies by introducing general conditions to demonstrate the contextuality of quantum mechanics from measurements of observables of arbitrary dimensions. Among the consequences of our results is the impossibility of having a maximal violation of contextuality in the Peres–Mermin scenario with discrete observables of odd dimensions. In addition, we show how to construct a large class of observables with a continuous spectrum enabling the realization of contextuality tests both in the Gaussian and non-Gaussian regimes.

In a non-contextual theory the measurement result of an observable depends only on the state of the system and the observable A being measured. In particular, it does not depend on the compatible observables that are measured alongside. This corresponds to the classical view in which the system is in a well defined state with well defined properties at all times. It was first shown by Kochen and Specker that quantum mechanics is a contextual theory and that it is impossible to assign pre-determined values to all (compatible) observables describing a system [1]. Later on, Peres and Mermin derived a simple quantitative argument [2, 3] to test contextuality that also lead to the discovery of state independent non-contextual inequalities [4].

Here, we briefly recall the principles of the Peres–Mermin scenario (PMS) and show how it can be generalized to arbitrary operators. Using this approach we show that in the discrete case a state independent violation of non-contextuality with qudits of odd dimension is impossible. In the continuous case we will prove the existence of a whole class of operators, Gaussian and Non-Gaussian enabling the demonstration of the contextuality of quantum mechanics.

To see that quantum mechanics is contextual, one can consider a set of 9 dichotomic observables A_{ij}, with outcome $+1$ and -1, such that each observables sharing a common subscript are compatible (commute). From this set of observables, one can form the quantity :

$$\langle X\rangle = \langle A_{11}A_{12}A_{13}\rangle + \langle A_{21}A_{22}A_{23}\rangle + \langle A_{31}A_{32}A_{33}\rangle + \langle A_{11}A_{21}A_{31}\rangle + \langle A_{12}A_{22}A_{32}\rangle - \langle A_{13}A_{23}A_{33}\rangle . \quad (1)$$

In a non-contextual theory, one is able to assign pre-determined values +1 or −1 to every observable in 1, thus yielding the noncontextual bound of 4. However, in the quatum case, it is possible to find combinations of Pauli measurements such that $\langle X\rangle_{QM} = 6$ for every quantum state [4].

Instead of considering observables for the A_{ij}, we can consider arbitrary unitary operators $U_{ij} = A^R_{ij} + iA^I_{ij}$ where A^R_{ij} and A^I_{ij} are the real and imaginary hermitian parts. In this case X is not an observable anymore and the quantity to consider is the real part of X, made of all A^R_{ij} and A^I_{ij}. Under the unitarity condition, it is possible to prove that this quantity is bounded by $3\sqrt{3}$ (see [5]). In order to construct a PMS, one needs that the observables sharing common subscripts are compatibles. Also, in order to maximally violate the inequality, one needs that the product of the observables in each bracket is either $\mathbb{1}$ or $-\mathbb{1}$ (see [6] for details). These two conditions lead to the constraints:

$$\{\hat{U}_i, \hat{U}_j\} = 2\delta_{ij}\hat{U}_i^2, \tag{2}$$

$$[\hat{U}_i, \hat{U}_j] = \pm 2i\epsilon_{ijk}\hat{U}_k^\dagger. \tag{3}$$

A point worth discussing, but that is out of the scope of the present Letter, is the case of state dependent inequalities, as the ones of the CHSH type [7] and the relation between the obtained condition and fault tolerant quantum computation with continuous variables [8]: we can notice that a common feature to all these results is the necessity of unitary operations defined in continuous variables to obey the commutation relations (3), providing insight on the relationship between contextuality and quantum computation.

In the discrete case these conditions can be satisfied only if the dimension of the system is even. Examples of operators satisfying these relations are given by the rotation operators :

$$\hat{R}_1 = e^{i\hat{S}_x\pi}, \quad \hat{R}_2 = e^{i\hat{S}_y\pi} \quad \text{and} \quad \hat{R}_3 = e^{i\hat{S}_z\pi} \tag{4}$$

where $\hat{S}_x$, $\hat{S}_y$ and $\hat{S}_z$ are the three components of a half integer spin S.

In the continuous case it has been shown [5] that it is possible to use displacement operators $\mathcal{D}(\alpha_i) = \exp(\alpha_i\hat{a}^\dagger - \alpha_i * \hat{a})$ where the α_i obey $\mathrm{Im}(\alpha_i\alpha_j^*) = \pm\pi/2$. We further show that any set of operators of the form $\hat{V}\mathcal{D}(\alpha_i)\hat{V}^\dagger$, where $\hat{V}$ is an arbitrary unitary operator, is suitable for a contextuality test. In particular, using unitaries that are polynomials of order higher than 3 in the canonical variables, we can create a family of non-Gaussian operators enabling the violation of non-contextuality.

We have shown that that it is possible to derive a general non-contextual inequality in terms of complex observables that can be applied to discrete and continuous variable measurements. This allowed us to derive general conditions that must be fulfilled by the operators in order to have a maximal violation of the contextual inequality. In particular, we showed that in the case of discrete system it is possible to have a maximal state independent violation of the contextual inequality only for system of even dimension. In the case of continuous variables we can derive a family of operators, Gaussian and non-Gaussian, suitable for contextuality tests.

References

[1] Kochen S and Specker E P 1967 *J Math. Mech.* **17** 59
[2] Peres A 1990 *Phys. Lett. A* **151** 3
[3] Mermin N D 1990 *Phys. Rev. Lett.* **65** 3373
[4] Cabello A 2008 *Phys. Rev. Lett.* **101** 210401
[5] Asadian A, Budroni C, Steinhoff F E S, Rabl P and Gühne O 2015 *Phys. Rev. Lett.* **114** 250403
[6] Laversanne-Finot A *et al* 2015 (*Preprint* `arXiv:1512.03334 [quant-ph]`)
[7] Ketterer A, Keller A, Coudreau T and Milman P 2015 *Phys. Rev. A* **91** 012106
[8] Gottesman D, Kitaev A and Preskill J 2001 *Phys. Rev. A* **64** 012310

Extending Bell's Theorem: Ruling out Paramater Independent Hidden Variable Theories

G J Leegwater

Faculty of Philosophy, Erasmus University Rotterdam, Burgemeester Oudlaan 50, 3062 PA Rotterdam, The Netherlands

E-mail: leegwater@fwb.eur.nl

Abstract. Bell's Theorem may well be the best known result in the foundations of quantum mechanics. Here, it is presented as stating that for any hidden variable theory the combination of the conditions Parameter Independence, Outcome Independence, Source Independence and Compatibility with Quantum Theory leads to a contradiction. Based on work by Roger Colbeck and Renato Renner, an extension of Bell's Theorem is considered. In this extension the theorem is strengthened by replacing Outcome Independence by a strictly weaker condition.

1. Introduction

Quantum mechanics provides limited predictions for measurement outcomes: generally, its predictions are probabilistic. It has been a long-standing question under what conditions quantum mechanics can be extended such that more precise predictions can be given. In such extensions, known as 'hidden variable theories', the quantum description of a system is supplemented with extra variables. A number of impossibility theorems have been proven, showing that large classes of hidden variable theories are incompatible with quantum mechanics. The best known example is Bell's Theorem [1, 2], which rules out hidden variable theories satisfying two locality conditions: Parameter Independence and Outcome Independence. Using recent work of Roger Colbeck and Renato Renner [3, 4], this result can be extended: Outcome Independence can be replaced by a strictly weaker condition, enlarging the class of hidden variable theories that is ruled out. In this extension, any hidden variable theory satisfying Parameter Independence is trivial, meaning that the probabilities of measurement outcomes are independent of the hidden variables, rendering such variables redundant.

2. Extending Bell's Theorem

2.1. *Bell's Theorem*

Bell's Theorem can be stated as the appearance of a contradiction when the following four conditions are combined:

- **ParInd: Parameter Independence**
  When considering measurements on subsystems of some larger system, ParInd guarantees that the outcome probabilities for a measurement on any subsystem are independent of what measurement is being performed on any other subsystem. Quantum mechanics itself satisfies ParInd. A violation of ParInd gives rise to tension with relativity theory.

- **OutInd: Outcome Independence**
  When considering measurements on subsystems of some larger system, OutInd guarantees that the outcome probabilities for a measurement on any subsystem are independent of the outcome of a measurement on any other subsystem. Quantum mechanics itself violates OutInd.
- **SrcInd: Source Independence**
  This condition guarantees that the distribution of the hidden variables is independent of what measurement is being performed. Since the choice of measurement can be based on anything an experimentator wishes, a violation of this condition is often seen as a 'conspiracy'.
- **CompQuant: Compatibility with Quantum Mechanics**
  According to this condition, when integrating out the hidden variables, the hidden variable theory gives the same probabilities for measurement outcomes as quantum mechanics.

2.2. Extension
Bell's Theorem can be strengthened by replacing OutInd by the following condition:

- **NonTriv: Nontriviality**
  Triviality means that for any measurement, outcome probabilities are independent of the hidden variables. This renders such variables redundant: with or without the hidden variables, the probabilities equal the probabilities given by quantum mechanics. Nontriviality is the opposite condition: there are measurements for which the outcome probabilities do depend on the hidden variables.

Note that, since quantum mechanics (without additional variables) violates OutInd, any hidden variable theory satisfying OutInd must predict outcome probabilities different from those of quantum mechanics. This means that the outcome probabilities depend on the hidden variables, redering such a theory nontrivial. So, OutInd implies NonTriv. Furthermore, NonTriv does not imply OutInd, since it is easy to construct a hidden variable theory which gives the same predictions as quantum mechanics for EPR experiments, thereby violating OutInd, while for some other measurement it gives nontrivial predictions, thereby satisfying NonTriv. This makes NonTriv a strictly weaker condition than OutInd.

3. Further reading
The above is explained in more detail in a forthcoming paper [5], where also the relation between this work and the work of Colbeck and Renner is elucidated.

Acknowledgments
I would like to thank the conference organizers for making possible the inspiring gathering that EMQM15 was.

This research is made possible by a grant from the Netherlands Organisation for Scientific Research (NWO) for the research programme 'The Structure of Reality and the Reality of Structure'.

References
[1] Bell J S 1964 *Physics* **1** 195–200
[2] Bell J S 1985 *Dialectica* **39** 86–96
[3] Colbeck R and Renner R 2011 *Nature Communications* **2** 411
[4] Colbeck R and Renner R 2012 (*Preprint* `arxiv:1208.4123 [quant-ph]`)
[5] Leegwater G J 2016 An impossibility theorem for parameter independent hidden variable theories *Manuscript submitted for publication*

The weak value of spin for atomic systems

V Monachello and R Flack

Dept. Physics and Astronomy, University College London, Gower Street, London WC1E 6BT

E-mail: `vincenzo.monachello.14@ucl.ac.uk`, `r.flack@ucl.ac.uk`

Abstract. A system is being designed and constructed in order to measure the weak value of spin for atomic systems. The experiment utilises spin-1 metastable helium atoms in the 2^3S_1 state. This paper outlines the experiment and its features.

1. "Weak measurement" process

The "weak measurement" process [1] describes a scenario in which a system is weakly coupled to a pointer between well-defined pre- and post-selected states. The spin of the pre-selected metastable helium (He*) atoms used in this experiment will be weakly coupled to its centre-of-mass (COM).

To measure the weak value of spin, two perpendicular inhomogeneous magnets are required, the first with the magnetic field gradient in the z-axis and the second in the x-axis. Figure 1 shows the variation on the double Stern–Gerlach (SG) setup consisting of both the weak and strong stages as detailed by Aharonov, Albert and Vaidman (AAV) [1] and later discussed by Duck, Stevenson and Sudarshan [2] and Pan and Matzkin [3].

The COM for He* can be modelled as a one-dimensional Gaussian, with the displacement between the initial $|\phi_i\rangle$ and final $|\phi_f\rangle$ COM states determining the weak value of spin. Due to the "weak measurement" process, the average COM shift for a spin-1 atom is proportional to the weak value as shown in Eq. (1)

$$\left(\hat{S}_z\right)_w = \frac{\langle S_x|\,\hat{\sigma}_z\,|S_i\rangle}{\langle S_x|\,S_i\rangle} = \tan\left(\frac{\theta}{2}\right), \tag{1}$$

where $|S_i\rangle$ and $|S_x\rangle$ are the initial (pre-) and final (post-selected) spin states of the system, $\hat{\sigma}_z$ is the spin operator and θ is the spin vector angle bound between the x- and z-axis. This is in contrast to the standard case yielding the quantum mechanical expectation value. Looking at Eq. (1) it can be seen that the weak value of the chosen observable is complex and can hold varying values depending on the chosen spin states. As the pre- and post-selected states approach orthogonality, the weak value becomes larger and eventually undefined, giving rise to the term weak value amplification.

2. Experimental setup

The final setup is shown in Figure 2. A pulsed supersonic beam of He* atoms in both the 2^3S_1 (triplet) and 2^1S_0 (singlet) states will be produced using an electron-seeded direct-current discharge method [4].

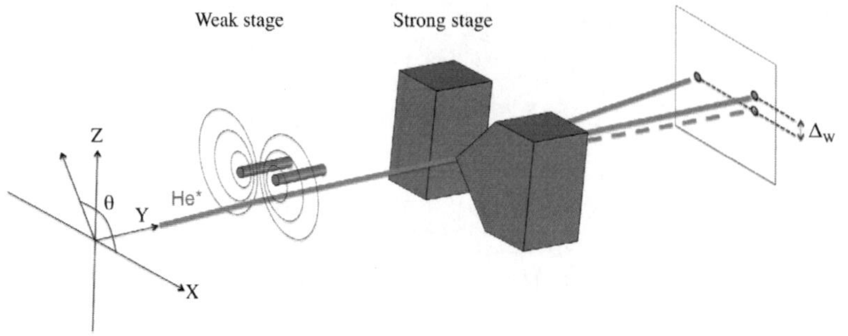

Figure 1. Schematic showing both the weak and strong stages. The weak stage produces a displacement Δ_W on the post-selected state, from this the weak value of spin can be determined.

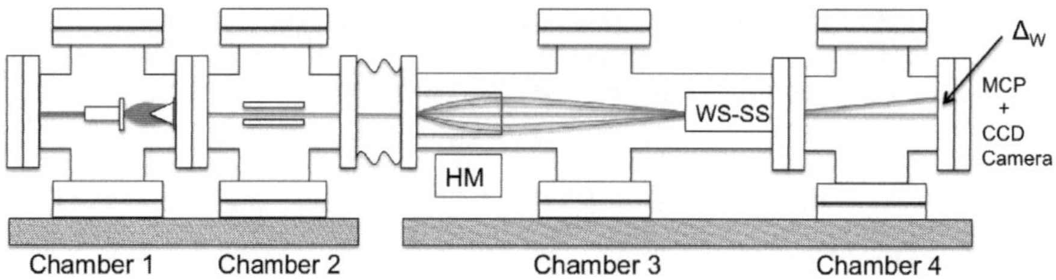

Figure 2. The experimental setup, including the location of the hexapole magnet (HM), the weak stage (WS) and the strong stage (SS) magnets.

Chamber 2 will house two metal plates producing an electric field perpendicular to the beam line, removing unwanted ions and electrons from the discharge. A 2060 nm LED is then used to remove neutral singlet atoms utilising the $2^1S \rightarrow 2^1P$ transition, decaying shortly after to the ground state.

Spin polarisation is achieved in chamber 3 by employing a permanent hexapole magnet. Spin up $m_s = +1$ atoms are focused to a point along the axis of propagation, spin down $m_s = -1$ atoms become defocused and are lost to the hexapole and chamber walls. He* atoms in the 2^3S_1 $m_s = 0$ state feel no force and pass through unhindered. A Helmholtz coil setup located shortly after the pre-selection phase will be used to set the angle θ.

The focal point of the hexapole magnet will coincide with an assembly containing both a slit of length $\approx 300\,\text{nm}$ and the weak stage consisting of a two-wire field. The small slit size ensures transverse coherence of the atomic beam, and maximises the magnetic field gradient while conforming to the weak limit set in Eq. (2)

$$\Delta \frac{\mu \frac{\partial B}{\partial z} \tau}{\bar{h}} \left(\hat{S}_z\right)_w \ll 1, \tag{2}$$

where Δ is the initial width of the wave packet, μ is the magnetic moment of the atom, $\frac{\partial B}{\partial z}$ and τ are the magnetic field gradient in the weak stage and the time the atom spends in this field respectively. The atomic beam will then pass through the strong stage, a permanent SG magnet producing a field gradient of $100\,\text{Tm}^{-1}$ over a 100 mm length. The deviation of the post-selected $m_s = +1$ state as measured using a micro-channel plate (MCP), phosphor screen and CCD camera setup will allow for the determination of the weak value of spin.

Journal of Physics: Conference Series **701** (2016) 012028 doi:10.1088/1742-6596/701/1/012028

References

[1] Aharonov Y, Albert D Z and Vaidman L 1988 *Phys. Rev. Lett.* **60** 1351–54
[2] Duck I M, Stevenson P M and Sudarshan E C G 1989 *Phys. Rev.* D **40** 2112–17
[3] Pan A K and Matzkin A 2012 *Phys. Rev.* A **85** 022122
[4] Halfmann T, Koensgen J and Bergmann K 2000 *Meas. Sci. Technol.* **11** 1510

About what can be witnessed by a Leggett–Garg inequality test: modeling its violation

Saulo V Moreira [1], Arne Keller[2], Thomas Coudreau[1] and Pérola Milman[1]

[1]Univ. Paris Diderot, Sorbonne Paris Cité, Laboratoire Matériaux et Phénomènes Quantiques, UMR 7162, CNRS, F-75205, Paris, France

[2]Univ. Paris Sud, Institut des Sciences Moléculaires d'Orsay (UMR 8214 CNRS), F-91405 Orsay, France

E-mail: saulo-vicente.moreira@univ-paris-diderot.fr

Abstract. The Leggett–Garg inequality is a widely used test of the "quantumness" of a system, and involves correlations between measurements realized at different times. According to its widespread interpretation, a violation of the Legget–Garg inequality disproves macroscopic realism and non-invasiveness. Nevertheless, recent results point out that macroscopic realism is a model dependent notion and that one should always be able to attribute to invasiveness a violation of a Legget–Garg inequality. This opens some natural questions: how to provide such an attribution in a systematic way? How can apparent macroscopic realism violation be recast into a dimensional independent invasiveness model? The present work answers these questions by introducing an operational model where the effects of invasiveness are controllable through a parameter associated with what is called the *measurability* of the physical system. Such a parameter leads to different generalized measurements that can be associated with the dimensionality of a system, to measurement errors or to back action.

1. Introduction

The Leggett–Garg inequality (LGI) was proposed by Leggett and Garg [1] in the 1980s as a witness to the "quantumness" in a macroscopic system. Its violation, thus, would imply nonclassicality, defined by the rejection of the two assumptions bellow:
(i) Macroscopic Realism (MR): a macroscopic system with two or more macroscopically distinct states available to it will at all times be in one of those states.
(ii) Noninvasive measurability (NIM): it is possible, in principle, to determine the state of the system with arbitrarily small perturbation to its subsequent dynamics.
This inequality involves measurements of observables that can be associated with macroscopic properties of a system at different times, defined here as $\hat{Q}(t_i)$. By defining $C_{kl} \equiv \langle \hat{Q}(t_k)\hat{Q}(t_l) \rangle$, we can write the following LGI [2]:

$$-2 \leq K_{LG} \equiv C_{12} + C_{23} + C_{34} - C_{14} \leq 2. \quad (1)$$

Nonetheless, there has been a recent debate in the literature [3, 5] about if a violation of the LGI allows one to reject both the assumptions. Maroney and Timpson [3] showed that what a violation of a LGI primarily witnesses is whether the measurement is invasive or not. MR is a model-dependent notion, as, for instance, the notion of quantum superposition is not incompatible with realism [4]. However, it is also discussed in Ref. [3] with the aid of the *ontic models* framework introduced by Spekkens [6] that despite the fact that what is most immediately witnessed by a violation of a LGI is invasiveness of measurements, this violation allows one to reject a specific notion of MR, the 'operational eigenstate mixture macrorealism'. This corresponds to a given preparation such that, for certain measurements, would return a given outcome with probability 1. As discussed by the authors, our experience tell us that when observation of macroscopic properties takes place, the system is put into an operational eigenstate of those properties, such that if a particular value for a macroscopic property is obtained, then we must obtain that same value if we look at the system immediately afterwards. Therefore, operational eigenstates can provide a characterisation of the behaviour of macroscopic systems when observed, and the violation of the LGI would permit us to reject MR in these terms.

2. Modeling invasiveness of measurements in LGIs

We propose, in [7], a model to test a LGI using positive operator valued measurements (POVM). In our model, invasiveness can be associated with the resolution and efficiency of a measurement apparatus. These parameters, according to the experimental situation, may, or not, be associated with the dimensionality of the system. The introduced POVMs provide a physically sound and operational interpretation of what is actually being tested by the LGI.

The system considered consists of a arbitrary spin j prepared in a initially maximally mixed state of the eigenstates m of the z-component, $\hat{J}_z$, which dynamics is given by a precession around the x-component, $\hat{J}_x$ - as discussed above, this state is not supposed to violate the LGI, what ensures that nonclassicality will arise from the system's dynamics and the measurement processes. In order to measure the correlations C_{kl} in Eq. 1, we define a two-valued POVM, which allows us to introduce parameters defined as the *measurability*, σ, and the *resolution*, Δm_μ, μ being the specific m's that can always be faithfully determined by the measurement apparatus. Therefore, for the particulars $m = \mu$, one is always able to associate with them a well-defined parity once they are perfectly determined by the measurement apparatus, whilst for $m \neq \mu$, the interplay between the parameters σ and Δm_μ will determine how well the correct parity can be associated with these outcomes. For a fixed resolution, Δm_μ, invasiveness can be associated with the measurability, σ, and this later will then control the violation of the LGI. We refer the reader to Ref. [7] for all the details.

This scheme is quite general and can be associated with different physical origins, for instance the increase of dimensionality of a system and the classical disturbance created by a the measurement process. The interpretation of each parameter depends on the physical system one uses to test a LGI, the role of each parameter clearly identified to invasiveness, whatever its physical origin is. Our model can be used to help understanding and interpretation of LGI tests and can be tested experimentally in a number of physical systems, such as Stern–Gerlach-like experiments with inhomogeneous fields and the orbital angular momentum of photons.

Acknowledgments

S. V. M. acknowledges financial support from the Brazilian agency CAPES.

References

[1] Leggett A J and Garg A 1985 *Phys. Rev. Lett.* **54** 857
[2] Emary C, Lambert N and Nori F 2014 *Rep. Prog. Phys.* **77** 016001
[3] Maroney O J E and Timpson C G 2014 (*Preprint*, `arXiv:1412.6139 [quant-ph]`)
[4] Bohm D and Hiley B J 1993 *The Undivided Universe* (London: Routledge)
[5] Clemente L and Kofler J 2015 *Phys. Rev. A* **91** 062103
[6] Spekkens R 2005 *Phys. Rev. A* **71** 052108
[7] Moreira S V, Keller A, Coudreau T and Milman P 2015 *Phys. Rev. A* **92** 062132

Measuring the weak value of momentum in a double slit atom interferometer

J Morley, P D Edmunds and P F Barker

Dept. Physics & Astronomy, University College London, Gower St, London WC1E 6BT, UK

E-mail: p.barker@ucl.ac.uk

Abstract. We describe the development of an experiment to measure the weak value of the transverse momentum operator (local momentum [1]) of cold atoms passing through a matter-wave interferometer. The results will be used to reconstruct the atom's average trajectories. We describe our progress towards this goal using laser cooled argon atoms.

1. Introduction

There is currently considerable interest in the use of weak measurements [2] to explore fundamental quantum processes and for quantum metrology. Of particular impact has been the experiment of Kocsis *et al.*, who employed weak measurements of photon momentum, to reconstruct average trajectories of photons [3]. Here, the polarisation of the photon was used as a pointer that weakly coupled to the transverse photon momentum. This type of experiment has only been performed on photons, for which a classical wave-optics interpretation can be applied. We describe the development of a two-slit, matter-wave interferometer that uses argon atoms to perform weak measurements of momentum.

2. Double slit interferometer

We are developing a matter wave interferometer using laser cooled metastable ($4s[3/2]_2$) argon atoms from a magneto-optical trap (MOT). Cold atoms from the MOT, shown in figure 1, drop approximately 2 cm onto double slits which have a separation of 6 μm. The slits, with a length of 1 mm, are formed on 1 μm thick membrane.

The atoms transmitted through the slits fall through the weak measurement stage formed by a tilted magnetic field. The matter wave interference pattern is produced 25 cm below the slits with a period of approximately 250 μm. This pattern is recorded on a micro-channel plate detector (MCP) as the atoms ionise when they strike its surface.

3. Measuring a weak value of momentum

The pointer that weakly couples to atomic momentum in our atom interferometer is the spin of the 5 Zeeman sub-levels (M') of this state. The weak stage is a magnetic field with a small tilt, θ, from the axis perpendicular to the falling atoms. As the atoms falling through the field undergo Larmor precession, they accumulate a phase shift that is proportional to their duration in the field. Due to the tilt in the field, the accumulated phase is dependent on the transverse momentum of the atoms and is given for small angles of θ by $\phi = \frac{g\mu_B M' w}{\hbar^{2}_{g}} p_x$ where p_x is the

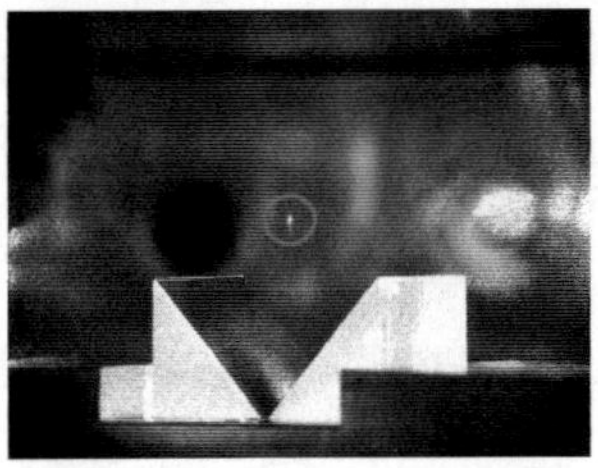

Figure 1. The laser cooled atoms trapped above the slits of the interferometer located below two prisms.

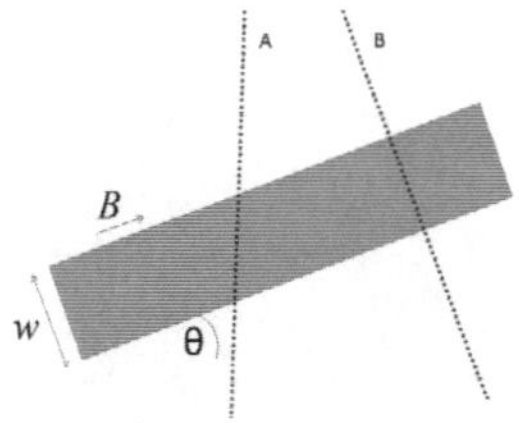

Figure 2. Atoms with different trajectories, A and B, undergo different phase shifts as they pass through a region of constant magnetic field, tilted by θ with respect to the axis of the interferometer.

transverse momentum, p_y is the momentum along the direction of propagation and w is the width of the region of constant magnetic field that constitutes the weak stage. To measure the phase shift, the atoms are first spin polarised in the state $|J = 2, M = 2\rangle$ before entering the weak stage. An abrupt 90° change in field direction induces a Majorana flip, which projects this initial state into a superposition of Zeeman sub-levels (M') as they enter the weak stage [6]. The M' states then accumulate the velocity dependent phase as described above. On leaving the field another 90° rotation projects the superposition into 5 M'' states. The induced phase shift in the weak stage, and therefore the atoms transverse momentum, is extracted by measuring the spatial modulation of the interference pattern for each M'' state on leaving the weak stage. For the $M'' = 0$ state this is given by $\frac{3}{8}\sin^4\phi$.

To measure the interference pattern for each M'' state they are split into separate interference fringes using a Stern–Gerlach field, produced by a wire. We can reconstruct the average trajectories of the atoms by calculating the transverse momentum at regular intervals below the slits. We have constructed the source of cold metastable argon atoms and the slits required for the interferometer. We are currently optimizing the geometry and field strength of the weak stage to obtain highly localised momentum measurements along the direction of propagation of the atoms.

4. References

[1] Berry M V 2013 *Eur. J. Phys* **34** 1337–48
[2] Aharonov Y, Albert D Z and Vaidman L 1988 Phys. Rev. Lett. 60 1351–54
[3] Kocsis S, Braverman B, Ravets S, Stevens M J, Mirin R P, Krister Shalm L and Steinberg A M 2011 *Science* **332** 1170–73
[4] Edmunds P D and Barker P F 2014 *Phys. Rev. Lett.* **113** 183001
[5] Gähler R, Golub R, Habicht K, Keller T and Felber J 1996 *Physica B* **229** 1–17
[6] Viaris de Lesegno B et al. 2003 *Eur. Phys. J. D* **23** 25

Regularization of derivatives on non-differentiable points

Dimiter Prodanov

Department of Environment, Health and Safety, Neuroscience Research Flanders, Imec, Leuven, Belgium

E-mail: dimiterpp@gmail.com; dimiter.prodanov@imec.be

Abstract. The notion of derivative is limited only to the idealization of linear growth. The paper presents a formal regularization procedure for the derivatives of functions using the fractional velocity, i.e. a local fractional derivation operation. The approach can be applied in strongly non-linear or fractal settings, such as singular functions, Brownian motion, quantum paths, etc.

1. Fractional approach to derivation

A central notion of physics is the rate of change. It can be argued that this perception inspired Newton and Leibniz to develop the apparatus of differential calculus. However, the notion of *derivative* is limited only to the idealization of linear growth. Classical physical variables, such as velocity or acceleration, are considered to be differentiable functions of position. On the other hand, quantum mechanical paths were found to be non-differentiable and stochastic in simulations [2]. The relaxation of the differentiability assumption could open new avenues in describing physical phenomena, for example, using the **scale relativity theory** developed by Nottale [1], which assumes fractality of quantum-mechanical trajectories. Cherbit [4] introduced the notion of *α fractional (fractal) velocity* as the limit of the fractional difference quotient. His main application was the study of fractal phenomena and physical processes for which the instantaneous velocity was not well defined. This concept can be extended to mixed-orders as follows [5, 6] :

Definition 1 (Mixed-order velocity). *Define the fractional velocity of mixed order $n+\beta$ of function $f(x) \in \mathbb{C}^n$ as*

$$v_{\pm}^{n+\beta} f(x) := (\pm 1)^{n+1}(n+1)! \lim_{\epsilon \to 0} \frac{f(x \pm \epsilon) - T_n(x, \pm\epsilon)}{\epsilon^{n+\beta}} \tag{1}$$

where $n \in \mathbb{N}$; $\epsilon > 0$, $0 < \beta \leq 1$ are real parameters and $T_n(x,\epsilon)$ is the Taylor polynomial $T_n(x,\epsilon) = f(x) + \sum\limits_{k=1}^{n} \frac{f^{(k)}(x)}{k!}\epsilon^k$.

Some of the most peculiar properties of α-velocity are that it takes discrete values, it is discontinuous and non-vanishing only at points where the usual derivative is unbounded[3, 5].

The **fractional velocity** is able to characterize growth of functions varying on fractal sets. This can be exemplified by its application to singular functions, such as the Cantor's function.

Example 1. *Cantor's function obeys the functional equation*

$$C(x) := \begin{cases} 0, & x = 0 \\ \frac{1}{2}C(3x), & 0 \le x < 1/3 \\ \frac{1}{2} & 1/3 \le x \le 2/3 \\ \frac{1}{2} + \frac{1}{2}C(3x-2), & 2/3 \le x \le 1 \\ 1, & x = 1 \end{cases}$$

Therefore $v_{+}^{\alpha}C(x) = 1$ *if* x *belongs to the Cantor set* $\mathbb{C}_{1,3}$ *and* $v_{+}^{\alpha}C(x) = 0$ *otherwise for* $\alpha = \frac{\log 2}{\log 3}$. *The* ***Cantor's set*** *in turn is given by* $\mathbb{C}_{1,3} = \left\{x : x = \overline{0.d_1 \ldots d_n},\ d \in \{0, 2\}\right\}$ *in ternary number representation.*

2. Regularized Taylor expansions

In other applications it transpires that **fractional velocity** acts as an auxiliary object with regard to integer-order derivatives. Paradoxically, the irregularity of the fractional velocity can be used to regularize the usual derivatives at singular points. This can be demonstrated in the regularization procedure for the derivatives of Hölder functions, which allows for removal of the weak singularity in the derivative caused by strong non-linearities:

Definition 2. *Regularized derivative of a function is defined as:*

$$\frac{\bar{d}^{\,\beta\pm}}{dx} f(x) := \lim_{\epsilon \to 0} \frac{\Delta_{\epsilon}^{\pm}[f](x) - v_{+}^{\beta} f(x)\, \epsilon^{\beta}}{\epsilon}$$

We will require as usual that the forward and backward regularized derivatives be equal for a uniformly continuous function.

Then the following statement holds: Let $f(t, w) \in \mathbb{C}^2$ be composition with $w(x)$, a Hölder function $\mathbb{H}^{1/q}$ at x, then

$$\frac{\bar{d}^{\,\pm}}{dx} f(x, w) = \frac{\partial f}{\partial x} + \frac{\bar{d}^{\,\pm}}{dx} w(x) \cdot \frac{\partial f}{\partial w} + \frac{(\pm 1)^q}{q!} [w^q]^{\pm} \cdot \frac{\partial^q f}{\partial w^q} \tag{2}$$

where

$$[w^q]^{\pm} = \lim_{\epsilon \to 0} \left(v_{1/q}^{\pm\epsilon} [w](x) \right)^q$$

is the fractal q-adic (co-)variation. Possible applications of presented approach are regularizations of quantum mechanical paths and Brownian motion trajectories, which are Holder 1/2.

Acknowledgments

The work has been supported in part by a grant from Research Fund - Flanders (FWO), contract number 0880.212.840.

References

[1] Nottale L 1989 *Int J. Mod. Phys. A* **4** 5047–5117
[2] Amir-Azizi S, Hey A J G and Morris T R 1987 *Complex Systems* **1** 923–938
[3] Adda F B and Cresson J 2001 *J. Math. Anal. Appl.* **263** 721–737
[4] Cherbit G 1991 *Fractals, Non-integral dimensions and applications* (Paris: John Wiley & Sons) chap Local dimension, momentum and trajectories, pp 231–238
[5] Prodanov D 2015 *Fract. Calc. Appl. Anal.* **18** 580–602
[6] Prodanov D 2016 *Fract. Calc. Appl. Anal.*, in press

Journal of Physics: Conference Series **701** (2016) 012032 doi:10.1088/1742-6596/701/1/012032

A process algebra model of QED

William Sulis

CI Lab, McMaster University, 255 Townline Rd E., Cayuga, Ontario, Canada N0A 1E0

E-mail: `sulisw@mcmaster.ca`

Abstract.
The process algebra approach to quantum mechanics posits a finite, discrete, determinate ontology of primitive events which are generated by processes (in the sense of Whitehead). In this ontology, primitive events serve as elements of an emergent space-time and of emergent fundamental particles and fields. Each process generates a set of primitive elements, using only local information, causally propagated as a discrete wave, forming a causal space termed a causal tapestry. Each causal tapestry forms a discrete and finite sampling of an emergent causal manifold (space-time) $\mathcal{M}$ and emergent wave function. Interactions between processes are described by a process algebra which possesses 8 commutative operations (sums and products) together with a non-commutative concatenation operator (transitions). The process algebra possesses a representation via nondeterministic combinatorial games. The process algebra connects to quantum mechanics through the set valued process and configuration space covering maps, which associate each causal tapestry with sets of wave functions over $\mathcal{M}$. Probabilities emerge from interactions between processes. The process algebra model has been shown to reproduce many features of the theory of non-relativistic scalar particles to a high degree of accuracy, without paradox or divergences. This paper extends the approach to a semi-classical form of quantum electrodynamics.

1. Process algebra

The goal of the process approach is to develop an emergent quantum mechanics which rests upon a model of reality that is discrete, finite, determinate (but not deterministic), and mostly non-contextual and local, in other words, as *realist* a reality as possible. The primitive elements of this ontology are *generated* by processes [1] and the usual entities of physics are emergent upon these primitives. The model reproduces the non-relativistic quantum mechanics (NRQM) of scalar particles to a high degree of accuracy and, it has been argued, without paradoxes [2, 3]. The present paper extends the model to a semi-classical version of quantum electrodynamics.

Formally, represent each primitive element as an *informon* $[n] < \mathbf{m}_n, \phi_n(\mathbf{z}), \Gamma_n, \mathbf{p}_n > \{G_n\}$, possessing *intrinsic* features (Γ_n-local process coupling constant; $\mathbf{p}_n$-vector of properties; G_n-causally ordered content set of *exclusively prior* informons, causally prior to n), and *extrinsic* features [interpretations by an observer] ($\mathbf{m}_n$-element of a causal space $\mathcal{M}$ [4]; $\phi_n(\mathbf{z})$-local contribution to a wave function in $\mathcal{H}(\mathcal{M})$, $\phi_n(\mathbf{m}_n) = \Gamma_n \hat{\phi}_n(\mathbf{m}_n) = \Gamma_n$). Moreover, $\Gamma_n = \sum_{k \in G_n} K(n,k)\Gamma_k$, for some propagator K. Each process generates R informons per each of N rounds, each of which is comprised of r short rounds, forming a *causal tapestry* $\mathcal{I}$, which embeds space-like into $\mathcal{M}$ while G_n-causal relations are preserved. $\mathcal{I}$ then becomes information for a new process generating a new tapestry. A process is *primitive* if $R = 1$. The informons of $\mathcal{I}$ generate, via an interpolation procedure [5], an emergent wave function on $\mathcal{M}$ of the form

$\Phi_{\mathcal{I}}(\mathbf{z}) = \sum_{n\in\mathcal{I}} \phi_n(\mathbf{z}) = \sum_{n\in\mathcal{I}}(\sum_{k\in G_n} K(n,k)\Gamma_k)\hat{\phi}_n(\mathbf{z})$. The physics plays out on the causal tapestry, independent of the interpretations. The wave function $\Phi_{\mathcal{I}}(\mathbf{z})$ represents a real discrete wave and a particular reality. In the case of NRQM it agrees closely with the usual NRQM wave function in the limit $N, r \to \infty$ and with Planck scale causal embeddings [2, 3].

Complex processes are formed from primitive processes using the process algebra which has 4 Abelian sums (sequential action of sub-processes, for states involving a single entity and superpositions), 4 Abelian products (simultaneous action of sub-processes, for states involving multiple entities), and concatenation which describes transitions between processes resulting from interactions with newly generated informons. Measurement is understood as an interaction between processes, triggered by the generation of individual informons and dependent upon their compatibility [6] (proportional to the local coupling constants), giving rise to an emergent contextual probability [7]. The process algebra possesses a useful heuristic representation as an algebra of non-deterministic combinatorial games [8]. Analysis requires the game tree [8], a heuristic and combinatorial tool (akin to configuration space) depicting all possible tapestries generated by the process, needed to generate set-valued process and configuration space covering maps, (defined on function space products (calculational) or co-products (ontological)) which link to NRQM [2, 3], enabling one to calculate correlations and measurement outcomes.

2. QED

A naive quantum electrodynamics is developed semi-classically following the idea of collective electrodynamics [9]. The local coupling constant is the local electrodynamic potential $(\phi_n, A_n^1, A_n^2, A_n^3)$. K is the classical electromagnetic propagator. A primitive process $\mathbb{P}_{\omega,\mathbf{k}}$ creates, sequentially, informons corresponding to a single photon of energy $E = \hbar\omega$ and momentum $\mathbf{p} = \hbar\mathbf{k}$. From a single informon, $\mathbb{P}_{\omega,\mathbf{k}}$ will generate a set of R informons embedding within a ball of radius Planck length in $\mathcal{M}$. Each of these nascent informons serves as a source during the next cycle of generation so the next generator takes the form $\hat{R}\mathbb{P}_{\omega,\mathbf{k}} = \overbrace{\mathbb{P}_{\omega,\mathbf{k}}\hat{\oplus}\cdots\hat{\oplus}\mathbb{P}_{\omega,\mathbf{k}}}^{R}$ where $\hat{\oplus}$ is the free sum which allows subprocesses to provide information to any other subprocess (the process analogue of superposition). A measurement will still always register single photons. Successive generations will result in the process $\hat{R}^n\mathbb{P}_{\omega,\mathbf{k}}$ as $n \to \infty$. The generated informons will spread out over an ever widening region of $\mathcal{M}$ giving the appearance of a discrete sampling of a field. The resulting (ontological) wave function will be

$$\Phi(\mathbf{z}) = \lim_{n\to\infty} \sum_{i=1}^{n} \Phi_i(\mathbf{z}) = \lim_{n\to\infty} \sum_{i=1}^{n} \sum_{k\in\mathcal{I}_i} (\phi_k, A_k^1, A_k^2, A_k^3)\hat{\phi}_k(\mathbf{z})$$

which gives an interpolation of the electrodynamic potential. Different plays of the process result in different $\Phi(\mathbf{z})$, expressed in the set-valued process covering map. Second quantization, needed for deriving the usual QED wave function, requires consideration of complex processes and the use of the more subtle and complicated configuration space process map.

References

[1] Eastman T and Keeton H 2004 *Physics and Whitehead: Quantum, Process and Experience* (Albany: SUNY)
[2] Sulis W H 2014 *A Process Model of Non-Relativistic Quantum Mechanics* Ph.D. Thesis University of Waterloo
[3] Sulis W H 2014 A Process model of quantum mechanics *J. Mod. Phys.* doi:10.4236/jmp.2014.516176
[4] Borchers H J and Sen R N 2006 *Mathematical Implications of Einstein-Weyl Causality* (New York: Springer)
[5] Zayed A I 1993 *Advances in Shannon's Sampling Theory* (Boca Raton: CRC Press)
[6] Trofimova I 2001 *Principles, concepts, and phenomena of ensembles with variable structure Nonlinear Dynamics in the Life and Social Sciences* ed W H Sulis and I N Trofimova (Amsterdam: IOS Press)
[7] Khrennikov A (2010) *Ubiquitous Quantum Structures* (Berlin: Springer)
[8] Conway J H (2001) *On Numbers and Games* (Natick: A.K. Peters)
[9] Mead C (2000) *Collective Electrodynamics: Quantum Foundations of Electromagnetism* (Boston: MIT Press)

Journal of Physics: Conference Series **701** (2016) 012033 doi:10.1088/1742-6596/701/1/012033

Classical gravity from certain models of emergent quantum mechanics

Ricardo Gallego Torromé

[1]Departamento de Matemática, Universidade Federal de São Carlos, Rodovia Washington Luís, km 235, SP, Brazil

E-mail: rigato39@gmail.com

Abstract. It is shown how an interaction that resembles classical gravity emerges in certain dynamical systems.

1. Hamilton–Randers theory

Hamilton-Randers systems, the class of dynamical systems that we consider in this contribution, are deterministic, local and relativistic dynamical models (named Hamilton–Randers systems in [1]). The configuration space for each Hamilton–Randers model is a product of the form

$$TM = \prod_{k=1}^{N} TM_4^k. \tag{1}$$

Here $\{M_4^k\}_{k=1}^N$ is a family of diffeomorphic four-manifolds to a given *model manifold* M_4. There is also a probability measure μ_P and a *topological asymmetric metric*.

In Hamilton–Randers theory there is assumed two independent dynamics, the $U(t)$-flow and the *usual Heisenberg dynamics* along the external s-time, $V(s)$. The $U(t)$-flow is a geometric flow in the co-tangent space T^*TM. Several properties that such flow must have are postulated. One of the fundamental property is that the flow must be composed of *fundamental cycles* that depend on the properties and characteristics of the $U(t)$-flow and physical system under consideration. One generic characteristic is that each cycle is composed by three dynamical regimes: an ergodic-contractive-expanding regime. Then fixed a system, each cycle associated to the system follows another cycle in a *sequence* defining any external time parameter s. The ontological degrees of freedom are described by points $u = (x, y) \in TM$. The Hamiltonian function is

$$H(u,p,t) = (1-\kappa(t))\left(\sum_{k=1}^{8N} \beta^k(u)p_k\right), \tag{2}$$

where κ is a scalar factor determined by the $U(t)$-flow.

It is postulated the existence in each of the fundamental cycle of a contractive regime such that the dynamics determined by the flow is 1-Lipschitz continuous. In such regime one has that

$$\lim_{t\to T} H = 0, \quad \lim_{t\to T}(1-\kappa(t)) = 0. \tag{3}$$

The Hamilton–Randers models are canonically quantized by using the quantization scheme

$$(u,p)\mapsto(\hat{u},\hat{p}),\quad [\hat{u}^i,\hat{u}^j]=0,\quad [\hat{p}_i,\hat{p}_j]=0,\quad [\hat{u},\hat{p}]=\delta^i_j,\quad i,j=1,...,8N. \tag{4}$$

The Hamiltonian operator $\widehat{H}$ is obtained by applying the Weyl prescription to (2). Also, one introduce an *ontological basis* $\{|u\rangle\}$ of eigenstates for the $\hat{u}$ operators that generate the Hilbert space H_p for the dynamical degrees of freedom at the Planck scale. Moreover, by property (3)

$$\lim_{t\to T}\widehat{H}(\hat{u},\hat{p},t)|\psi\rangle=0 \tag{5}$$

holds good.

In terms of the basis $\{|x_k,y_k\rangle\}$ a generic quantum state is of the form

$$\psi(x)=\sum_{k=1}^{N}\int_{T_xM_4^k}d^4z_k\,e^{\imath\vartheta_k(x_k,y_k)}\frac{1}{\sqrt{N}}n_k(x_k,y_k)\,|x_k,y_k\rangle. \tag{6}$$

Then if the locality condition

$$\int_{T_xM_4^k}d^4z_k\,e^{\imath\vartheta_k(x_k,y_k)}\,e^{-\imath\vartheta_k(x'_k,y'_k)}\longrightarrow\delta(x_k-x'_k)\delta(y_k-y'_k) \tag{7}$$

is imposed, then there is defined a scalar product in the space of vectors of the form (6), there is a well defined Heisenberg image for the dynamics and the Born rule holds good [1].

2. Concentration of measure and the origin of classical gravity

Concentration of measure is the fact that *in a measure metric space of large dimension, every real 1-Lipschitz function of many variables is almost constant almost everywhere.* This principle applies to Hamilton–Randers models, that attach to each quantum system (as an electron, for example) $N\gg1$ degrees of freedom. Then we can argue [1] that the following concentration

$$\mu_p\left(|f-M_f|>8\,N\sigma_f\right)\leq\frac{1}{2}\exp(-32N^2) \tag{8}$$

holds for any 1-Lipschitz function $f:T^*TM\to\mathbf{R}$; M_f is the median and σ_f the variance of f.

Let $X^\mu(S_i)$, $i=A,B$ be the macroscopic observable μ-coordinate associated to the system S_i. In the domain $t\to(2n+1)T$ the functions $X^\mu((2n+1)T,s)=X^\mu(s)$ are 1-Lipschitz [1]. We apply concentration to the systems A and B when $M^\mu_i(s=0))$ and σ_i are the same for i=A,B,

$$\mu_p(\left(\frac{1}{\sigma_{X^\mu}}\,|X^\mu(S_i(s))-M^\mu(s)|>\rho\right)\leq C_1\,\exp\left(-C_2\,\frac{\rho^2}{2\rho_p^2}\right). \tag{9}$$

Thus a form universality of free fall is proved by geometric and measure theoretical arguments.

Collecting all the properties of the $U(t)$-flow in the equilibrium domain we have that **1**. The diffeomorphism invariance (5) holds good, **2**. A weak equivalence principle holds, as consequence of the concentration (9), **3**. It is a classical interaction (because it happens in the concentration domain), **4**. There is a maximal speed for the fundamental degrees of freedom [1]. Moreover, Newtonian gravity is a 1-Lipschitz interaction at scales corresponding to the standard model domain [1]. Therefore, such hypothetical interaction should correspond to classical gravity.

Acknowledgments

This work was supported by PNPD-CAPES n. 2265/2011, Brazil.

References

[1] Gallego Torromé R 2015 A theory of emergent quantum mechanics (*Preprint* `arXiv:1402.5070v7 [math-ph]`)
[2] Gromov M 1999 Riemannian Structures for Riemannian and Non-Riemannian Spaces *PM* **152**, Ed. LaFontaine J and Pansu P (Boston: Birkhäuser)

Quantum mechanics emerging from stochastic dynamics of virtual particles

Roumen Tsekov

Department of Physical Chemistry, University of Sofia, 1164 Sofia, Bulgaria

E-mail: Tsekov@chem.uni-sofia.bg

Abstract. It is shown how quantum mechanics emerges from the stochastic dynamics of force carriers. It is demonstrated that the Moyal equation corresponds to dynamic correlations between the real particle momentum and the virtual particle position, which are not present in classical mechanics. This new concept throws light on the physical meaning of quantum theory, showing that the Planck constant square is a second-second position-momentum cross-cumulant.

According to modern physics, the Newtonian interactions in classical mechanics occur via exchange of force carriers [1]. These virtual particles transmit the long-range interactions among the real particles and interact only locally with the latter. Generally, it is expected that the force carriers dynamics is stochastic, which will result in random Newtonian potentials. Thus, the stochastic motion of the virtual particles can cause quantum dynamics [2]. Stochastic electrodynamics is an important example, for instance, which is already proposed as an origin of quantum mechanics [3]. In the present paper it is shown how quantum mechanics emerges from the stochastic dynamics of force carriers [4]. It is also demonstrated that the quantum Moyal equation [5] corresponds to statistical correlations between the real particle momentum and the virtual particle position, not present in classical mechanics. The model throws light on the physical meaning of the Planck constant, a quintessence of quantum theory.

Let us consider first a real and a virtual particles, located at positions r and R, respectively. Since by definition there is no long-range forces, their interaction is a delta-potential $\varepsilon\delta(r-R)$. The specific parameter ε accounts for the type of interaction, e.g. electrostatic, gravitational, etc. In general, the position of the virtual particle is expected to be random and, hence, the interaction potential above is a stochastic one [4]. The corresponding force, acting on the real particle, reads $-\varepsilon\partial_r\delta(r-R)$. Introducing the real particle phase-space distribution density $W(p,r,t)$, one can write its dynamics in the form

$$\partial_t W + p\cdot\partial_r W/m = \partial_p \cdot \int_{-\infty}^{\infty} \varepsilon\partial_r\delta(r-R)F(R,p,r,t)dR = \partial_p\cdot(\varepsilon\partial_R F)_{R=r} \tag{1}$$

The interaction between the real and virtual particles is expressed in Eq. (1) via a collision integral, where $F(R,p,r,t)$ is the joint probability density for the virtual particle to occupy the position R and for the real particle to have momentum p and coordinate r at time t. The real particle marginal probability density W can be derived from F via a simple integration over R

$$W(p,r,t) \equiv \int_{-\infty}^{\infty} F(R,p,r,t)dR \qquad \rho(R,t) \equiv \int_{-\infty}^{\infty}\int_{-\infty}^{\infty} F(R,p,r,t)dpdr \tag{2}$$

while $\rho(R,t)$ is the local density of force carriers. A macro-particle interacts with many force carriers at once and thus stochastic forces cancel each other. This results in classical mechanics, where the virtual particle distribution is independent of the real particle. Hence, the joint probability density factorizes in a product $F(R,p,r,t) = \rho(R,t)W(p,r,t)$ of the virtual particle density and the real particle probability density. In this case Eq. (1) reduces straightforward to the Liouville equation from classical dynamics

$$\partial_t W + p \cdot \partial_r W/m = \partial_r U \cdot \partial_P W \qquad U(r,t) = \int_{-\infty}^{\infty} \varepsilon\delta(r-R)\rho(R,t)dR = \varepsilon\rho(r,t) \tag{3}$$

The corresponding Newtonian potential U follows exactly the distribution of the virtual particles in the system, which is not perturbed by the motion of the real particle. There are indications in quantum mechanics that the real particle motion affects the distribution of virtual particles. The Liouville equation (3) changes to the Wigner–Liouville or Moyal equation

$$\partial_t W + p \cdot \partial_r W/m = \sum_{n=0}^{\infty} \frac{(i\hbar/2)^{2n}}{(2n+1)!} \partial_r^{2n+1} U \cdot \partial_p^{2n+1} W \tag{4}$$

Keeping in mind the definition of the Newtonian potential $U = \varepsilon\rho$, the juxtaposition of the Wigner–Liouville equation (4) and Eq. (1) unveils a possible expression for the joint distribution density

$$F(R,p,r,t) = \rho W + \sum_{n=1}^{\infty} \frac{(i\hbar/2)^{2n}}{(2n+1)!} \partial_R^{2n} \rho \cdot \partial_p^{2n} W \tag{5}$$

As is seen, the last sum represents a correlation function between the virtual and real particles, which vanishes in the classical limit at . The analysis of the probability density (5) is easier in the Fourier space, where the corresponding characteristic function is given by

$$\tilde{F}(K,q,k,t) = \int_{-\infty}^{\infty}\int_{-\infty}^{\infty}\int_{-\infty}^{\infty} \exp(iK \cdot R + iq \cdot p + ik \cdot r)F(R,p,r,t)dRdpdr \tag{6}$$

Substituting here Eq. (5) yields the characteristic function

$$\tilde{F}(K,q,k,t) = \tilde{\rho}(K,t)\frac{\sin(\hbar K \cdot q/2)}{\bar{h}K \cdot q/2}\tilde{W}(q,k,t) \tag{7}$$

expressed by the characteristic functions of the marginal distributions (2). The corresponding generating function of the cross-cumulant of the real particle momentum and the force carrier position reads

$$\Phi \equiv \ln \frac{\tilde{F}(K,q,k,t)}{\tilde{\rho}(K,t)\tilde{W}(q,k,t)} = \ln \frac{\sin(\hbar K \cdot q/2)}{\bar{h}K \cdot q/2} \tag{8}$$

As is seen, Φ is independent of time and of the real particle position. Hence, quantum mechanics is due to disturbances of the virtual particle sea caused by the real particle motion. The only parameter in Φ is the Planck constant $\hbar$ and Eq. (8) can elucidate its physical meaning. The expansion in power series $\Phi = -(\hbar K \cdot q)^2/24 + \cdots$ shows that the second-second cross-cumulant of the virtual particle position and the real particle momentum equals to $\kappa_{22} \equiv < (r \cdot p)^2 > - < R \otimes R >:< p \otimes p >= -\bar{\hbar}/2$. It indicates a firm anti-correlation between the virtual particle position and the real particle momentum. In contrast to the famous Heisenberg inequality $\sigma_r \sigma_p \geq \bar{h}/2$, restricting the real particle momentum and position$\hbar^2 = -2\kappa_{22}$ is an equality which could be considered as the Planck constant definition. A new Heisenberg relation $\sigma_{R^2}\sigma_{p^2} \geq \bar{\hbar}/2$ follows from the Cauchy-Schwarz inequality $\kappa_{22} \geq -\sigma_{R^2}\sigma_{p^2}$.

References

[1] Zee A 2010 *Quantum Field Theory in a Nutshell.* (New Jersey: Princeton University Press)
[2] de la Torre A C 2015 *Preprint* `arXiv:1505.04326 [quant-ph]`
[3] de la Peña L, Cetto A M and Valdés-Hernández A 2015 *The Emerging Quantum: The Physics Behind Quantum Mechanics* (New York: Springer)
[4] Tsekov R 2012 *Ann. Univ. Sofia, Fac. Phys.* **SE** 112
[5] Groenewold H J 1946 *Physica* **12** 405; Moyal J E 1949 *Math. Proc. Cambridge Phil. Soc.* **45** 99; Cherbit G 1991 *Fractals, Non-integral Dimensions and Applications* (New York: Wiley)

Journal of Physics: Conference Series **701** (2016) 012035 doi:10.1088/1742-6596/701/1/012035

Gravity-matter entanglement in Regge quantum gravity

Nikola Paunković[1] and Marko Vojinović[2]

[1]SQIG — Security and Quantum Information Group, Instituto de Telecomunicacões, and Departamento de Matemática, Instituto Superior Técnico, Universidade de Lisboa, Avenida Rovisco Pais 1049-001, Lisboa, Portugal

[2]Grupo de Física Matemática, Faculdade de Ciências da Universidade de Lisboa, Campo Grande, Edifício C6, 1749-016 Lisboa, Portugal

E-mail: npaunkov@math.ist.utl.pt, vmarko@ipb.ac.rs

Abstract.
We argue that Hartle–Hawking states in the Regge quantum gravity model generically contain non-trivial entanglement between gravity and matter fields. Generic impossibility to talk about "matter in a point of space" is in line with the idea of an emergent spacetime, and as such could be taken as a possible candidate for a criterion for a plausible theory of quantum gravity. Finally, this new entanglement could be seen as an additional "effective interaction", which could possibly bring corrections to the weak equivalence principle.

Introduction. The unsolved problems of interpreting quantum mechanics (QM) and formulating quantum theory of gravity (QG) are arguably the two most prominent ones of the twentieth century theoretical physics. Up to date, most of the efforts to solve the two were taken independently. Indeed, the majority of the interpretations of QM do not involve explicit dynamical effects (with notable exceptions of the spontaneous collapse and the de Broglie–Bohm theories), while the researchers from the QG community often adopt the many-world interpretation of QM. Nevertheless, the two problems share a number of similar unsolved questions and counter-intuitive features, such as nonlocality: entanglement-based quantum nonlocality, as well as the anticipated explicit dynamical nonlocality in QG (a consequence of quantum superpositions of different gravitational fields, i.e., different spacetimes and their respective causal orders). We analyse the generic entanglement between gravitational and matter fields in the Regge model of quantum gravity, and its possible impact to the fundamental questions regarding QM and QG.

Regge quantum gravity model. A simple toy model of quantum gravity with matter fields is the Regge quantum gravity with one real scalar field, whose construction can be motivated by the Loop Quantum Gravity research program [1, 2]. The path integral of the model is

$$Z_T = \int \mathcal{D}L \int \mathcal{D}\Phi \ \exp\left[iS_{\text{Regge}}(L) + iS_{\text{matter}}(L,\Phi)\right], \tag{1}$$

where L are the lengths of the edges of the triangulation T of a 4-manifold $\mathcal{M}_4$, and Φ are the values of the scalar field in 4-simplices of T. The measure terms $\mathcal{D}L$ and $\mathcal{D}\Phi$ are defined via discretization induced by T. The actions S_{Regge} and S_{matter} represent lattice discretizations of the Einstein–Hilbert action for gravity and an action of the scalar field coupled to gravity, respectively. See [3, 4] for details.

A generic state of the system, belonging to the kinematical Hilbert space $\mathcal{H}_G \otimes \mathcal{H}_M$, is

$$|\Psi\rangle = \int \mathcal{D}l \int \mathcal{D}\phi \;\; \Psi(l,\phi) \;\; |l\rangle|\phi\rangle \,. \tag{2}$$

However, since gravity is a theory with constraints, not every kinematical state is allowed, so we must choose the coefficients $\Psi(l,\phi)$ such that $|\Psi\rangle$ is an element of the physical Hilbert space $\mathcal{H}_{\text{phys}} \subset \mathcal{H}_G \otimes \mathcal{H}_M$. One such class of states are the Hartle–Hawking (HH) states [5], defined by the following choice of the coefficients, for a given triangulation T:

$$\Psi(l,\phi) = \Psi_{\text{HH}}(l,\phi) \equiv \int \mathcal{D}L \int \mathcal{D}\Phi \;\; \exp\left[iS_{\text{Regge}}(L,l) + iS_{\text{matter}}(L,\Phi,l,\phi) \right] . \tag{3}$$

This expression differs from (1) in that the triangulation T is now assumed to have a nontrivial 3-dimensional boundary ∂T, and that the variables l, ϕ living on the boundary are not integrated over, in contrast to the bulk variables L and Φ.

Using (2) and (3), one can calculate the reduced density matrix of the Hartle–Hawking state,

$$\hat{\rho}_M = \text{Tr}_G |\Psi\rangle\langle\Psi| = \int \mathcal{D}\phi \int \mathcal{D}\phi' \;\; \left[\int \mathcal{D}l \;\; \Psi_{\text{HH}}(l,\phi)\Psi^*_{\text{HH}}(l,\phi') \right] \;\; |\phi\rangle\langle\phi'| \,,$$

where the integral in the brackets can be denoted as $Z_{T\cup\bar{T}}(\phi,\phi')$. The resulting density matrix can then be tested for entanglement by checking if the trace of its square equals one [6]. In the Regge quantum gravity model the path integrals reduce to a finite number of ordinary integrals, which can then in principle be evaluated. For a generic triangulation, we obtain

$$\text{Tr}_M \, \hat{\rho}_M^2 = \int \mathcal{D}\phi \int \mathcal{D}\phi' \;\; \left| Z_{T\cup\bar{T}}(\phi,\phi') \right|^2 \neq 1 \,,$$

i.e., the gravitational and scalar degrees of freedom in the generic HH state are entangled.

Discussion. In [7] Penrose argues that gravity-matter entanglement is at odds with (classical) spacetime, seen as a (four-dimensional) differentiable manifold. In light of this, our result could be seen as a quantitative indicator that in quantum gravity one cannot talk of "matter in a point of space", i.e., this result could be seen as a confirmation of a "spacetime as an emergent phenomenon".

Thus, generic gravity-matter entanglement could be seen as a possible candidate for a criterion for a plausible theory of quantum gravity.

Entanglement is in standard quantum mechanics a generic consequence of the interaction. This new entanglement can be regarded as a consequence of an effective interaction (such as the "exchange interactions", which are a consequence of quantum statistics). This additional "effective interaction" can potentially lead to corrections to the weak equivalence principle.

Acknowledgments

NP was partially supported, under the CV-Quantum internal project at IT, by FCT PEst-OE/EEI/LA0008/2013 and UID/EEA/50008/2013. MV was supported by the FCT grant SFRH/BPD/46376/2008, the FCT project PEst-OE/MAT/UI0208/2013, and partially by the project ON171031 of the Ministry of Education, Science and Technological Development, Serbia.

Journal of Physics: Conference Series **701** (2016) 012035 doi:10.1088/1742-6596/701/1/012035

References

[1] Rovelli C 2004 *Quantum Gravity* (Cambridge: Cambridge University Press)
[2] Rovelli C and Vidotto F 2014 *Covariant Loop Quantum Gravity* (Cambridge: Cambridge University Press)
[3] Miković A and Vojinović M 2012 *Class. Quant. Grav.* **29** 165003
[4] Miković A 2013 *Rev. Math. Phys.* **25** 1343008
[5] Hartle J B and Hawking S W 1983 *Phys. Rev. D* **28** 2960
[6] Nielsen M A and Chuang I L 2000 *Quantum Computation and Quantum Information* (Cambridge: Cambridge University Press)
[7] Penrose R 1996 *Gen. Relativ. Gravit.* **28** 581

Journal of Physics: Conference Series **701** (2016) 012036 doi:10.1088/1742-6596/701/1/012036

Emergent quantum mechanics without wavefunctions

J Mesa Pascasio[1,2], S Fussy[1], H Schwabl[1], G Grössing[1]

[1] Austrian Institute for Nonlinear Studies, Akademiehof, Friedrichstr. 10, 1010 Vienna, Austria
[2] Atominstitut, TU Wien, Operng. 9, 1040 Vienna, Austria

E-mail: ains@chello.at

Abstract. We present our model of an Emergent Quantum Mechanics which can be characterized by "realism without pre-determination". This is illustrated by our analytic description and corresponding computer simulations of Bohmian-like "surreal" trajectories, which are obtained classically, i.e. without the use of any quantum mechanical tool such as wavefunctions. However, these trajectories do not necessarily represent ontological paths of particles but rather mappings of the probability density flux in a hydrodynamical sense. Modelling emergent quantum mechanics in a high-low intesity double slit scenario gives rise to the "quantum sweeper effect" with a characteristic intensity pattern. This phenomenon should be experimentally testable via weak measurement techniques.

We have studied beam attenuation experiments in neutron interferometry as introduced by Summhammer *et al.* [1] for extreme attenuation values. We use the *transmission factor a* as the beam's transmission probability [2], in the cases of a (deterministic) chopper wheel with $a = t_{\rm open}/t_{\rm open}+t_{\rm closed}$ given by the temporal open-to-closed ratio, and a (stochastic) semitransparent material with $a = I/I_0$ being the relation of the intensity I with absorption compared to the intensity I_0 without. Thus the beam modulation behind the interferometer is obtained in two ways. With φ denoting the phase difference, the intensities are given for respective cases by

$$I_{\rm det} \propto 1 + a + 2a\cos\varphi \tag{1}$$
$$I_{\rm stoch} \propto 1 + a + 2\sqrt{a}\cos\varphi \tag{2}$$

Although the same number of particles is observed in both cases, in Eq. (1) the contrast of the interference pattern is proportional to a, whereas in Eq. (2) it is proportional to $\sqrt{a}$. In our calculations we assume stochastic type attenuation.

The probability density distribution for, say, the right slit of a double-slit system exhibits marked signs of interference effects due to the compressed wave superpositions within the bunching area caused by the sweeper effect. For transmission factors below $a \lesssim 10^{-4}$, new effects appear which are not taken into account in linear extrapolation of expectations based on higher-valued transmission factors.

With ever lower values of a, one can see a steadily growing tendency for the attenuated beam to become swept aside. In our model, this phenomenology is explained by processes of diffusion, due to the presence of accumulated vacuum heat (i.e. kinetic energy) mainly in the "strong" beam. The sweeper effect is thus the result of the vacuum heat sweeping aside the very

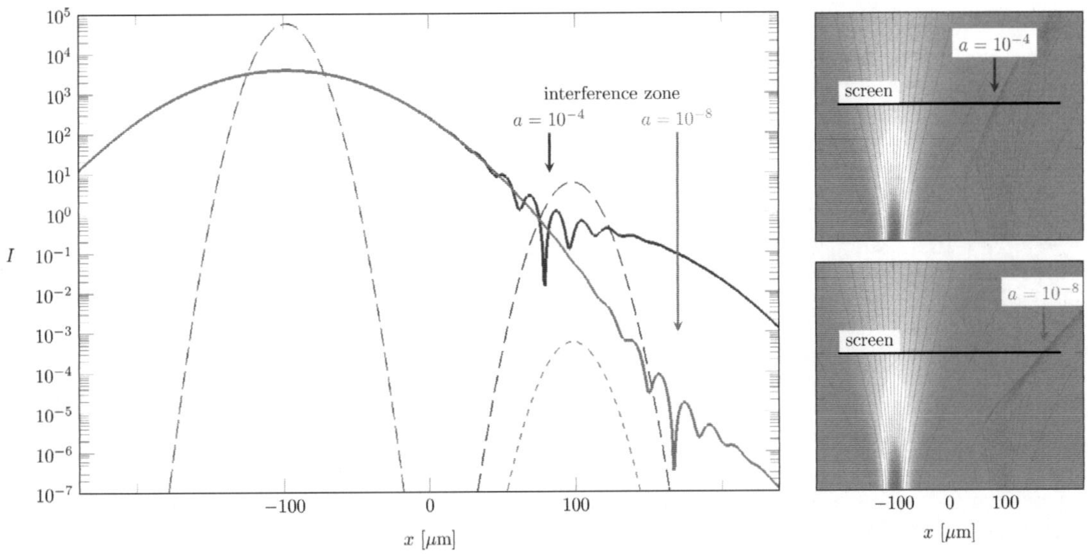

Figure 1. (Left) Dashed initial distributions for the cases of $a = 10^{-4}$ (blue) and $a = 10^{-8}$ (red), respectively, develop into distributions clearly showing interference phenomena which have been "swept aside" far to the right. This sweeper effect is due to the explicit appearance of the nonlinear structure of the probability density current in these domains for very low values of a. (Right) Probability density distributions I emanating from the double slit with transmission factor $a = 10^{-4}$ (top) and $a = 10^{-8}$ (bottom), respectively.

low intensity beam, with a sharp boundary defined by the balancing of the osmotic momenta coming from the two beams, respectively. Due to the different a-dependencies in Eqs. (1) and (2), we observe different shifts of the intensity patterns and correspondingly different angles of the above-mentioned boundary.

As was already pointed out by Luis and Sanz [3], the resulting trajectories such as those of Fig. 1 can be understood as a nonlinear effect that is not usually considered in standard quantum mechanics, but explainable in the Bohmian picture. There, it is the structure of the velocity field which is genuinely nonlinear and therefore allows for the emergence of the type of trajectory behavior observed. Moreover, in our approach, the emergence of the trajectories of Fig. 1 can be traced back to the non-vanishing of the entangling current $\frac{\hbar}{m}(R_1\nabla R_2 - R_2\nabla R_1)$ [4].

Acknowledgments

We thank Maurice de Gosson and Jan Walleczek for fruitful discussions and the Fetzer Franklin Fund of the John E. Fetzer Memorial Trust for partial support of the current work.

References

[1] Summhammer J, Rauch H and Tuppinger D 1987 Stochastic and deterministic absorption in neutron-interference experiments *Phys. Rev.* A **36**(9) 4447–4455

[2] Grössing G, Fussy S, Mesa Pascasio J and Schwabl H 2015 Extreme beam attenuation in double-slit experiments: Quantum and subquantum scenarios *Ann. Phys.* **353** 271–281 (*Preprint* arXiv:1406.1346 [quant-ph])

[3] Luis A and Sanz Á S 2015 What dynamics can be expected for mixed states in two-slit experiments? *Ann. Phys.* **357** 95–107 (*Preprint* arXiv:1311.2612 [quant-ph])

[4] Grössing G, Fussy S, Mesa Pascasio J and Schwabl H 2015 The quantum sweeper effect *J. Phys.: Conf. Ser.* **626**(1) 012017 (*Preprint* arXiv:1502.04034 [quant-ph])